# VIBRATION DAMPING

# VIBRATION DAMPING

**AHID D. NASHIF**

*President, Anatrol Corporation*

**DAVID I. G. JONES**

*Air Force Wright Aeronautical Laboratories*

**JOHN P. HENDERSON**

*Air Force Wright Aeronautical Laboratories*

**A Wiley-Interscience Publication**

## JOHN WILEY & SONS

New York    Chichester    Brisbane    Toronto    Singapore

*Library of Congress Cataloging in Publication Data:*

Nashif, Ahid D.
  Vibration damping.

"A Wiley-Interscience publication."
  Includes index.
  1. Vibration.  2. Damping (Mechanics)  I. Jones, David
I. G.  II. Henderson, John P. (John Phillips).
III. Title.

TA355.N26  1985      620.3′7      84-17247
ISBN 0-471-86772-1

Printed in the United States of America

10 9 8 7 6 5 4 3

# PREFACE

This book is written for individuals responsible for solving vibration and noise problems in a wide variety of industrial structures and machines. Practicing engineers in the automotive, aerospace, naval, and other fields of engineering will find here much valuable practical information, as well as a strong theoretical framework forming a basis for application of damping technology in new and unfamiliar situations. Vibration control by the use of viscoelastic damping materials has, in recent years, grown from a specialized approach for the solution of difficult, expensive, vibration problems in certain military aerospace systems into a widely used, often low cost, element in the package of structural and functional changes particularly needed to solve noise and vibration control problems in the field of general engineering such as automotive, diesel engine, office, and computer machinery and transportation system production. The authors have been in the forefront in many aspects of this transition, and much of the information in this book is distilled from their original research in the area and their in-depth experience of applying the damping technology to realworld problems.

The material in this book assumes a basic understanding of mechanical engineering, including vibration theory, acoustics, and engineering mathematics. With this as a basis, the book provides a systematic phenomenological development of the theory of dynamic behavior of damping materials, as well as guidelines for the application of the theory in practical situations where data are often not available at the time needed, or with the accuracy needed, and where deadlines often require rapid decisions based on a deep understanding of the phenomena and variables involved. It is this, rather than extensive analysis constructed without the pressures of deadlines, which the reader will gain most from careful reading of this book.

The first chapter presents an elementary review of the basic theory of vibration and the role of damping in this analysis. Chapters 2 and 3 introduce the basic concepts of damping phenomenology, with greatest emphasis on viscoelastic damping behavior of polymeric, elastomeric, and glassy materials. Chapter 4 describes the effect of viscoelastic damping on response of structures, with the greatest emphasis being on the particularly fruitful case of the single degree of freedom system. Chapter 5 follows this by studying the effect of discrete damping devices, such as tuned dampers, on structural response. Chapter 6 describes the effect of the other great class of damping treatments, namely the layered surface, or laminated, systems on structural response. The last chapter consists of data sheets of complex modulus data for a large number of useful damping materials. In each chapter illustrations, examples, and case histories are incorporated to show the reader how the theory and the data sheets can be used during the solution of practical problems in noise and vibration control. It is hoped that this book will be of help to practicing engineers and designers concerned with noise and vibration problems in many diverse industries.

The authors would be remiss if they did not acknowledge the debt they owed to their predecessors and contemporaries in this diverse field. The authors were each introduced to the field of vibration control technology by means of viscoelastic damping materials during the early 1960s when they participated in a major damping development program under the direction of Mr. Walter J. Trapp, Chief of the Strength and Dynamics Branch, at what was then the Air Force Materials Laboratory at Wright-Patterson Air Force Base in Dayton, Ohio. The initial impulse given by Mr. Trapp has never dissipated, though it certainly has dispersed as the paths of the authors and the other individuals at the Materials Laboratory diverged, and it can with some justice be said that both this book and many of the multifaceted applications of damping now in use around the world, owe their very existence to Mr. Trapp. Along with his early encouragement, the authors also owe much inspiration to the leading practitioners of the early 1960s and later, including but not limited to Professors B. J. Lazan and R. Plunkett at the University of Minnesota, Dr. Hermann Oberst at Farbwercke-Hoechst AG in West Germany, Dr. Eric Ungar at Bolt, Beraneck and Newman, Inc. in Cambridge, Massachusetts, Professor D. J. Mead at the University of Southampton, England, and more recently, Dr. L. C. Rogers of the Flight Dynamics Laboratory, WPAFB. In more recent times the number of colleagues and friends who work with or have worked with the authors to advance further the field of vibration damping technology has become far too great to encompass in a few pages, and the authors must therefore confine themselves to thanking them en masse, rather than individually, for their contributions, solicited or not, to this book. Of course the authors' opinions remain their own, and none of our colleagues should be blamed for any shortcomings of this book. Last, but by no means least, the authors owe a great debt to Donna Phillips for her patient and uncomplaining efforts to put this book into shape, despite the worst efforts of the authors.

AHID D. NASHIF
DAVID I. G. JONES
JOHN P. HENDERSON

*Cincinnati, Ohio*
*Dayton, Ohio*
*Dayton, Ohio*
*December 1984*

# CONTENTS

## 4. MODELING OF STRUCTURAL RESPONSE OF DAMPED SYSTEMS                                                                 117

# NOMENCLATURE: STANDARD AND GENERIC SYMBOLS

| | |
|---|---|
| $a$ | length (Chapters 1, 6) |
| $b$ | breadth (Chapter 1) |
| $A, B, C \dots$ | nondimensional parameters (Chapters 1, 4) |
| $c$ | velocity of sound |
| $C$ | viscous damping coefficient |
| $C_c$ | coefficient of critical viscous damping |
| $D$ | energy dissipated per cycle, per unit volume, or neutral axis location (according to context) |
| $E, E'$ | Real part of Young's modulus |
| $E''$ | imaginary or quadrature part of Young's modulus ($E'' = E'\eta$) |
| $e^x$ | exponential function of argument $x$ |
| $F$ | amplitude of exciting force |
| $\vec{F}$ | force vector |
| $G, G'$ | shear modulus (real part) |
| $g$ | shear parameter (non-dimensional) |
| $G''$ | imaginary or quadrature part of shear modulus |
| $H$ | thickness |
| $I$ | second moment of area |
| $i$ | $\sqrt{-1}$ |
| $i, j$ | integers ($i, j = 1, 2, \dots$) |
| $k$ | stiffness |

| | |
|---|---|
| $l, L$ | length |
| $\ln(x)$ | natural logarithm of argument $x$ |
| $m, M$ | masses |
| $\vec{\mathbf{M}}$ | moment vector |
| $m, n$ | integers ($m, n = 1, 2, \ldots$), including mode number |
| $N$ | integer ($N = 1, 2, \ldots$), including number of terms in series or number of cycles |
| $p$ | dynamic pressure (relative to atmospheric) |
| $r$ | radius or length dimension |
| $Q$ | quality factor |
| $S$ | area |
| $t$ | time |
| $T$ | temperature (absolute) |
| $U$ | energy stored per unit volume during cycle |
| $w$ | displacement |
| $W$ | displacement amplitude |
| $x, y, z$ | coordinate axes (Cartesian) |
| $\alpha$ | nondimensional parameter (including receptance) |
| $\beta$ | nondimensional parameter |
| $\Gamma$ | nondimensional stiffness parameter |
| $\delta$ | Kronecker or Dirac delta, logarithmic decrement or phase angle, according to context |
| $\Delta$ | determinant or dimensionless length parameter |
| $\varepsilon$ | nondimensional parameter including phase angle and dynamic strain |
| $\phi$ | modal function |
| $\psi$ | nondimensional parameter |
| $\eta$ | loss factor |
| $\mu$ | coefficient of friction |
| $\kappa$ | dynamic stiffness |
| $\lambda$ | nondimensional parameter |
| $\pi$ | ratio of circumference of circle to diameter (3.14159 ...) |
| $\rho$ | density |
| $\sigma$ | stress |
| $\theta$ | angle of rotation |
| $\xi$ | nondimensional frequency parameter |
| $\omega$ | circular frequency |

# VIBRATION DAMPING

# 1

# FUNDAMENTAL CONCEPTS IN STRUCTURAL DYNAMICS

## ADDITIONAL SYMBOLS

| | |
|---|---|
| $a$ | crack length |
| $A_1 \cdots A_4$ | constants of integration |
| $B_1 \cdots B_4$ | constants of integration |
| $\dfrac{da}{dN}$ | crack growth rate |
| $f(x)$ | sectionally continuous function |
| $F_D$ | force on structure from point damper (amplitude) |
| $F_1 \cdots F_n$ | discrete exciting forces |
| $\vec{F}(t)$ | force vector |
| $F_x, F_y, F_z$ | Cartesian components of force vector (amplitudes) |
| $k_1 \cdots k_n$ | discrete stiffness elements |
| $l$ | position of internal boundary ($l < L$) |
| $m_1 \cdots m_n$ | discrete mass elements |
| $M_{mn}$ | modal mass |
| $\vec{M}(t)$ | moment vector |
| $M_x, M_y, M_z$ | Cartesian components of moment vector (amplitudes) |

| | |
|---|---|
| $n$ | mode number |
| $T$ | time interval |
| $\vec{w}(t)$ | displacement vector |
| $w'$ | $dw/dx$ |
| $w''$ | $d^2w/dx^2$ |
| $w'''$ | $d^3w/dx^3$ |
| $\dot{w}$ | velocity ($dw/dt$) |
| $\ddot{w}$ | acceleration ($d^2w/dt^2$) |
| $W_D$ | amplitude of direct component of displacement |
| $W_Q$ | amplitude of quadrature component of displacement |
| $W_x, W_y, W_z$ | Cartesian components of response vector (amplitudes) |
| $x_j, y_j, z_j$ | coordinates of point $j$ ($j = 1, 2, \ldots$) |
| $\alpha_1, \alpha_2$ | nondimensional functions |
| $\alpha, \beta$ | nondimensional parameters for damped beam |
| $\alpha_{12}(\omega)$ | transfer receptance between points 1 and 2 |
| $\delta(x)$ | Dirac delta function of argument $x$ |
| $\delta_{nm}$ | Kronecker delta |
| $|\Delta|$ | determinant |
| $\Delta K$ | stress intensity factor |
| $\phi_n$ | $n$th modal function |
| $\eta, \eta_1, \eta_2$ | loss factors |
| $\kappa_{12}(\omega)$ | transfer dynamic stiffness between points 1 and 2 |
| $\sigma_D$ | stress amplitude (fatigue load) |
| $\sigma_s$ | static stress (fatigue load) |
| $\xi_n$ | eigenvalues ($n = 1, 2, \ldots, \infty$) |
| $\omega_D$ | frequency of oscillatory stress load |
| $\omega_j$ | resonant frequency ($j = 1, 2, \ldots$) |

## 1.1.  INTRODUCTION

In this book we attempt to describe the fundamentals of the theory, prediction, and practice of vibration control by means of viscoelastic materials. We present theory in sufficient detail, and with sufficient reference material, for the determined reader to be able to understand many of the main developments of the past several decades. Those readers seeking to apply their understanding to the solution of practical problems in engineering will have a framework on which to base their efforts. Although the theory may not be directly applicable to all the complex problems arising in the real world, it does always serve as a guide for understanding the results of experimental or semiempirical approaches, thereby eliminating many needless cut and try cycles.

## 1.2.  METHODS OF PREDICTING RESPONSE

### 1.2.1.  Introductory Remarks

A necessary prerequisite to the control of the vibrations of a mechanical system is an understanding of the detailed dynamic behavior of the system under excitation by forces at various points. Many approaches toward this task have been taken, including direct measurement of the required information, mathematical modeling and exact solution of the resulting partial differential equation(s) of motion, discrete finite element modeling and solution of the resulting large array of second-order differential equations, energy methods, and the joining together of solutions corresponding to subsets of the total system. All these approaches have advantages and disadvantages, and no one method is always best. The approach should be tailored to the resources and time available and the experience and skills of the individuals facing each specific task, for no mathematical or experimental methods are better than the people using them. This caution may be more important than ever in a time when computers can be used to perform rapidly prodigious calculations because ill-conceived or erroneous results can be achieved with as little or as much effort as well-conceived and accurate ones. There is just no substitute for care, common sense, and the checking of solutions in as many ways as possible. Dynamic analysis is no area for blind trust in anything, whether it be the infallibility of humans or machines! So in this section we shall briefly explore the area of dynamic response of structures, highlighting the features of several methods and some of the main theorems on which the science is based.

Real structures and machines are made up of components possessing finite levels of rigidity and mass, and nonperfect energy transfer characteristics. As a result the imposition of external or internal loads, during operation of the structure or machine, will lead to nonzero deformations at all times and will under certain conditions lead to very high amplitude vibrations at specific frequencies, or to instabilities of the static or dynamic deformations. It is very important in modern engineering practice to be able accurately to predict these deflections, instabilities, or high amplitude vibrations and to exercise some optimization during design and manufacture so that control can be exercised over the static and/or dynamic stress levels, response amplitude levels, and transmitted or radiated noise levels in accordance with the needs of the particular application.

Consider the conceptual structure shown in Figure 1.1. A general three-dimensional structure can be characterized by physical properties such as the Young's modulus, shear modulus, bulk modulus, and mass distribution. If a force vector $\vec{F}(t)$ is applied at an arbitrary point $1(x_1, y_1, z_1)$, it will generate a response vector $\vec{w}$ at an arbitrary point $2(x_2, y_2, z_2)$. The magnitude of $\vec{w}$ will be proportional to the magnitude of $\vec{F}$ for a linear system, but the direction of $\vec{w}$ will depend on the physical properties of the structure and the three components of $\vec{F}(F_x, F_y, F_z)$. Similarly a moment $\vec{M}(t)$, comprising three components

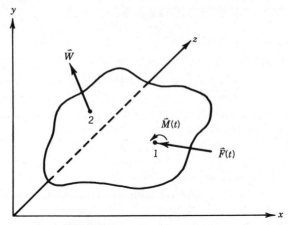

**FIGURE 1.1.**   Loads on structure.

$M_x$, $M_y$, $M_z$, will have a corresponding response vector $\vec{w}$. The principle of superposition applies for linear systems so that responses from two or more inputs may be added vectorially. The usual aim of vibration analysis is to predict $\vec{w}$ given $\vec{F}$ and/or $\vec{M}$. For stationary structures, which do not rotate, the response will always be finite for finite values of $\vec{F}$ or $\vec{M}$ except that if $\vec{F}$ or $\vec{M}$ contain harmonic components of the form $\sin(\omega t + \varepsilon)$, then at certain frequencies the mass and stiffness characteristics of the structure exactly cancel out each other's influence, and the amplitude of the harmonic response $\vec{w}(t)$ increases, ultimately reaching infinity unless other factors come into play to limit the amplitude. Such factors include nonlinear constitutive behavior i.e., nonlinear relationships between stress and strain, failure of the structure—another type of nonlinearity—or finite levels of energy dissipation, at one or more points in the structure, which increase with the second or greater power of the amplitude, the *damping*. For low levels of damping the specific frequencies at which high resonant amplitudes occur are called the natural frequencies of the structure, and these frequencies and the corresponding distributions of amplitude are global properties, which do not depend on the exact point(s) at which $\vec{F}$ (or $\vec{M}$) are applied. The amplitude distributions at the natural frequencies are known as the natural modes of vibration. Once the natural frequencies and modes of vibration are known, it is possible to use this information to solve many problems of forced ($\vec{F} \neq 0$) or free ($\vec{F} = 0$) vibration. Referring again to Figure 1.1, if a force $\vec{F}(t)$, of form $\vec{F} = F_x \sin \omega t$, is applied at any point 1, then the response vector $\vec{w}$ at any point 2 will typically look something like the sketch in Figure 1.2. The peak amplitudes at frequencies $\omega_1$, $\omega_2, \ldots$ will be infinite in the absence of damping but in reality will always be finite, even if usually large. The frequencies $\omega_{\text{I}}$, $\omega_{\text{III}}, \ldots$ correspond to zero response, where the point 2 is an "antiresonance point." Points such as $\omega_{\text{II}}$ also occur, for which the response is a minimum but does not reach zero. Each specific pattern of behavior depends on the points 1 and 2, the structure

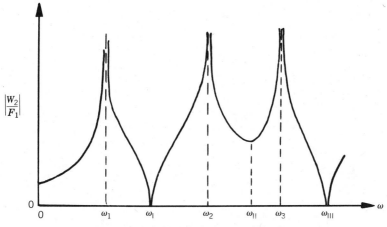

**FIGURE 1.2.** Typical response behavior.

geometry, and its mass and stiffness characteristics. For some structures the natural frequencies will be well separated, whereas for others they will be very close, as illustrated in Figure 1.3.

If a structure is rotating, as, for example, the rotor in a turbine engine illustrated schematically in Figure 1.4., other forces come into play to modify the behavior, sometimes drastically. These are related to centrifugal and/or Coriolis type accelerations and can modify not only modes of vibration and natural frequencies but also lead to instabilities such as whirl of rotating shafts. These instabilities occur when the forces and moments arising from these rotary accelerations feed energy from the motive power which drives the system into one or more of the natural modes of vibration. Such instabilities are very destructive when they occur, since there is no way of limiting the amplitudes. For this reason dynamics of rotating structures is a discipline subject to intensive study.

Another type of instability that can occur in rotating or stationary structures is known as flutter. This occurs when some relative movement takes place between the structure and its surrounding medium (e.g., air). The presence of vibration in the structure then causes disturbances of the medium, and under some conditions these disturbances give rise to forces back on the structure which reinforce the vibration that started them so that the structure becomes self-excited.

### 1.2.2.  Direct Classical Method of Predicting Response

The classical methods of predicting dynamic response of systems have adopted the viewpoint that one should first obtain a differential equation of motion (exact within the limits of the original physical assumptions) and then obtain an exact mathematical solution [1.1–1.10]. Clearly this is possible only in a

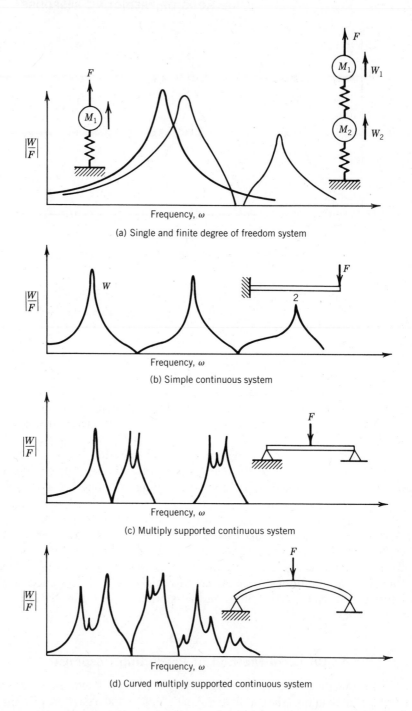

(a) Single and finite degree of freedom system

(b) Simple continuous system

(c) Multiply supported continuous system

(d) Curved multiply supported continuous system

**FIGURE 1.3.** Response of typical types of structures.

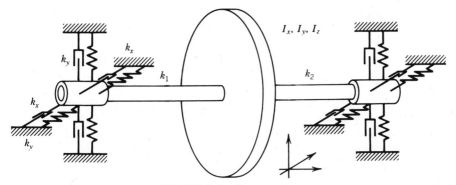

**FIGURE 1.4.** Rotor system.

limited number of cases, so today the most useful function of the classical methods is to give physical insight into what is happening and also a standard of comparison for checking out the currently more fashionable and useful discrete methods. No analyst should venture to utilize modern finite element techniques without feeling the occasional need to check his or her models for accuracy, stability, uniqueness, and reasonableness. Far too many errors are due simply to neglect of this necessary task of constant self-checking. Figure 1.5 shows just a few of the configurations for which classical methods work well. We shall address only one example, namely the pinned-pinned beam. The well-known Euler–Bernoulli equation of motion for the beam is

$$EI \frac{\partial^4 w}{\partial x^4} + \rho bH \frac{\partial^2 w}{\partial t^2} = F(x, t) \tag{1.1}$$

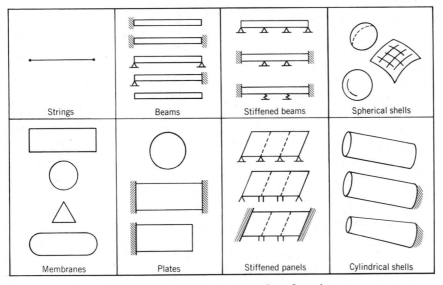

**FIGURE 1.5.** Classical structural configurations.

with the boundary conditions

$$w\left(\pm\frac{L}{2}\right) = w''\left(\pm\frac{L}{2}\right) = 0 \tag{1.2}$$

where $w'' = d^2w/dx^2$, $x$ is measured from the center of the beam, and $L$ is the length of the beam. The direct solution of equation (1.1) is obtained by adding a particular solution to the general solution of the homogeneous equation ($F = 0$) and imposing the boundary conditions. For example, if $F(x, t) = Fe^{i\omega t}$, it can readily be shown that

$$w(x, t) = \frac{FL^4}{2EI\xi^4}\left[\frac{\cos\left(\dfrac{\xi x}{L}\right)}{\cos\left(\dfrac{\xi}{2}\right)} + \frac{\cosh\left(\dfrac{\xi x}{L}\right)}{\cosh\left(\dfrac{\xi}{2}\right)} - 2\right]e^{i\omega t} \tag{1.3}$$

where

$$\xi^4 = \rho b H \frac{\omega^2 L^4}{EI} \tag{1.4}$$

Clearly equation (1.3) represents an exact, closed form, solution. Errors may arise from the fact that several approximations are inherent in the Euler–Bernoulli equation, such as the neglect of rotatory inertia and shear deflection, both of which would have to be allowed for if the beam were thick relative to its length ($H/L > 0.1$ or so) and/or if one were to consider response far beyond the first few modes of vibration, but no approximations are involved in the mathematical analysis itself. Similar results, albeit more complicated, can be obtained for many, though not all, of the configurations shown in Figure 1.5, and the many other configurations studied over the past several centuries. The ability to obtain closed form solutions to ideal problems is very important, but unfortunately it is neither possible nor cost-effective to obtain such solutions for real-world problems. One reason is the general refusal of real problems to fit into ideal categories; circular plates invariably seem to have holes in them at odd points, for example. Another is that the mathematical functions needed to describe the solution are often neither well tabulated nor available to a sufficient degree of precision in computer subroutines, or the numerical values of the mathematical functions grow too large for computers to handle with precision. For example, in equation (1.3), the $\cosh(\xi x/L)$ and $\cosh(\xi/2)$ functions become very large indeed as $\xi$ exceeds 10, and this means that equation (1.3) can in real life deal with only relatively few resonant modes before most unusual and expensive measures must be taken to rescale the problem. It can be done of course, but other methods eventually become equally or more attractive, even when exact solutions are possible.

Another problem arising in the direct classical approach is the modeling of damping. If the structure is made up of homogeneous material, one approach is to replace $E$ in equation (1.1) by a complex modulus $E(1 + i\eta)$ [1.11–1.13], as will be discussed in Chapter 2, but this is effective only if the material has linear damping characteristics, which may or may not be functions of frequency. If damping is applied at a joint, support, bearing, or other assembly, one must introduce damping forces and/or moments which are obtained from experiment or analytical methods. This particular difficulty really has nothing whatever to do with the classical methods of solution but is equally a problem for all methods of solution. For complex modulus damping of a pinned-pinned beam, the equation of motion now becomes

$$EI(1 + i\eta)\frac{\partial^4 w}{\partial x^4} + \rho b H \frac{\partial^2 w}{\partial t^2} = F e^{i\omega t} \tag{1.5}$$

with, as before,

$$w\left(\pm\frac{L}{2}\right) = w''\left(\pm\frac{L}{2}\right) = 0 \tag{1.6}$$

The solution becomes, by direct analogy with equation (1.3),

$$w(x, t) = \frac{FL^4}{2EI\xi^4}\left[\frac{\cos\{(\xi x/L)(1 + i\eta)^{-1/4}\}}{\cos\{(\xi/2)(1 + i\eta)^{-1/4}\}} + \frac{\cosh\{(\xi x/L)(1 + i\eta)^{-1/4}\}}{\cosh\{(\xi/2)(1 + i\eta)^{-1/4}\}} - 2\right]e^{i\omega t} \tag{1.7}$$

One difficulty now becomes immediately apparent, namely the fact that we must determine circular and hyperbolic functions of complex arguments. This is not unduly difficult for this particular problem but is certainly difficult in many other cases. For example, circular plates require Bessel functions to describe the solution, and Bessel functions of complex arguments do not make for elementary mathematical and computer operations. At any rate it is clearly possible to obtain exact solutions to some ideal problems, and one should not minimize the importance of this. When one performs the necessary algebraic manipulations, equation (1.7) can be written in the form

$$w(x, t) = \frac{FL^4}{2EI\xi^4}(W_D + iW_Q)e^{i\omega t} \tag{1.8}$$

where

$$W_D = \frac{AC + BD}{C^2 + D^2} + \frac{EG + FH}{G^2 + H^2} - 2 \tag{1.9}$$

$$W_Q = \frac{AD - BC}{C^2 + D^2} + \frac{FG - EH}{G^2 + H^2} \tag{1.10}$$

and $A$, $B$, $C$, and so forth, are simple expressions involving circular and hyperbolic functions of $\alpha = \xi(1 + \eta^2)^{-1/8} \cos \varepsilon/2$ and $\beta = \xi(1 + \eta^2)^{-1/8} \sin \varepsilon/2$, with $\tan \varepsilon = \eta$, namely

$$A = \cos\left(\frac{\alpha x}{L}\right) \cosh\left(\frac{\beta x}{L}\right)$$

$$B = \sin\left(\frac{\alpha x}{L}\right) \sinh\left(\frac{\beta x}{L}\right)$$

$$C = \cos\left(\frac{\alpha}{2}\right) \cosh\left(\frac{\beta}{2}\right)$$

$$D = \sin\left(\frac{\alpha}{2}\right) \sinh\left(\frac{\beta}{2}\right)$$

$$E = \cosh\left(\frac{\alpha x}{L}\right) \cos\left(\frac{\beta x}{L}\right)$$

$$F = \sinh\left(\frac{\alpha x}{L}\right) \sin\left(\frac{\beta x}{L}\right)$$

$$G = \cosh\left(\frac{\alpha}{2}\right) \cos\left(\frac{\beta}{2}\right)$$

$$H = \sinh\left(\frac{\alpha}{2}\right) \sin\left(\frac{\beta}{2}\right)$$

If one instead introduces the damping at a particular point on the structure (already this is another idealization), then as far as the classical direct method is concerned, one has just introduced another set of boundary conditions. If, in the beam considered hitherto, this point is at $x = l$, then the new boundary conditions for equation (1.1), which is not changed, are

$$w\left(\pm \frac{L}{2}\right) = w''\left(\pm \frac{L}{2}\right) = 0 \tag{1.11}$$

$$\lim_{\Delta \to 0} w(l - \Delta) = \lim_{\Delta \to 0} w(l + \Delta) \tag{1.12}$$

$$\lim_{\Delta \to 0} w'(l - \Delta) = \lim_{\Delta \to 0} w'(l + \Delta) \tag{1.13}$$

$$\lim_{\Delta \to 0} w''(l - \Delta) = \lim_{\Delta \to 0} w''(l + \Delta) \tag{1.14}$$

$$\lim_{\Delta \to 0} EIw'''(l + \Delta) = \lim_{\Delta \to 0} EIw'''(l - \Delta) + F_D \tag{1.15}$$

where $F_D$ is the amplitude of the harmonic force due to the damping mechanism at the point $x = l$, which is opposed by the change in shear force across the point of application of the damping force. For a damped, linear, complex spring $F_D = k(1 + i\eta)w(l)$, and such a description of the damping force would allow one to complete the set of boundary conditions in equations (1.11) to (1.15). Note that in this case eight boundary conditions arise and that four new boundary conditions are introduced for each new point damping device. The problem soon reaches the point where other methods of solution become more attractive.

### 1.2.3.  Classical Normal Mode Method

The classical normal mode method starts from the same differential equation of motion as the direct method, but differs in that an approximate solution is sought in terms of a truncated infinite series [1.14–1.18]. The only approximation lies in the truncation of the series, so that this method differs from the normal mode method as it applies to discrete models, where the equations of motion are also approximate or, more precisely, the physical model of the structure is approximated by a finite set of masses and stiffnesses describable by linear algebraic equations in the spatial variables rather than by differential equations. The method whereby the form of the infinite series is derived is basically similar to that used in the direct method. The solution of the homogeneous equation of motion is first obtained by letting $F(x, t) = 0$. As for the direct method, one assumes a solution of the form $w(x, t) = A_1 e^{\lambda x/L} e^{i\omega t}$ and seeks values of $\lambda$ for which $A_1 \neq 0$ (i.e., the solution is nontrivial). This can happen only if

$$\lambda^4 - \xi^4 = 0 \tag{1.16}$$

$$\lambda = \pm\xi \quad \text{or} \quad \pm i\xi \tag{1.17}$$

$$\therefore \quad W = A_1 e^{\xi x/L} + A_2 e^{-\xi x/L} + A_3 e^{i\xi x/L} + A_4 e^{-i\xi x/L} \tag{1.18}$$

Putting the boundary conditions $w(0) = w(L) = w'(0) = w'(L) = 0$, where $x$ is now measured from one end of the pinned-pinned beam, gives

$$\begin{bmatrix} 1 & 1 & 1 & 1 \\ e^{\xi} & e^{-\xi} & e^{i\xi} & e^{-i\xi} \\ 1 & 1 & -1 & -1 \\ e^{\xi} & e^{-\xi} & -e^{i\xi} & -e^{-i\xi} \end{bmatrix} \begin{Bmatrix} A_1 \\ A_2 \\ A_3 \\ A_4 \end{Bmatrix} = \begin{Bmatrix} 0 \\ 0 \\ 0 \\ 0 \end{Bmatrix} \tag{1.19}$$

This can only have a nontrivial solution if

$$|\Delta| = \begin{vmatrix} 1 & 1 & 1 & 1 \\ e^{\xi} & e^{-\xi} & e^{i\xi} & e^{-i\xi} \\ 1 & 1 & -1 & -1 \\ e^{\xi} & e^{-\xi} & -e^{i\xi} & -e^{-i\xi} \end{vmatrix} = 0 \tag{1.20}$$

Equally well, one could utilize the relationships between the exponential and hyperbolic functions and between the complex exponential and circular functions to write $W$ in the alternate form:

$$W = B_1 \sinh\left(\frac{\xi x}{L}\right) + B_2 \cosh\left(\frac{\xi x}{L}\right) + B_3 \sin\left(\frac{\xi x}{L}\right) + B_4 \cos\left(\frac{\xi x}{L}\right) \quad (1.21)$$

For this particular problem the boundary conditions are more readily introduced, and $|\Delta|$ now takes the form

$$|\Delta| = \begin{vmatrix} 0 & 1 & 0 & 1 \\ \sinh\xi & \cosh\xi & \sin\xi & \cos\xi \\ 0 & 1 & 0 & -1 \\ \sinh\xi & \cosh\xi & -\sin\xi & -\cos\xi \end{vmatrix} = 0 \quad (1.22)$$

Equations (1.20) and (1.22) give identical solutions, from which it can readily be shown that an infinite number of values of $\xi_n$ exist, for which $|\Delta| = 0$, namely

$$\xi_n = n\pi \quad (1.23)$$

and the corresponding expression for $W_n(x)$, normalized to give a maximum value of 1.0 is

$$\phi_n = \frac{W_n(x/L)}{W_n(x/L)_{\text{max}}} = \sin\left(\frac{n\pi x}{L}\right) \quad (1.24)$$

The discrete numbers $\xi_n$ are known as the eigenvalues and are directly related to the natural frequencies of the structure, and the functions $\phi_n(x/L)$ are known as the eigenfunctions or normal modes. Since they represent solutions of the undamped, homogeneous, equation, it is seen that any normal mode, once set in existence, will persist indefinitely at its natural frequency $\omega_n$.

Having defined the normal modes of a system as the discrete (infinite) set of functions that satisfy the homogeneous equation of motion, and the eigenvalues in terms of the set of discrete frequencies for which these motions can exist, we must now look at some useful properties of normal modes and eigenvalues. First, looking again at the equation of motion, we see that

$$\frac{d^4\phi_n}{d(x/L)^4} = \xi_n^4\phi_n \quad (1.25)$$

It can also be shown that

$$\int_0^L \rho b H \phi_n \phi_m \, dx = M_{nm}\delta_{nm} \quad (1.26)$$

where $\delta_{nm} = 0(n \neq m)$ or $1(n = m)$ and $\delta_{nm}$ is the Kronecker delta. These two properties are evident for the pinned-pinned beam, but they are also true for most nondissipative boundary conditions. Equation (1.26) expresses a particularly important property of the normal modes, namely that of orthogonality. $M_{nm}$ is known as a generalized mass since it has the dimensions of mass. Note also that discrete changes of mass or cross section, in the most general case, can be allowed for in equation (1.26).

One further property of the normal modes now completes the basis for the normal mode method. This is the fact that for any motion of the system such as described by equation (1.1) and boundary conditions (1.2), or any other set of equations and boundary conditions, the motion can always be expressed as an infinite series of the corresponding normal modes, that is,

$$w(x, t) = \sum_{n=1}^{\infty} \phi_n\left(\frac{x}{L}\right)[W_n e^{i\omega t} + W_n^* e^{i\omega_{nt}}] \tag{1.27}$$

Here the term $W_n e^{i\omega t}$ represents the forced motion, and $W_n^* e^{i\omega_{nt}}$ the transient motion, often assumed to be damped out in steady state vibration analysis. If we now go back to equation (1.1), expand $W(x)$ as a series of the form described by equation (1.27), with $W_n^* \equiv 0$, and invoke equation (1.25) to eliminate the fourth-order spatial derivatives, then we can write

$$\sum_{n=1}^{\infty} \left(\frac{EI\xi_n^4}{L^4} - \rho H b\omega^2\right) W_n \phi_n e^{i\omega t} = F(x)e^{i\omega t} \tag{1.28}$$

Now factor equation (1.28) by $\phi_m(x/L)$, and integrate from 0 to $L$, to get the matrix equation

$$\begin{bmatrix} \alpha_1\left(\dfrac{EI\xi_1^4}{L^4} - \rho b H\omega^2\right) & 0 & 0 & \cdots \\ 0 & \alpha_2\left(\dfrac{EI\xi_2^4}{L^4} - \rho b H\omega^2\right) & 0 & \cdots \\ 0 & 0 & 0 & \cdots \\ 0 & 0 & 0 & \cdots \\ \vdots & \vdots & & \vdots \\ 0 & 0 & \alpha_n\left(\dfrac{EI\xi_n^4}{L^4} - \rho b H\omega^2\right) \end{bmatrix} \begin{Bmatrix} W_1 \\ W_2 \\ \vdots \\ W_n \end{Bmatrix}$$

$$= \begin{Bmatrix} F_1 \\ F_2 \\ \vdots \\ F_n \end{Bmatrix} \tag{1.29}$$

which can be written in the simpler form

$$[A_{ij}\delta_{ij}]\{W_j\} = \{F_j\} \tag{1.30}$$

where $\alpha_n = \int_0^L \phi_n^2 \, dx$ and $F_n = \int_0^L F(x)\phi_n(x) \, dx$, and $A_{ij} = \alpha_i(EI\xi_j^4/L^4 - \rho bH\omega^2)$. Note that the use of normal modes at once diagonalizes the matrix describing the solution, so that the $n$th term of the series can be obtained directly:

$$w = \sum_{n=1}^{\infty} \frac{F_n L^4 \phi_n e^{i\omega t}}{EI(\xi_n^4 - \rho bH\omega^2 L^4/EI)\alpha_n} \tag{1.31}$$

For the pinned-pinned beam $\xi_n = \pi n$ and $F(x) = F$, so

$$\alpha_n = \int_0^L \sin^2\left(\frac{n\pi x}{L}\right) dx = \frac{L}{2}$$

$$F_n = \int_0^L F(x) \sin\left(\frac{n\pi x}{L}\right) dx = \frac{2F}{\pi n}$$

$$\therefore \quad w = \frac{4FL^4}{\pi EI} \sum_{n=1}^{\infty} \frac{\sin(n\pi x/L)e^{i\omega t}}{n(n^4\pi^4 - \xi^4)} \tag{1.32}$$

Equations (1.31) and (1.32) represent exact solutions in the limit as $n \to \infty$ but are approximate in the sense that, in practice, one must truncate the series. Clearly the point at which truncation is effected will depend on how many modes of vibration are of interest. It can be shown analytically and numerically that solutions (1.3) and (1.32) are identical. Figure 1.6 shows a comparison between equations (1.3) and (1.32) when truncated at 1, 3, and 5 terms, demonstrating how the series solution improves as more terms are added.

Introducing linear damping through the normal mode method is now quite simple. For example, if we wish to replace $E$ by $E(1 + i\eta)$ in the equation of motion, as before, nothing in the solution process is changed, and $E$ can be replaced by $E(1 + i\eta)$ wherever it appears in the series and at any stage. The solution (1.31) therefore becomes

$$w = \sum_{n=1}^{\infty} \frac{F_n L^4 \phi_n e^{i\omega t}}{EI[(1 + i\eta)\xi_n^4 - \rho bH\omega^2 L^4/EI]\alpha_n} \tag{1.33}$$

which is much simpler than equation (1.7), though some manipulations are necessary to calculate the real (or direct) and imaginary (or quadrature) parts of $W$, namely $W_D$ and $W_Q$. Even when the damping is applied at a point, such as a complex spring $k(1 + i\eta)$ linked to ground at $x = l$, as considered before,

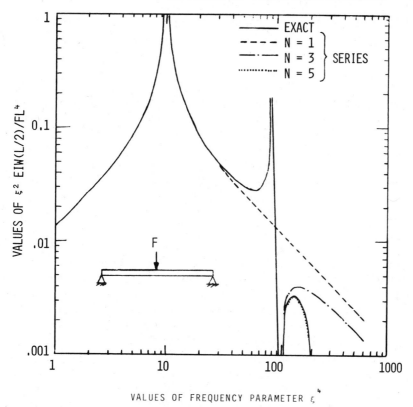

**FIGURE 1.6.**   Response of pinned-pinned beam: exact and series solutions.

the solution is not too difficult to formulate. In this case the equation of motion is written

$$EI \frac{d^2 W}{dx^4} - \rho b H \omega^2 W = F e^{i\omega t} - k(1 + i\eta) W \delta(x - l) \qquad (1.34)$$

where $\delta(x - l)$ is the Dirac delta function, being zero when $x \neq l$, infinite at $x = l$, but having the finite property

$$\int_0^L \delta(x - l) f(x) \, dx = f(l) \qquad (1.35)$$

for any sectionally continuous finite function $f(x)$. Since the mode shapes $\phi_n$ satisfy these conditions, the same procedure as before leads to the matrix equation

$$[B_{ij} + A_{ij}\delta_{ij}]\{W_j\} = \{F_j\} \qquad (1.36)$$

with

$$B_{ij} = k(1 + i\eta)\phi_i\left(\frac{l}{L}\right)\phi_j\left(\frac{l}{L}\right) \quad \text{and} \quad A_{ij} = \alpha_i\left(\frac{EI\xi_j^4}{L^4} - \rho b H\omega^2\right)$$

Note that in this case the matrix is complex and nondiagonal, although the effect of the off-diagonal terms will often be small in comparison with the diagonal terms. Again the series must be truncated, but now the most efficient method of solution would be to use a computer code to solve the set of complex matrix equations. We shall not do this here since our aim is to illustrate what can and cannot be done, rather than solve this particular problem in detail. One important consideration to note is that the order of the problem has not been increased as a result of allowing for this discrete term at $x = l$, which is in contrast to what happens for the direct method. The ease with which these solutions can be obtained illustrates the important fact that it is not the mathematical analysis that makes it difficult to allow for off-diagonal terms in the solution matrix (through this certainly can happen on occasion) but rather that in many problems neither the mechanism of the damping nor the points of application are well known, and working back from measurements of modal response to create a model of the physical mechanisms (the exact opposite of the process described earlier) is extremely difficult at best and impossible at worst. For many of the elastomeric, polymeric, or vitreous materials discussed in this book, however, a rational quantitative mathematical description is not only possible but has been accomplished, and it can be used for prediction of the effects of damping treatments or damping devices, using such materials, on structural response, noise transmission, or noise emission. The same can be said for some nonlinear damping systems, such as high damping metals or dry friction damping, although the mathematical difficulties are greatly increased as a result of the nonlinearity.

### 1.2.4.  Discrete Methods

For relatively simple structures, in the first few modes of vibration, it is quite possible to model the response in terms of a mass-spring system having relatively few degrees of freedom, as illustrated in Figure 1.7a [1.19–1.24]. There is an advantage in this if one intends to use the model so created for other purposes, such as a substructure in a larger calculation, or for determining the qualitative effect of varying some parameters of interest. However, substantiation of the validity of the model depends on the acquisition of supporting data, such as the results of exact analyses, finite element analyses, or experiments. To illustrate the approach, let us examine once again the pinned-pinned beam and consider its modeling in terms of a three degree of freedom system, as illustrated in Figure 1.7b. As far as correctly modeling the system is concerned, many points of view have been expressed, but one way or another one must determine values for the flexural stiffnesses $k_1 \cdots k_5$, $m_1 \cdots m_3$, and $F_1$ through $F_3$ which best reproduce the response of the actual structure in the first three modes of vibration. If the mass of the continuous beam is divided

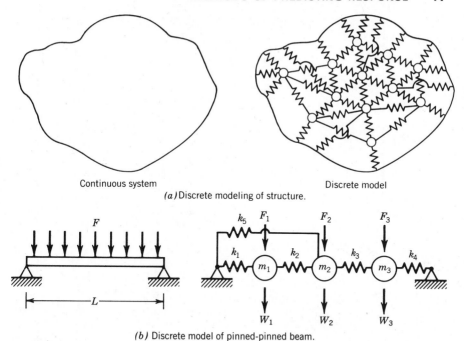

Continuous system                    Discrete model

*(a)* Discrete modeling of structure.

*(b)* Discrete model of pinned-pinned beam.

**FIGURE 1.7.**  Discrete modeling. (*a*) Discrete modeling of structure. (*b*) Discrete model of pinned-pinned beam.

into three equal parts, then $k_1$ through $k_4$ remain difficult to specify because the boundary conditions for each subelement are not identical. One might be able to model the system quite well by making some arbitrary, if reasonable, estimates of $k_1$ to $k_4$ based on the length of each segment into which the continuous beam was divided. For example, one might let $m_1 = m_2 = m_3 = \rho b H L / 4$ and $k_1 = k_2 = k_3 = k_4 = 3EI/(L/4)^3$ and $k_5 = 0$, based on an arbitrary splitting of the beam into four parts. It remains true, however, that the accuracy of this particular identification of parameters can be verified only by comparing natural frequencies, mode shapes, and forced response of the model with that of the original structure, and this is not easily done for a very complicated structure. In general, it is probably best to let $m_1$, $m_2$, $m_3$, $k_1$ through $k_5$ be arbitrary and adjust them until one obtains the best fit with whatever analytical or experimental data are available for the particular structure being modeled. The equations of motion of the discrete system can be written in the form

$$
\begin{bmatrix}
-m_1\omega^2 + k_1 + k_2 & -k_2 & 0 \\
-k_2 & -m_2\omega^2 + k_2 + k_3 + k_5 & -k_3 \\
0 & -k_3 & -m_3\omega^2 + k_3 + k_4
\end{bmatrix}
\begin{Bmatrix} W_1 \\ W_2 \\ W_3 \end{Bmatrix}
$$

$$
= \begin{Bmatrix} F_1 \\ F_2 \\ F_3 \end{Bmatrix} \quad (1.37)
$$

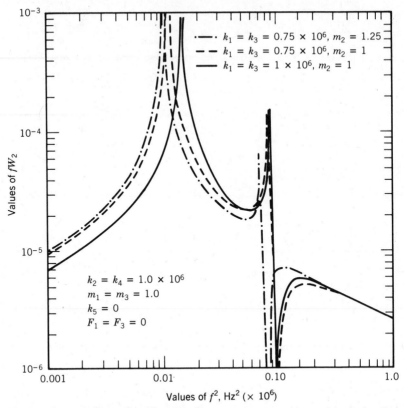

**FIGURE 1.8.**   Predicted response of discrete structure representing model of pinned-pinned beam.

Note that the mass-stiffness matrix is not diagonal although it can be diagonalized by proper matrix algebra operations. These equations are readily solved for $W_1$, $W_2$, and $W_3$. One must then introduce specific numerical values of $m_1$ through $m_3$ and $k_1$ through $k_5$ and seek those values that best match the exact solution at the points $x = L/4$, $L/2$, and $3L/4$. Figure 1.8 shows some comparisons between the exact and approximate solutions for specific values of the mass and stiffness parameters. It is seen that though a perfect fit is difficult to achieve, the first one or two modes can be modeled with some measure of accuracy. In this way one can seek guidelines for choosing the best values of the parameters and hence "calibrate" the model.

**Introduction of damping.**   Damping by a point damper at points 1, 2, or 3 can be added simply as another element in the equations of motion. For example, if a viscoelastic spring $k(1 + i\eta)$ is connected between mass $m_1$ and

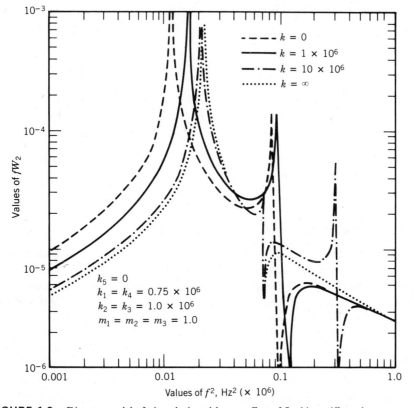

**FIGURE 1.9.**   Discrete model of pinned-pinned beam; effect of flanking stiffness $k$ on response.

ground, an additional force $k(1 + i\eta)w_1$ acts on $m_1$, and the matrix equation becomes

$$\begin{bmatrix} -m_1\omega^2 + k_1 + k_2 + k(1 + i\eta) & -k_2 & 0 \\ -k_2 & -m_2\omega^2 + k_2 + k_3 + k_5 & -k_3 \\ 0 & -k_3 & -m_3\omega^2 + k_3 + k_4 \end{bmatrix}$$

$$\times \begin{Bmatrix} W_1 \\ W_2 \\ W_3 \end{Bmatrix} = \begin{Bmatrix} F_1 \\ F_2 \\ F_3 \end{Bmatrix} \quad (1.38)$$

This can again be solved without difficulty. The stiffness $k$ will change the natural frequencies and modes of vibration, but no change is involved in the method of solution, or in the order of the problem. Figure 1.9 illustrates some effects of changing $k$ on response.

If, on the other hand, the damping is distributed throughout the structure, so that $k_1, k_2, \ldots$ are all replaced by complex stiffnesses, then equation (1.38) becomes instead

$$\begin{bmatrix} \kappa_{11} & -k_2(1+i\eta) & 0 \\ -k_2(1+i\eta) & \kappa_{22} & -k_3(1+i\eta) \\ 0 & -k_3(1+i\eta) & \kappa_{33} \end{bmatrix} \begin{Bmatrix} W_1 \\ W_2 \\ W_3 \end{Bmatrix} = \begin{Bmatrix} F_1 \\ F_2 \\ F_3 \end{Bmatrix} \tag{1.39}$$

with $\kappa_{11} = (k_1 + k_2)(1+i\eta) - m_1\omega^2$, $\kappa_{22} = (k_2 + k_3 + k_5)(1+i\eta) - m_2\omega^2$, and $\kappa_{33} = (k_3 + k_4)(1+i\eta) - m_3\omega^2$. This can be solved just as readily. The modes for equation (1.38) are complex, whereas those for the distributed damping case, equation (1.39), are real.

### 1.2.5. Receptance/Impedance Methods

Receptance methods of predicting structural response have been described by various users as "impedance methods," "mobility methods," "compliance methods," and so on, but the common underlying principle far outweighs differences of interpretation or specific technique [1.25–1.29]. Essentially the receptance techniques do not start from an equation of motion as such but utilize solutions of certain problems, whether obtained by classical, discrete, or experimental methods, in combination, to solve other problems. Essentially for any structural system, as illustrated in Figure 1.10, having any boundary conditions within some limits, the response vector at any point 1 resulting from any force vector at point 2 can be determined as a function of the frequency $\omega$, by analytical or experimental methods, as just stated, that is,

$$\vec{W}_1 = \vec{\alpha}_{12}(\omega) \wedge \vec{F}_2 \tag{1.40}$$

where $\vec{\alpha}_{12} \wedge \vec{F}_2$ ($\wedge$ is a symbol denoting the vector product) is the vector product of $\vec{F}_2(F_i, F_j, F_k)$ and $\vec{\alpha}_{12}(\alpha_{12i}, \alpha_{12j}, \alpha_{12k})$ and $i, j, k$ are three orthogonal axes in space. For a simple plane structure, such as a beam or plate, $\vec{W}$, $\vec{F}$, and $\alpha_{12}$ can be treated as scalars, because only one direction is important, so that

$$W_1 = \alpha_{12}(\omega)F_2 \tag{1.41}$$

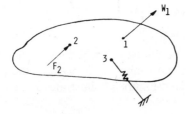

**FIGURE 1.10.** Structure with loads and displacements.

What is important in the receptance method is that $\alpha_{12}$ encompasses all the boundary conditions (internal and external) once and for all, and the response of the system to more than one force can be determined by simple summation, as long as the system is linear; that is, the response to forces $F_3$, $F_4$,... at points 3, 4,... will be, by superposition,

$$W_1 = \alpha_{12}(\omega)F_2 + \alpha_{13}(\omega)F_3 + \alpha_{14}(\omega)F_4 + \cdots \tag{1.42}$$

To illustrate the application of the receptance method, let us once more return to the single span pinned-pinned beam discussed earlier. If a force $F_2 e^{i\omega t}$ is applied at point $2(x = x_2)$ and the response is required at point $1(x = x_1)$, then application of the normal mode method can be shown to give the receptance or compliance:

$$\alpha_{12} = \frac{\partial W(x_1)}{\partial F_2} = \frac{2L^3}{EI} \sum_{n=1}^{\infty} \frac{\sin(n\pi x_1/L) \sin(n\pi x_2/L)}{\xi_n^4(1 + i\eta_n) - \xi^4} \tag{1.43}$$

$$= \sum_{n=1}^{\infty} \frac{\sin(n\pi x_1/L) \sin(n\pi x_2/L)}{M_{nn}[\omega_n^2(1 + i\eta_n) - \omega^2]}$$

where $\eta_n$ is the damping at each eigenvalue $\xi_n$, and is not necessarily constant, and $M_{nn} = \rho H b L/2$ is a modal mass parameter. If then a spring connection $k(1 + i\eta)$ to ground is added at another point 3, the additional force due to this influence is $F_3 = k(1 + i\eta)W_3$, and hence the response at point 1 is now

$$W_1 = \alpha_{12}F_2 + k(1 + i\eta)\alpha_{13}W_3 \tag{1.44}$$

Furthermore at point 3,

$$W_3 = \alpha_{32}F_2 + k(1 + i\eta)\alpha_{33}W_3$$

$$\therefore \quad W_3 = \frac{\alpha_{32}F_2}{1 + k(1 + i\eta)\alpha_{33}} \tag{1.45}$$

This can be substituted back in (1.44) to give

$$W_1 = \alpha_{12}F_2 + \frac{k(1 + i\eta)\alpha_{32}\alpha_{12}F_2}{1 + k(1 + i\eta)\alpha_{33}}$$

$$= \alpha_{12}\left[1 + \frac{k(1 + i\eta)\alpha_{13}}{1 + k(1 + i\eta)\alpha_{33}}\right]F_2 \tag{1.46}$$

Hence the effect of the element $k(1 + i\eta)$ can readily be calculated.

**Relationship between modal and discrete models.** In general, the response of a system can be described by the modal expression

$$\alpha_{12} = \sum_{n=1}^{\infty} \frac{\phi_n(x_1)\phi_n(x_2)}{M_{nn}[\omega_n^2(1 + i\eta_n) - \omega^2]} \qquad (1.47)$$

where $\phi_n(x_i)$ is the $n$th modal function at point $x_i$, $M_{nn}$ is the $n$th modal mass, and $\omega_n$ is the $n$th resonant frequency. Usually the series must be truncated, and the modal expansion is often obtained from identification of measured response data. If accurate measured values of $\alpha_{12}$ are available, for all frequencies within the range $0 \leq \omega \leq \omega_{max}$, then from equation (1.47) at $\omega = \omega_n$

$$|\alpha_{12}|_{max} \doteq \frac{\phi_n(x_1)\phi_n(x_2)}{M_{nn}\omega_n^2\eta_n} \qquad (1.48)$$

$\omega_n$ is measured from the maxima of $|\alpha_{12}|$, and $\eta_n$ is measured by the half-power bandwidth method to be discussed in later chapters. Therefore $M_{nn}/\phi_n(x_1)\phi_n(x_2)$ can be calculated for all pairs of points 1, 2:

$$\frac{M_{nn}}{\phi_n(x_i)\phi_n(x_j)} = \frac{1}{\omega_n^2\eta_n|\alpha_{12}|_{max}} \qquad (1.49)$$

Since $M_{nn}$ is supposed to be constant for all points 1, 2, if one fixes point 2 and varies point 1 over the entire structure, one can find the point 3, for each $n$, for which $M_{nn}/\phi_n(x_1)\phi_n(x_2)$ is a minimum, and this is the point at which each mode is normalized, that is, $\phi_n(x_3) = 1$ and $x_1 = x_2 = x_3$.

$$\therefore \quad M_{nn} = \frac{1}{\omega_n^2\eta_n|\alpha_{33}|_{max}} \qquad (1.50)$$

With $M_{nn}$, $\eta_n$, $\omega_n$, and $\phi_n(x_i)$ known for each value of $n$ within the range $0 \leq w \leq \omega_{max}$, equation (1.47) completely specifies the response of the structure at any point 1 to an oscillating force at any point 2. It is sometimes convenient to replace this modal model by an equivalent discrete model. For example, if one is looking at the first two modes of the system, then only two discrete masses should be needed to model the system. A set of equations is now postulated for excitation at point 1 (mass $m_1$) as follows:

$$\kappa_{11}W_1 + \kappa_{12}W_2 = F_1 \qquad (1.51)$$

$$\kappa_{21}W_1 + \kappa_{22}W_2 = 0 \qquad (1.52)$$

These equations can be interpreted in terms of the unknown dynamic stiffnesses $\kappa_{11}$, through $\kappa_{22}$, and the known receptances $\alpha_{11} = W_1/F_1$ and

$\alpha_{21} = W_2/F_1$. Equally one can consider the same hypothetical system excited at point 2 by a force $F_2$:

$$\kappa_{11}W_1 + \kappa_{12}W_2 = 0 \tag{1.53}$$

$$\kappa_{21}W_1 + \kappa_{22}W_2 = F_2 \tag{1.54}$$

which relates $\kappa_{11}$ through $\kappa_{22}$ to the receptances $\alpha_{12} = W_1/F_2$ and $\alpha_{22} = W_2/F_2$. We therefore have four equations of the form

$$\begin{bmatrix} \alpha_{11} & \alpha_{21} & 0 & 0 \\ \alpha_{12} & \alpha_{22} & 0 & 0 \\ 0 & 0 & \alpha_{11} & \alpha_{21} \\ 0 & 0 & \alpha_{12} & a_{22} \end{bmatrix} \begin{Bmatrix} \kappa_{11} \\ \kappa_{12} \\ \kappa_{21} \\ \kappa_{22} \end{Bmatrix} = \begin{Bmatrix} 1 \\ 0 \\ 0 \\ 1 \end{Bmatrix} \tag{1.55}$$

from which $\kappa_{11}$, $\kappa_{12}$, $\kappa_{21}$, and $\kappa_{22}$ can be determined. Note that for a damped system $\kappa_{ij}$ and $\alpha_{ij}$ are complex, so one must compare real (direct) and imaginary (quadrature) parts to solve the problem completely. The behavior of $\kappa_{ij}$ with frequency for each value of $i,j$ allows one to construct the model.

### 1.2.6.  Transfer Matrix Methods

The third of the "exact" methods of solution is not usually regarded as a "classical" technique, yet it is exact in the same sense that solutions can be obtained to specific types of problems without any further approximations beyond the physical approximations involved in deriving the original equations of motion [1.30–1.39]. One limitation is that the structure being analyzed must be essentially one-dimensional. For example, the method is readily applicable to beams with many internal boundary conditions, such as beams on many supports, but is applicable to complex plate geometries only if one set of parallel boundaries is pinned, as shown in Figure 1.11. In this system, if the response in the $y$ direction can be represented adequately by a series of the form $\sum_{n=1}^{N} W_n(x) \sin(n\pi y/l)$, then the two-dimensional Euler–Bernoulli, or any

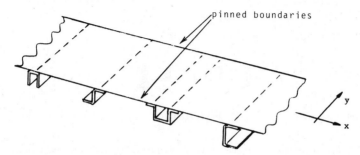

**FIGURE 1.11.**   One-dimensional skin-stringer structure.

other, plate equation can be reduced to one involving the $x$ and $t$ variables only. The essence of the transfer matrix method is to solve the equations of motion step by step using a matrix notation, regarding the values of displacement, slope, shear, and bending moment as variables to be determined simultaneously in the solution process, not separately as in the direct classical or normal mode methods. The transfer matrix technique is discussed further in Chapters 4 and 5.

### 1.2.7. Finite Element Methods

The true finite element methods differ from the lumped mass approach mainly in that when the structure is divided up, the element stiffnesses are determined from classical static stress analysis of the elements themselves rather than from a process of structural identification [1.40–1.46]. Figure 1.12a illustrates some

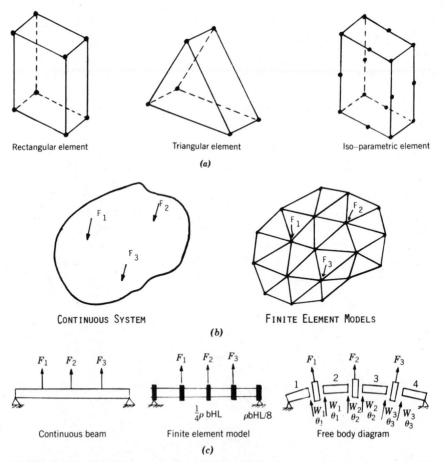

FIGURE 1.12. Finite element modeling. (a) Typical finite elements. (b) Finite element modeling of continuous structure. (c) Finite modeling of pinned-pinned beam.

of the types of elements which are commonly used. Each element is defined usually by 6, 8, 16, or 20 points, or "nodes," where compatibility of displacements and loads with adjacent elements is enforced. The motions of the nodes, in three dimensions, are the primary variables, and equations of motion are written down for these variables using usually some variational approach. Strain energies are calculated for the elements in terms of all the nodal displacements, and the mass is often lumped at the nodes, so the kinetic energy is readily described in terms of nodal velocities. Because the elements are formed directly from the structure geometry, the process of identifying stiffnesses is eliminated, and corresponding terms are calculated directly from the geometry of each element. A large number of nodes is required to represent adequately a complex structure, so the main tasks in finite element analysis are (1) to set up the "geometry file," that is, the element geometry matrix, and (2) to solve a large number of second-order differential equations (in the time domain) to predict the dynamic response. Many general and special programs have been developed in which the procedures for setting up the geometry file and solving the equations of motion are handled in an interactive manner with a computer.

## 1.3.  TECHNIQUES FOR VIBRATION CONTROL

### 1.3.1.  Introductory Remarks

In engineering practice, controlling vibration and noise in structures and machines is partly art and partly science. This is because, though one can in principle obtain from analysis or experiment the data needed to develop and optimize the appropriate control measures, in practice one is constrained by factors of time, equipment, and economics and is often obliged to make decisions concerning the control measures without having complete information. This means that guesses must be made, past experience must be drawn on, and less than optimum measures must be sought which do the job without necessarily being the most perfect solution. The discussion in this section will address several of the measures available to engineers for vibration and/or noise control, showing how they work and some of the basic parameters involved.

One of the least understood aspects of the design of a damping treatment is determination of how much damping is really needed to deal with a given problem. This question cannot be resolved unless there exists an understanding of how much damping there is in the original structure. The definition of this initial damping is very important, because all improvements are related to that value. To illustrate this matter, consider the difference in structural response between so-called "built-up" structures and welded or integrally machined structures.

Built-up structures are joined together or assembled by mechanical fasteners, such as rivets, bolts, and screws. Examples of such structures are the riveted skin-stringer structures used in aircraft fuselage constructions, and complex diesel engine blocks. Examples of integrally machined or welded structures are muffler shells and blades. Built-up structures usually have high initial structural damping, which could give a loss factor as high as 0.05. Such values are considerably higher than those found in welded or integrally machined structures because the joint damping is then minimal, and the measured structural loss factor is comparable to that for the material, on the order of $10^{-4}$ or $10^{-5}$ for steel or aluminum structures. Thus increasing the built-up structure damping by a factor of, say, 10 is considerably more difficult than for integrally machined or welded structures. Different damping treatments and approaches have to be taken for different applications, depending on the initial structural damping values.

### 1.3.2. Effects of Mass, Stiffness, and Damping

When a structure suffers from excessive vibration or noise, it is being subjected to undesirable excitations which can sometimes be measured. The excitation can be random or deterministic, or a combination of both, as illustrated in Figure 1.13a. The excitation signal in the time domain may be difficult to

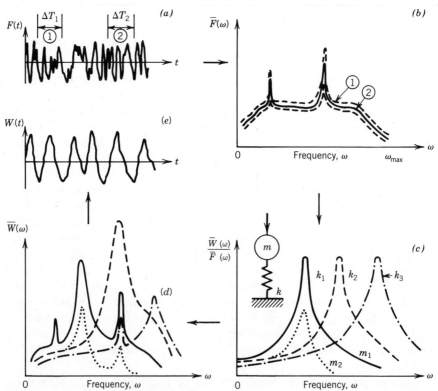

**FIGURE 1.13.** Random response behavior. (a) Excitation signal. (b) Spectrum of excitation. (c) Transfer function. (d) Response signal.

interpret, but if a Fourier transform is performed over a time window such as $\Delta T_1(\Delta T_1 > 1/\omega_{min})$, then the spectrum of the transformed signal might look as in Figure 1.13$b$, consisting of a broadband random part, along with some discrete periodic signals, such as those corresponding to blade passage frequencies. If the Fourier transform operation is repeated for several intervals $\Delta T_2$, $\Delta T_3, \ldots$, each spectrum will be different, but all will cluster around an average, and a root mean square (rms) spectrum level can be defined, for both the random and discrete parts, provided that the signal is stationary. If it is not stationary, then the spectrum will change with time, and analysis is difficult unless the change over each interval $\Delta T_n$ is small.

If the structure responds in one or more modes of vibration, as illustrated in Figure 1.13$c$, to any imposed excitation, then the combination of the receptance spectrum of the structure, which may itself be random for an array of similar structures, and the excitation spectrum, can give many possibilities as shown in the figure. For example, if the stiffness and mass characteristics are so chosen, the resonance peak may be right on top of one of the discrete excitation peaks, giving especially large responses. Figure 1.13$c$ shows possible effects of varying stiffness ($k_1 \rightarrow k_2 \rightarrow k_3$) and mass ($m_1 \rightarrow m_2$), that increasing stiffness can reduce response unless one moves the resonance to a peak of the excitation; increasing the mass can do the same thing. Increasing the damping reduces response near each resonance but cannot reduce the effect of the discrete excitation "spikes" unless the resonance frequency lies atop one of those spikes, which is not desirable anyway. Modifying the excitation to reduce the spikes and/or the broadband spectrum is effective in reducing response amplitudes but is not always easy to accomplish.

### 1.3.3.  Criteria for Damping

The usual approach in the application of damping treatments to structures is to optimize the system for maximum damping only. Such an approach, although correct from the point of view of damping optimization, neglects the fact that other modal parameters can change whenever a damping treatment is applied or incorporated into a structure. Therefore it is frequently essential to consider the variation of all three parameters—damping, mass, and stiffness—and try to optimize the damping treatment for all three parameters and not merely for one. Depending on the nature of the problem, and on the response of the structure, different parameters need to be optimized. This point will be illustrated for two cases, in one of which a structure is subjected to boundary excitation and in the other to force excitation. For simplicity, a single degree of freedom system will be considered. However, the argument is unchanged for complicated structures.

**Boundary excitation.**   Consider a structure excited at its boundary and vibrating in its fundamental mode of vibration. Such a structure can be represented by a single degree of freedom system as shown in Figure 1.14. The

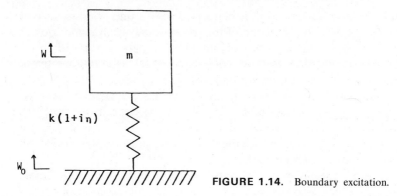

**FIGURE 1.14.** Boundary excitation.

equation of motion of such a system can be written in the form:

$$m\ddot{w} + k(1 + i\eta)(w - w_0) = 0 \qquad (1.56)$$

where $k$, $m$, and $\eta$ are the modal stiffness, mass, and damping values, respectively. The ratio of the response of the mass to the input vibration, $|W/W_0|$, is illustrated in Figures 1.15 and 1.16 for two different natural frequencies, $\omega_0 = \sqrt{k/m}$. The selected natural frequencies in the figure are $\omega_0 = 1$ rad/sec

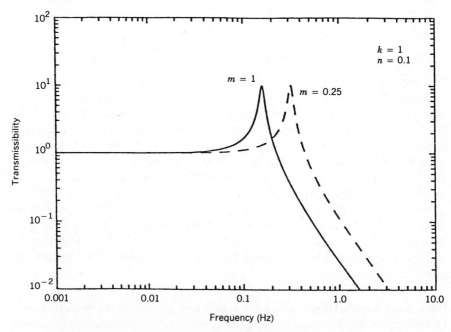

**FIGURE 1.15.** Effect of mass changes for the boundary excitation case.

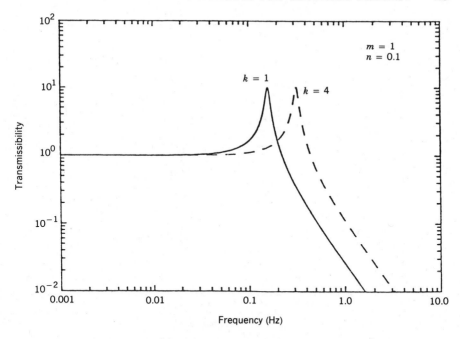

**FIGURE 1.16.**   Effect of stiffness changes for the boundary excitation case.

and $\omega_0 = 2$ rad/sec. These two cases illustrate the effect of two levels of stiffness and/or mass modifications for the structure, without change of the damping in the system. As usual it is possible to think of the plots in Figures 1.15 and 1.16 in terms of three different zones, one below the natural frequency, one at the natural frequency, and one above the natural frequency. The effect of changing the level of damping from a loss factor of 0.1 to 0.4 is illustrated in Figure 1.17. It is seen that damping helps only near the resonant zone and increases the amplitude above resonance. Below resonance the response is essentially unchanged. Figures 1.15 and 1.16 illustrate the effect of changes to the mass and stiffness, respectively. The two cases plotted in the figures represent different levels of stiffness and/or mass for the structure. Consider first the case for which $\omega_0 = 0.3$ Hz, assuming that this is the baseline structure and that a vibration reduction is being sought above the natural frequency. Lowering the natural frequency of the system to $\omega_0 = 0.15$ Hz, either by changing its mass or its stiffness will considerably reduce the high frequency response. Thus to separate the structure from the exciting force in rubberlike materials, we are using the fundamental frequency of the system to provide the necessary isolation at high frequency. It should be emphasized that lowering the natural frequency of the system could line it up with another discrete exciting force, so that undesirably high response could result for the modified system. Below the resonant frequency the mass and stiffness modifications result in virtually no improvements. This is because the mass and the foundation are vibrating in

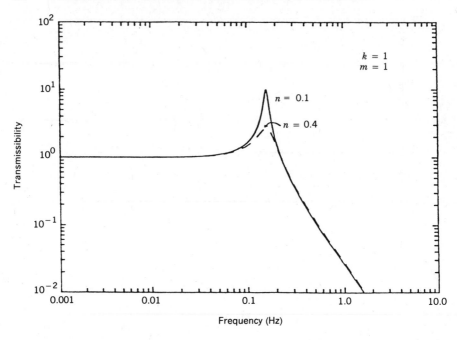

**FIGURE 1.17.** Effect of damping changes for the boundary excitation case.

phase, with the spring acting as a rigid link at frequencies below the fundamental frequency, and therefore the ratio of the two amplitudes is always unity. It is evident from this discussion that the stiffness and mass modifications affect the frequency range above resonance, whereas damping affects the response near resonance, and no important changes occur below the resonant frequency from changes in damping, mass or stiffness.

To illustrate further the overall combined effects of stiffness, mass, and damping on the single degree of freedom system, consider now the random response. For random response the rms value $\bar{W}/\bar{W}_0$ is shown in Chapter 4 to be given by

$$\left(\frac{\bar{W}}{\bar{W}_0}\right) = \frac{\pi\sqrt{\omega_0}}{2\sqrt{2\eta}}[1 + \sqrt{1 + \eta^2}] \tag{1.57}$$

For small to medium modal damping values the rms response $\bar{W}/\bar{W}_0$ is proportional to $\sqrt{\omega_0/\eta}$. Thus the random response $\bar{W}/\bar{W}_0$ is minimized not only by increasing the modal damping but also by reducing the resonant frequency.

**Forced excitation.** In this case it is assumed that the structural component is excited by forces at one or more points of the structure. Consider in

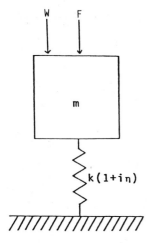

**FIGURE 1.18.**  Forced excitation.

Figure 1.18 a single degree of freedom system in which a force $Fe^{i\omega t}$ is applied to the mass and its response $w$ is observed. The equation of motion for this system is

$$m\ddot{w} + k(1 + i\eta)w = Fe^{i\omega t}$$

The response measured at the mass, in terms of frequency, is illustrated in Figure 1.19 for two levels of damping, $\eta = 0.1$ and 0.4. It is seen that the

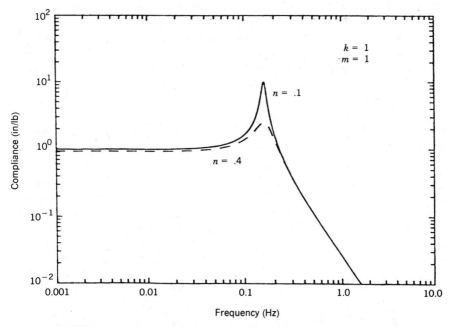

**FIGURE 1.19.**  Effect of damping changes for the forced excitation case.

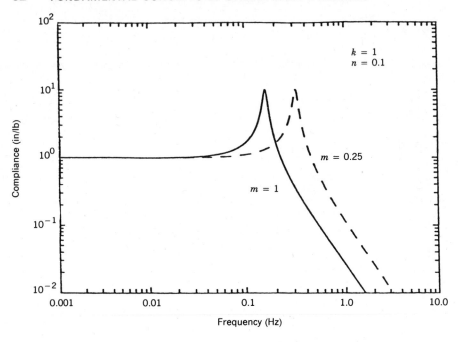

**FIGURE 1.20.**   Effect of mass changes for the forced excitation case.

increased damping level affects the response at resonance only, without significant changes to either high frequency or low response. Consider the two plots shown in Figures 1.20 and 1.21 for the cases $\omega_0 = 0.15$ and $\omega_0 = 0.3$ Hz. These two plots are for a loss factor of 0.1. It is assumed that for the baseline structure $\omega_0 = 0.15$ Hz. It is evident that, by stiffening the spring to make $\omega_0 = 0.3$ Hz, considerable improvement can be achieved below the resonant frequency of the system. This reduction in amplitude is directly proportional to the increase in stiffness of the system, in contrast to the base excited case. It is emphasized that before a structure is stiffened, the nature of the excitation forces must be known so that the resonant frequency is not shifted to a place where it might encounter exciting forces with higher amplitude. Also it can be noted from this figure that the response at high frequency is essentially unchanged and that the stiffness modification helps most below and at the resonant frequency. The effect of mass modification on this system is shown in Figure 1.20, and it is seen that it has little effect below resonance. The combined effects of damping, stiffness, and mass changes are best illustrated again by considering the rms response of the system, which can be written in the form

$$\left(\frac{\overline{W}}{\overline{F}}\right) = \frac{\pi\sqrt{\omega_0}}{2\sqrt{2\eta}\,k}\left[\frac{1 + \sqrt{1 + \eta^2}}{1 + \eta^2}\right]^{1/2}$$

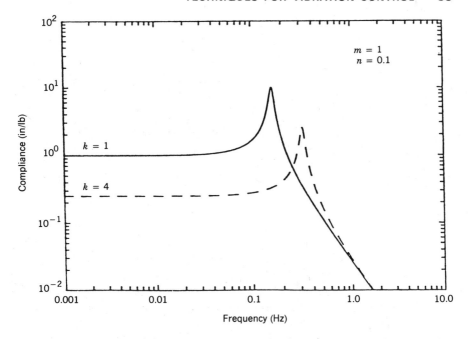

**FIGURE 1.21.**   Effect of stiffness changes for the forced excitation case.

For small to medium modal damping values the rms response $\overline{W}/\overline{F}$ is proportional to $\sqrt{\omega_0/k^2\eta}$. It can be seen from this expression that the response can be minimized not only by increasing the loss factor of the system but also by changing its modal mass and stiffness values.

### 1.3.4.   Isolation

Figure 1.22 illustrates a typical instance where isolation may be applied. If a structure is firmly attached to ground and excited by a driving force, it may be that excessive response or noise emission encountered can be reduced by (1) placing soft springs between the structure and ground, the springs being softer than both structure and ground at the points of attachment, and/or (2) by placing a soft spring system between the excitation force and the structure. In both cases the isolator springs may be damped or not, as the particular problem requires. To illustrate some of the effects of isolation, consider the simple structure shown in Figure 1.23, consisting of a mass $m_1$ on a spring $k_1(1 + i\eta_1)$, a mass $m_2$ being attached rigidly to ground, and a mass $m_3$ rigidly to $m_1$. The response of the system is given, for force excitation, by

$$\frac{W_1}{F} = \frac{1}{k_1(1 + i\eta_1) - (m_1 + m_3)\omega^2} \tag{1.58}$$

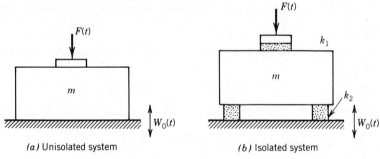

(a) Unisolated system

(b) Isolated system

**FIGURE 1.22.**   Unisolated and isolated system.

and, for base excitation,

$$\frac{W_1}{W_0} = \frac{k_1(1 + i\eta_1)}{k_1(1 + i\eta_1) - (m_1 + m_3)\omega^3} \tag{1.59}$$

If, on the other hand, the mass $m_2$ is separated from the ground by a complex stiffness $k_2(1 + i\eta_2)$ and the mass $m_3$ is separated from $m_1$ by a spring $k_3(1 + i\eta_3)$, then the response is that of a three degree of freedom system and is easily calculated. For force excitation the equations are

$$\begin{bmatrix} -m_3\omega^2 + k_3^* & -k_3^* & 0 \\ -k_3^* & -m_1\omega^2 + k_1^* + k_3^* & -k_1^* \\ 0 & -k_1^* & -m_2\omega^2 + k_1^* + k_2^* \end{bmatrix} \begin{Bmatrix} W_3 \\ W_1 \\ W_2 \end{Bmatrix} = \begin{Bmatrix} F_1 \\ 0 \\ 0 \end{Bmatrix} \tag{1.60}$$

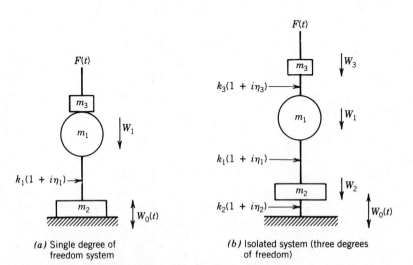

(a) Single degree of
freedom system

(b) Isolated system (three degrees
of freedom)

**FIGURE 1.23.**   Single degree of freedom system. (a) Without isolation. (b) With isolation.

where $k_j^* = k_j(1 + i\eta_j)$, $j = 1, 2, 3$. For base excitation the equations are

$$
\begin{bmatrix}
-m_3\omega^2 + k_3^* & -k_3^* & 0 \\
-k_3^* & -m_1\omega^2 + k_1^* + k_3^* & -k_1^* \\
0 & -k_1^* & -m_2\omega^2 + k_1^* + k_2^*
\end{bmatrix}
\begin{Bmatrix} W_3 \\ W_1 \\ W_2 \end{Bmatrix}
=
\begin{Bmatrix} 0 \\ 0 \\ k_2^* W_0 \end{Bmatrix}
$$

(1.61)

From these equations one may readily determine the effect of the additional parameters $m_2$, $m_3$, $k_2$, $k_3$, and $\eta_1$, $\eta_3$ on the response of the system.

## 1.3.5. Enclosures and Barriers

From the point of view of noise control a heavy, stiff enclosure placed completely around an offending machine or structure is used quite frequently in stationary industrial applications where weight is of no special concern and is used occasionally in nonstationary applications such as diesel engines on trucks where the weight penalty is not always considered unacceptable. However, questions of expense are rarely neglected these days, and barriers are not particularly effective on a combined weight, convenience, accessibility, and economic impact basis. If the enclosure is made lighter and more flexible, and hence cheaper, the surfaces of the barrier will vibrate under the excitation from the radiated noise of the structure and will in turn transmit some of the noise, hence reducing its effectiveness.

Similarly walls or barriers can be effective in blocking some of the radiated noise from selected areas, as illustrated in Figure 1.24c. However, gaps around the barrier can lead to reflection or refraction of some of the noise around the sides of the barriers, again reducing its efficiency. The enclosure and barrier approaches are certainly effective ways of controlling noise transmission in many cases and are recommended for consideration where conditions merit this approach. However, noise control at the source, by modification of the structure, isolation of the structure, introduction of damping into the structure, and/or reduction of the excitation are equally a part of the picture and

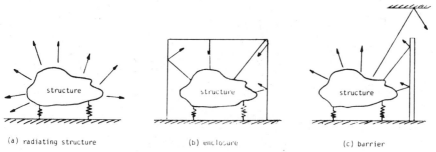

(a) radiating structure          (b) enclosure          (c) barrier

**FIGURE 1.24.**    Some soundproofing methods. (a) Radiating structure. (b) Enclosure. (c) Barrier.

should always be given equal attention until the evidence available from examination of each specific problem indicates the most cost-effective approach. It is terribly inefficient to put a barrier around a noisy machine when, for example, a set of simple isolators placed under it can reduce the noise by the required amount or damping of a single excessively vibrating subcomponent can do the same.

### 1.3.6.   Noise in Structures and Machines

Up to now we have concentrated on the mechanical elements involved in the response of structures. In most cases structures do not exist in isolation but are attached to the surface of the earth or are surrounded by a fluid medium such as water or air. Since sound waves are transmitted in the fluid and elastic waves in the earth, some coupling with these media is to be expected. For example when a structure vibrates, it will excite sound waves in the air which will be audible if sufficiently intense and within the frequency range of sensitivity of the ear. These sound waves will also in turn impinge back on the structure and affect its response. Similarly, when sound waves from one source such as a vibrating surface strike another flexible surface, they will generate oscillatory pressure loads on the surface that will cause it to vibrate and in turn reradiate sound waves, thereby causing transmission and reflection of the incident sound wave as illustrated in Figure 1.25. In principle the phenomena of structure-sound interaction can be predicted through the equations of motion of the structure and of the fluid medium. Yet, because of the complex

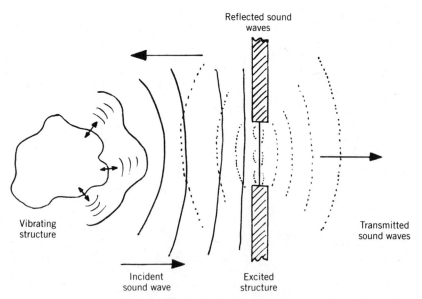

Reflected sound waves

Vibrating structure

Transmitted sound waves

Incident sound wave

Excited structure

**FIGURE 1.25.**   Sound-structure interactions.

geometries of real structures and multiple reflections of the sound waves, this is not at all easy in practice, and usually only very simple, ideal, problems can be solved with any degree of accuracy. These simple, classical solutions, however, can be a significant aid in understanding the phenomena involved and in interpreting the results of experimental measurements or very detailed computational predictions. This is especially important to help engineers understand the role of the various noise and vibration control measures available to them and the effects of varying the various parameters involved.

Without such experience it is a very difficult task, indeed, to seek and optimize measures for reducing practical machinery noise and vibration control problems in the real world. Some general references on noise control [1.47–1.52] might be of interest to the reader who is beginning the study of this subject.

**Prediction of sound pressure levels.** Referring again to Figure 1.25, one notes that the vibratory motion of each element of a structure or machine surface touching the surrounding fluid medium (e.g., air) causes sound waves to be emitted. These sound waves impinge on other parts of the structure, including the point of origin, and the excitation forces so created in turn cause further motion of the structure. The sound waves can be reflected by other surfaces in the vicinity, which in turn may be set in motion. In general, the problem so set may be impossible to solve in practice, although the general principles involved are quite well understood.

The basis for many calculations of noise fields created by vibrating structures is the celebrated equation, first made known by Lord Rayleigh [1.1] for the acoustic pressure $p$ at a point some distance from a vibrating surface in the absence of all other sound sources, namely

$$p = \frac{-i\omega\rho'}{2\pi} \int \frac{\dot{w}}{r} e^{i\omega(t-r/c)} \, dS$$

where $dS$ is the area of an element of the surface, a distance $r$ from the point where $p$ is being predicted and $\dot{w}$ is the local transverse velocity amplitude of the element. Note that this equation represents a summation of an infinite number (in the limit) of traveling waves originating at each element $dS$ and contributing individually to the total pressure at the point of interest. In this particular equation we are dealing with harmonic oscillation of frequency $\omega$.

### 1.3.7. Fatigue of Materials

When members of structures are subjected to time-varying loads, they can initiate and propagate cracks that eventually may lead to failure. Many parameters govern fatigue behavior of a member, including its microstructural characteristics (a function of composition and processing), the stress level and distribution, the relative levels of static and variable stresses, the nature and

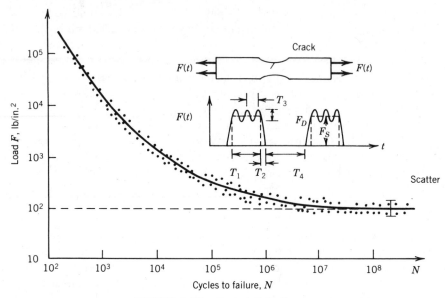

**FIGURE 1.26.**  Typical $F$-$N$ diagram.

distribution of initial flaws which can give rise to stress concentrations and local plastic deformation, temperature, local environmental conditions, and surface conditions. Clearly it will be quite impossible to summarize even a significant part of the fatigue-related work of the past half century, and we shall not attempt to do this since only a few aspects relate directly to the main subject of this book.

In the past the phenomenological approach to fatigue has concentrated on the utilization of large numbers of controlled tests on standardized specimens to relate fatigue life in cycles (since cyclic loading is the most commonly encountered fatigue situation) to the load amplitude, as illustrated in Figure 1.26. In such tests one can vary the dynamic stress amplitude $\sigma_D$, the frequency $\omega = 2\pi/T_3$, the superimposed static load $\sigma_s$, the dwell time $T_1$, the rest time $T_4$, as well as the crack length in an initially cracked specimen. For combined high frequency cyclic loading and quasi-static loading, the main feature of interest is often the cyclic stress level corresponding to a chosen number of cycles to failure (often $10^7$ cycles), and this forms the basis of the Goodman diagram which is merely a summary of fatigue data for a given material in terms of combinations of static and dynamic stresses giving lives of $10^7$ cycles, as illustrated in Figure 1.27.

The Goodman diagram serves, when adequate data are available, is a very useful design tool. If the actual loading of a component is estimated to correspond to point $A$, for example, one might have good reason to believe that it will not fail in less than $10^7$ cycles. Similarly point $B$ will clearly fail in far less

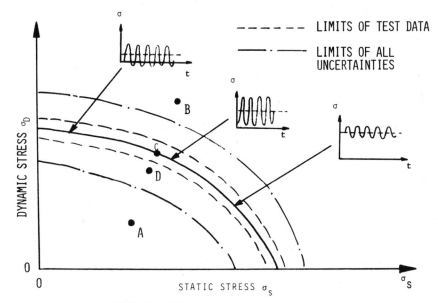

**FIGURE 1.27.** Idealized Goodman diagram.

than $10^7$ cycles. Points $C$ and $D$ are borderline, and in view of the usual scatter of fatigue data one could not give a categorical judgment of design adequacy especially when all other uncertainties are taken into account. For this reason the Goodman diagram is often useful to help designers develop a clearly conservative component (design point $A$) but cannot be used with any degree of reliability to design less conservatively (e.g., point $D$). This is reflected in Figure 1.27 by the two sets of limits: the inner limits correspond to uncertainties of the particular set of fatigue tests used to generate the Goodman diagram (point $D$ lies outside these limits); the outer limits correspond to the additional effects of all other uncertainties (point $D$ lies within these boundaries). In summary, the Goodman diagram is useful for very conservative design but not useful for the most efficient possible design.

In recent years efforts have been made to apply the science of fracture mechanics toward prediction of crack growth rates in specimens and/or components, and some success has been achieved for low cycle fatigue ($N < 10^4$ cycles or so). Some current work is now being concentrated on studies of combined high cycle–low cycle fatigue. In the fracture mechanics approach the crack growth rate is measured as a function of a "stress intensity factor." For simple standard specimens used for test purposes, it is not too difficult to define this stress intensity factor which is a measure of the stress distribution within the plastic region around the crack tip, but in real components subjected to varying stress fields it is not at all simple. Figure 1.28 shows a typical type of test specimen, and a typical form of measured crack growth rate

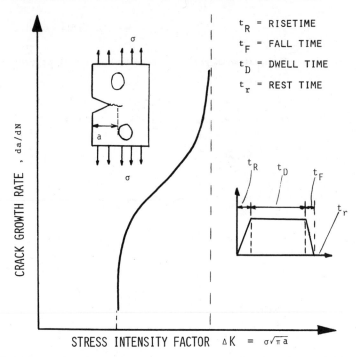

**FIGURE 1.28.**   Typical crack growth rate behavior as a function of stress intensity factor.

versus stress intensity factor data. The sinh shaped curve of $da/dn$ versus $\Delta K(\sigma\sqrt{\pi a})$, where $a$ is the crack length, is evidence of the very low initial rate of crack growth followed by a zone of linearly changing rate and, finally, an acceleration as $\Delta K$ becomes high and failure approaches. In the intermediate $\Delta K$ region $da/dn$ is quite well defined and can be used as the basis of an estimate of the remaining life of a cracked component. This capability depends on an ability to detect small cracks in the component during routine inspection and the establishment of a critical crack length (above the detection limit) below which size the component will survive and operate safely until the next inspection. If cracks are detected above the critical size, even though well below the immediate failure point, the component is rejected. This "retirement for cause" approach promises to improve greatly the economics of jet engine component replacement since the current philosophy is usually to reject components after a specific number of cycles of operation, regardless of condition. The readers who are interested in further study of fatigue, fracture mechanics, or retirement for cause will find an extensive literature in existence for their perusal. Since these are not the main subjects of this book, the references will be limited to a few general papers [1.53–1.56].

# REFERENCES

1.1. W. C. Strutt (Lord Rayleigh), *The Theory of Sound*, Vols. 1 and 2, 2nd ed., Dover, New York, 1945.

1.2. A. W. Leissa, "Plate vibration research, 1976–1980," *Shock Vib. Digest*, **13**(9), 11–22 (1981).

1.3. M. Lalanne, P. Berthier, and J. Der Hagopian, *Mechanical Vibrations for Engineers*, Wiley, New York, 1983.

1.4. A. W. Leissa, "The relative complexities of plate and shell vibrations," *Shock Vib. Bull.*, **50**(3), 1–9 (1980).

1.5. A. W. Leissa and M. S. Ewing, "Comparisons of beam and shell theories for the vibrations of thin turbomachinery blades," ASME Paper 82-GT-223, 1982.

1.6. P. M. Morse, *Vibrations and Sound*, McGraw-Hill, New York, 1948.

1.7. W. R. Callahan, "On the flexural vibrations of circular and elliptical plates," *Quart. Appl. Math.*, **13**, 371–380 (1955).

1.8. E. Reissner, "On vibrations of shallow-spherical shells," *J. Appl. Phys.*, **17**, 1038–1042 (1946).

1.9. C. M. Harris and C. E. Crede (eds.), *Shock and Vibration Handbook*, Vols. 1, 2, and 3, McGraw-Hill, New York, 1961.

1.10. R. E. D. Bishop and D. C. Johnson, *The Mechanics of Vibration*, Cambridge Univ. Press, England, 1960.

1.11. R. E. D. Bishop, "The General Theory of hysteretic damping," *Aero. Quart.*, **7** (Feb. 1956).

1.12. R. E. D. Bishop, "The treatment of damping forces in vibration theory," *J. Roy. Aero. Soc.*, **59, 738, 1955.

1.13. N. O. Myklestad, "The concept of complex damping," *J. Appl. Mech.*, **19**, 284 (1952).

1.14. K. A. Foss, "Coordinates which uncouple the equations of motion of damped linear systems," *J. Appl. Mech.*, **25**, 361 (1958).

1.15. A. G. J. MacFarlane, *Dynamical System Models*, George G. Harrap, London, 1970.

1.16. E. J. Richards and D. J. Mead (eds.), *Noise and Acoustic Fatigue in Aeronautics*, Wiley, London, 1968.

1.17. E. H. Dowell, "Component mode analysis of a simple non-linear, non-conservative system," *J. Sound Vib.*, **80**(2), 233–246 (1982).

1.18. A. W. Leissa and K. M. Iyer, "Modal response of circular cylindrical shells with structural damping," *J. Sound Vib.*, **77**(1) 1–10 (1981).

1.19. M. T. Soifer and A. W. Bell, "Simplifying a lumped parameter model," *Shock Vib. Bull.*, **39**, Pt 3, 153–160 (1969).

1.20. D. J. Halter, *Matrix Computer Methods of Vibration Analysis*, Wiley, New York, 1973.

1.21. E. C. Pestel and F. A. Leckie, *Matrix Methods in Elasto-Mechanics*, McGraw-Hill, New York, 1963.

1.22. M. T. Soifer and A. W. Bell, "Reducing the number of mass points in a lumped parameter system," *Shock Vib. Bull.*, **38**(2), 23–22 (1968).

1.23. J. P. Raney, "Identification of complex structures using near-resonance testing," *Shock Vib. Bull.*, **38**(2), 23–22 (1968).

1.24. N. Miramand, J. F. Billand, F. Leleux, and J. P. Kernevez, "Identification of structural modal parameters by dynamic tests at a single point," *Shock Vib. Bull.*, **46**(5), 197–212 (1976).

1.25. D. J. Ewins, "Why's and wherefores of modal testing," *SEE Jl*, **18**(3) (1979).

1.26. S. Mahalingam, "The synthesis of vibrating systems by use of internal harmonic receptances," *J. Sound Vib.*, **40**(3), 337–350 (1975).

1.27. R. J. Allemang, "Experimental modal analysis bibliography," Proc. 1st International Modal Analysis Conf., Orlando, Fla. (Sponsored by Union College, Schenectady, N.Y.), pp. 714–726, 1982.

1.28. H. G. H. Goyder, "Methods and applications of structural modeling from measured structural frequency response data," *J. Sound Vib.*, **68**(2), 209–230 (1980).

1.29. A. L. Klosterman and J. R. Lemon, "Dynamic design analysis via the building block approach," *Shock Vib. Bull.*, **42**, 97–104 (1972).

1.30. Y. K. Lin, "Response of multi-spanned beam and panel systems under noise excitation," Air Force Materials Laboratory report AFML-TR-64-348, Part 1, 1965.

1.31. C. A. Mercer and C. Seavey, "Prediction of natural frequencies and normal modes of skin-stringer panel rows," *J. Sound Vib.*, **6**(1), 149–162 (1967).

1.32. T. J. McDaniel, "Dynamics of circular periodic structures," *J. Aircraft*, **8**(3), 143–149 (1971).

1.33. J. P. Henderson and T. J. McDaniel, "The analysis of curved multi-span structures," *J. Sound Vib.*, **8**(2), 203–219 (1971).

1.34. F. Leckie and E. Pestel, "Transfer-matrix fundamentals," *J. Mech. Sci.*, **2**, 137–167 (1960).

1.35. Y. K. Lin, "Free vibrations of continuous skin-stringer panels," *J. Appl. Mech.*, **27**(4), 669–676 (1960).

1.36. Y. K. Lin and B. K. Donaldson, "A brief survey of transfer matrix techniques with special reference to the analysis of aircraft panels," *J. Sound Vib.*, **10**(1), 103–143 (1969).

1.37. R. E. D. Bishop, G. L. M. Gladwell, and S. Michaelson, *The matrix analysis of structures*, Cambridge Univ. Press, England, 1965.

1.38. C. A. Mercer, "Response of a multi-supported beam to a random pressure field," *J. Sound Vib.* **2, (1965).**

1.39. T. J. McDaniel and K. J. Chang, "Dynamics of rotationally periodic large space structures," *J. Sound Vib.*, **68**(3), 351–368 (1980).

1.40. V. H. Neufest and H. Lee, "Finite beam elements for dynamic analysis," *Shock Vib. Bull.*, **41**(7), 51–60 (1970).

1.41. S. H. Crandall and R. B. McCalley, Jr., "Numerical methods of analysis," *Shock and Vibration Handbook*, Vol. 2, McGraw-Hill, New York, 1961.

1.42. K. Bathe, and E. Wilson, *Numerical Methods in Finite Element Analysis*, Prentice-Hall, Englewood Cliffs, N. J., 1976.

1.43. R. Cook, *Concepts and Applications of Finite Element Analysis*, Wiley, New York, 1974.

1.44. O. C. Zienkiewicz, *The Finite Element Method in Engineering Science*, McGraw-Hill, New York, 1977.

1.45. Y. P. Lu and G. C. Everstine, "More on finite element modeling of damped composite systems," *J. Sound Vib.*, **69**(2), 199–205 (1980).

1.46. L. C. Rogers, C. D. Johnson, and D. A. Keinholz, "The modal strain energy finite element analysis method and its application to damped laminated beams," *Shock Vib. Bull.*, **51** (1981).

1.47. A. D. Nashif and W. G. Halvorsen, "Design evaluation of layered viscoelastic damping treatments," *Sound Vib.*, 12–15 (July 1978).

1.48. A. D. Nashif, "Control of noise and vibration with damping materials," *Sound Vib.*, 28–36 (July 1983).

1.49. A. D. Nashif, "Application of damping for noise control in diesel engine components," *Damping Applications for Vibration Control*, ASME, AMD-Vol. 38, pp. 133–143, 1980.

1.50. L. L. Faulkner (ed.), *Handbook of Industrial Noise Control*, Industrial Press, New York, 1976.

1.51. B. L. Clarkson, "The design of structures to resist jet noise fatigue," *J. Roy. Aero. Soc.*, **66**(622), 603–616 (1962)

1.52. R. Prybutok, "Materials for noise control," SAE paper 810859, *Controlling Truck Noise*, pp. 94ff., 1981.

1.53. H. Liebowitz, *Fracture*, Academic, New York, 1968.

1.54. G. C. Sih and S. R. Valuro (eds.), *Fracture mechanics in engineering applications*, Sijthoff and Noordhoff, The Netherlands, 1979.

1.55. R. S. Shane (chairman), *Testing for prediction of material performance in structures and components*, ASTM Special Tech. Publ. 515, 1971.

1.56. R. J. Hill, W. H. Reimann, and J. S. Ogg, "A retirement for cause study of an engine turbine disk," AFWAL-TR-81-2094, Air Force Wright Aeronautical Lab., WPAFB, 1981.

**2**

# CHARACTERIZATION OF DAMPING IN STRUCTURES AND MATERIALS

## ADDITIONAL SYMBOLS

| | |
|---|---|
| $D^\alpha$ | fractional derivative ($\alpha < 1$) |
| $f(t), F(\omega)$ | Fourier transform pair (forces) |
| $F(t)$ | force |
| $n$ | nondimensional parameter in description of hysteresis loop |
| $N$ | normal load |
| $\rho'$ | atmospheric pressure |
| $r', r_0, R$ | radii |
| $S_L$ | leak area |
| $t, \tau$ | time |
| $T$ | temperature |
| $U$ | energy stored |
| $V_0$ | initial volume |
| $W_n, W_{n+1}$ | peak amplitudes at $n$th and $n + 1$th cycles, respectively |
| $\gamma$ | ratio of specific heats of compressible fluids |
| $\varepsilon, \varepsilon_0$ | strains |
| $\varepsilon'$ | phase angle |

**44**

$\psi$          velocity potential

$v$          Poisson's ratio; also nondimensional parameter

$\theta, \theta_0$        angles


## 2.1.  EFFECTS OF DAMPING

During the course of daily life we only occasionally encounter structures that have little or no damping, and hence we rarely have the opportunity to observe the absence of a phenomenon that we take almost completely for granted. When a structure possesses no damping, no mechanism exists to remove the vibrational energy in it, implying that any vibratory motion, once set up, will continue for ever. Clearly this can never happen in the real world, but close approximations do occasionally occur. The first example that comes to mind is the bell, a structure designed through ancient experience to vibrate for a long time at a frequency (or frequencies) selected for its pleasing sound effect. The bell is designed in such a way that movement at the support location is very small, hence minimizing one group of energy dissipation mechanisms, and only one mode of vibration is strongly excited when the bell is impacted. Another well-known structure that possesses very little damping is the golf ball. In this case the highly elastic inner core of the ball is designed to absorb a great deal of energy, as it is deformed by the impact of the club, and then release almost all of it in a very short time as it recovers and is propelled, hopefully a large distance. In this case, as for the bell, any significant amount of damping would detract from the usefulness of the golf ball. A bell whose vibration died away too quickly would be perceived as "dead," and a golf ball that dissipated a significant fraction of the stored elastic energy would not travel as far as it should; indeed, "trick" golf balls have been produced to annoy golfers in just this manner!

So much for lack of damping. Most of the time structures built by humans or nature have so many mechanisms built in for dissipating vibrational energy, that the presence of extraneously excited vibrations is rarely noticed. It is for this reason that the need for damping is often not recognized; it has been there often enough to get us out of trouble nearly all of the time. Nowadays, as we continue to build ever more efficient and economical structures for various purposes and increase the demands we place on these structures, we also tend to eliminate many of the sources of damping which, though without full recognition, helped such structures to survive their service environments in the past. So more and more frequently we must take special efforts to return the damping we took away, and sometimes add more, in a reliable, safe, durable, and cost-effective way. To do this, we must understand what damping is all about. As in the study of most physical phenomena, damping is frequently best understood in terms of what it does; that is, how its presence, or absence, will affect the vibratory motion of the structure.

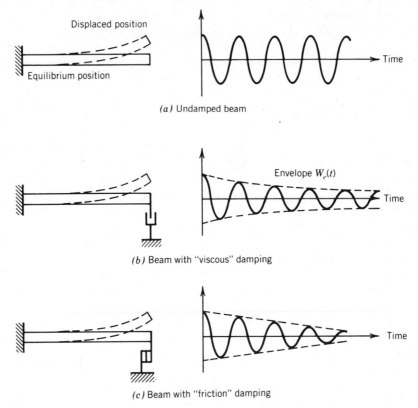

FIGURE 2.1. Effect of damping on free response.

—The first, easily observed effect of damping on structural response is that associated with free vibration. If, as an illustration, we deform a cantilever beam and then release it from rest, it will begin to oscillate regularly, and the amplitude of each successive oscillation will be smaller than the one before, the ratio of successive amplitudes being a measure of the amount of damping. Figure 2.1 illustrates this matter. If the beam is undamped, the oscillation, once set up, will continue indefinitely as in Figure 2.1a. If the beam tip is joined to ground by a viscous damper such as an automotive "shock absorber," as in Figure 2.1b, for which the force applied to the beam is directly proportional to and in opposition to the instantaneous velocity, then the response will die away with time, slowly for a "light" damper and quickly for a "heavy" damper. Theoretically the motion will take an infinite time to die away completely, but in practice other mechanisms of damping will eventually be greater than the velocity-dependent contributions of the damper and will bring the system to rest in a finite time. One such mechanism, illustrated in Figure 2.1c, is dry friction in which the damping force is constant but changes sign each half cycle so as to always oppose the velocity at each instant. Such damping brings the system to rest in a finite time. Clearly the vibrations of damped structures will normally die away with time at a rate that may be used as a measure of the amount of damping. In fact the well-known measure of damping so defined, known as the logarithmic decrement, is related to the ratio of the $n$th to the $n + N$th cycle amplitudes by:

$$\delta = \frac{1}{N} \ln \frac{W_n}{W_{n+N}} \tag{2.1}$$

where $W_n$ is the amplitude of the $n$th cycle and $W_{n+N}$ is the amplitude of the $n + N$th cycle. It is seen at once that only if the envelope of the decaying oscillation is an exponential curve of the form $W_e(t)e^{-\delta t}$ can this definition of damping be unique. This is the case for viscous damping and also for hysteretic damping describable by a complex modulus, at least as an approximation. It is certainly not the case for friction damping.

The next equally easily observed effect of damping is its influence on the final steady state amplitude attained by a structure when excited by a harmonically oscillating force (forced vibration). If the oscillating force is applied at time $t = 0$, the response will rapidly build up until the system is in dynamic equilibrium, as illustrated in Figure 2.2. At most frequencies the influence of damping will not be too great. At low frequencies the stiffness of the structure will provide the dominant restoring force, whereas at high frequencies it will be the inertial force. Somewhere in between, depending on the precise values of mass and stiffness, a resonance will occur. At this precise frequency, in the absence of damping, no dynamic equilibrium can be attained, and the system will be subjected to ever-increasing oscillation amplitudes. In practice of course some damping will always exist and will bring the system to equilibrium at higher or lower amplitudes, depending on the magnitude of the damping forces. Again a measure of damping is suggested by the observed behavior, namely

$$\eta = \frac{\omega_2 - \omega_1}{\omega_0} \tag{2.2}$$

where $\omega_0$ is the resonant frequency and $\omega_1$ and $\omega_2$ are the frequencies on either side of $\omega_0$ for which the amplitude is $1/\sqrt{2}$ times the resonant amplitude. $\eta$ is known as the system loss factor and increases as the damping increases. Again this measure of damping will be unique for certain types of damping, such as viscous or hysteretic, but will depend on the amplitude for other types of damping, such as dry friction, and must therefore be used with some caution, as indeed is the case for all measures of damping.

The third readily observable feature of damped harmonic vibration is the finite amount of energy that has to be expended in each cycle to sustain the motion. If the applied force on the structure is described as $f(t) = F \sin \omega t$,

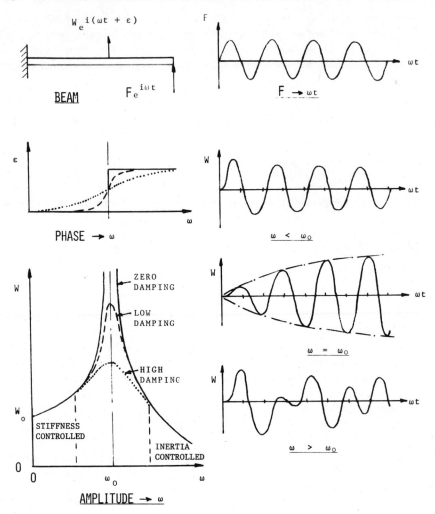

**FIGURE 2.2.** Effect of damping on steady state response.

then the response $w$ will be describable as $w(t) = W \sin(\omega t + \varepsilon)$. In the complete absence of damping $\varepsilon$ will be 0 or 180 degrees, with a jump at resonance as illustrated in Figure 2.2. When the structure is damped, no matter by what physical mechanism, $\varepsilon$ will deviate from these values by, sometimes, a substantial amount. The work done per cycle is

$$D = \int_0^{2\pi/\omega} F(t)\,(dw/dt)\,dt$$

$$= \int_0^{2\pi/\omega} (F \sin \omega t)[\omega w \cos(\omega t + \varepsilon)]\,dt$$

$$= + \pi W F \sin \varepsilon \qquad (2.3)$$

which is positive when $\varepsilon$ is positive. The strain energy in the system can be defined in many ways for a damped system, but the simplest and most consistent definition is half the product of the maximum displacement and the corresponding instantaneous value of the force, that is,

$$U = \left(-\frac{1}{2}\right)\left[W \sin\left(\frac{\pi}{2}\right)\right]\left[-F \sin\left(\frac{\pi}{2} - \varepsilon\right)\right]$$

$$= \left(\frac{1}{2}\right)FW \cos \varepsilon$$

Therefore from (2.3)

$$\tan \varepsilon = \frac{D}{2\pi U} \tag{2.4}$$

Clearly therefore the deviation of the phase angle $\varepsilon$ from 0 or 180 degrees and the ratio of energy dissipated to energy stored are strongly related and can be used to define yet another measure of damping, namely

$$\eta = \frac{D}{2\pi U} = \tan \varepsilon \tag{2.5}$$

Note again, however, that some difficulties can arise if one uses these measures of damping without care; since if the damping is high, $\tan \varepsilon$ will not be small, and the loss factor defined in terms of $D/2\pi U$ will not be identical to the loss factor defined by other measures described earlier. This inconsistency is not a problem in the sense that one cannot predict the response of a system if one knows the dynamic behavior characteristics of its members but only in the sense that one cannot readily compare data measured in one system, by one measure, with data on another system, by a different measure, and this is very important to note when one is comparing test data obtained by different test methods. One simply must compare data obtained by the same measures when trying to determine the effectiveness of a damping treatment, for example. The use of different measures will lead to confusion and error.

So far we have looked at damping from a purely phenomenological point of view, that is, with respect to its effect on structural response rather than in terms of the actual physical mechanisms that introduce the damping forces to the structure. One of the earliest attempts to introduce a realizable physical mechanism is the concept of the viscous damper, which forms the basis of most courses on damping even to this day. Essentially the approach is to introduce into the system a "device" whose damping force is proportional to relative velocity, as for the single degree of freedom system illustrated in Figure 2.3. The mass spring-dashpot (viscous damper) system could actually be made and probably has been for many laboratory demonstrations. The advantage of the concept is its physical and mathematical simplicity, which allows one to obtain solutions using elementary mathematical methods without any

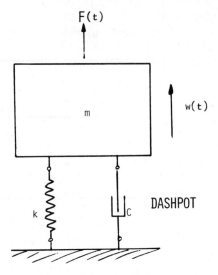

**FIGURE 2.3.** Single degree of freedom system with viscous damping.

of the apparent paradoxes or mathematical difficulties that other models of damping can lead to. Its disadvantage is that it is a contrived model, which only very rarely represents what happens in the real world. The equation of motion of the model shown in Figure 2.3 is

$$m\frac{d^2w}{dt^2} + C\frac{dw}{dt} + kw = F(t) \tag{2.6}$$

which has the advantage of being linear, having constant coefficients with respect to time, and being simple. As a means of illustrating many of the features of a damped system in a lecture room environment, it is ideal but does not accurately represent damping in the real world. Various combinations of springs and dashpots have been devised as a means of improving this model; some of these are illustrated in Figure 2.4. Certainly, by increasing the com-

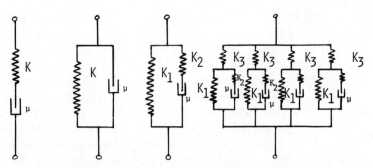

**FIGURE 2.4.** Models of viscoelastic behavior.

plexity of the damping model, one can improve the agreement between model and reality but only at the expense of increased complexity and the need to identify the numerical values of a large number of spring and dashpot parameters, presumably from tests on the damping material or device being modeled.

## 2.2. NONMATERIAL DAMPING

### 2.2.1. Acoustic Radiation Damping

The vibrational response of a structure will always couple with the surrounding fluid medium, which may commonly be air, water, oil, or other gases or liquids, in such a way as to modify the response characteristics [2.1–2.4]. In some cases this modification can be considerable, leading to noise emission into the fluid medium and, on occasion, significant changes of the natural frequencies and mode shapes. The damping effect of the fluid medium depends on many factors, including the density of the medium, the velocity of sound waves within it, and the mass and stiffness characteristics of the structure itself. To illustrate the principles involved, let us first examine a very simple system, illustrated in Figure 2.5, namely a mass supported by springs acting as a single degree of freedom coupled on each end to an acoustic medium. The equation of motion of the mass $m$ is

$$m \frac{d^2 w}{dt^2} + kw = F(t) - F_a \tag{2.7}$$

where $F_a$ is the force acting on the mass due to the acoustic medium, which is determined from the solution of the equation of motion of the acoustic medium, bounded on one side by the oscillating mass having velocity $\dot{w}(t)$ and

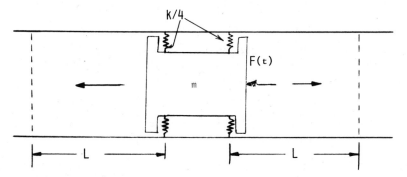

**FIGURE 2.5.**  Transmission of sound through one degree of freedom system.

open on the other end. The equation of motion to be satisfied is the one-dimensional wave equation:

$$\frac{d^2\psi}{dx^2} - \frac{1}{c^2}\frac{d^2\psi}{dt^2} = 0 \qquad (2.8)$$

In this equation $\psi$ is the velocity potential in the fluid medium, which is related to pressure increment $p$ and acoustical velocity $V$ by the following relations:

$$p = -\frac{\rho'\partial\psi}{\partial t}$$

$$V = \frac{\partial\psi}{\partial x}$$

If the velocity of the mass $m$ is $\dot{w}(t) = i\omega w e^{i\omega t}$, then we may assume that $\psi(x, t)$ is of the form $\psi(x)e^{i\omega t}$, so that equation (2.8) becomes

$$\frac{d^2\psi}{dx^2} + \frac{\omega^2}{c^2}\psi = 0 \qquad (2.9)$$

Now assume a solution in the space variable of the form $\psi(x) = \psi_0 e^{\lambda x}$, so that

$$\left(\lambda^2 + \frac{\omega^2}{c^2}\right)\psi_0 e^{i\omega t} = 0$$

that is,

$$\lambda = \frac{\pm i\omega}{c}$$

$$\therefore \quad \psi = A_1 e^{i\omega(t + x/c)} + B_1 e^{i\omega(t - x/c)} \qquad (2.10)$$

which clearly consists of two traveling waves, one moving to the right, the other to the left. If the fluid medium has no boundaries, as $L \to +\infty$, then the solution must be of the form $\psi = B_1 e^{i\omega(t - x/c)}$, but on the left side, as $L \to -\infty$, $\psi = A_1 e^{i\omega(t + x/c)}$. At the right-hand side, at $x = 0$

$$\dot{w} = \frac{\partial\psi}{\partial x} = -B_1\frac{i\omega}{c}$$

$$\therefore \quad B_1 = \frac{ic\dot{w}}{\omega}$$

Similarly $A_1 = -ic\dot{w}/\omega$. Hence the acoustic pressure acting on the right-hand surface of the mass is

$$p_R = -\rho' i\omega \frac{(ic\dot{w})}{\omega} = \rho' c\dot{w}$$

Similarly on the left side

$$p_L = -\rho' c\dot{w}$$

$$\therefore \quad F_a = \pi R^2 (p_R - p_L) = 2\pi R^2 \rho' c\dot{w}$$

Therefore equation (2.7) becomes

$$m\frac{d^2w}{dt^2} + 2\pi R^2 \rho' c\frac{dw}{dt} + kw = F(t)$$

If now $F(t) = Fe^{i\omega t}$, and $w(t) = We^{i\omega t}$, then

$$W = \frac{F}{k - m\omega^2 + (2\pi R^2 \rho' c\omega)i}$$

$$\therefore \quad \left|\frac{W}{F}\right| = \frac{1}{\sqrt{(k - m\omega^2)^2 + (2\pi R^2 \rho' c\omega)^2}} \tag{2.11}$$

This elementary analysis does not, as usual, give us the solution to any really practical engineering problem, but it does give us considerable insight into the main parameters involved and the order of magnitude of the acoustic damping effect. In fact equation (2.11) can be rewritten in the form

$$\left|\frac{W}{F}\right| = \frac{1}{|k(1 + i\eta_e) - m\omega^2|} \tag{2.12}$$

where

$$\eta_e = 2\pi R^2 \rho' c\omega/k \tag{2.13}$$

is the effective loss factor of the single degree of freedom system. Note especially that $\eta_e$ is proportional to $\omega$, so this type of damping is far more effective for higher frequencies than for low. Note also that $\eta_e$ is proportional to $\rho'$, so the acoustic damping will be much higher in water or oil, for example, than in air.

**Acoustically damped plate.** The problem of predicting the effect of a surrounding acoustic medium on the response of a plate is more complicated than for a simple piston in a tube. If we examine the plate illustrated in Figure 2.6, each vibrating element, such as that at point $X$, will set the acoustic medium in motion and will generate sound waves which will create pressures at any other point such as $Y$. The basis for predicting these pressures, in the absence of any reflecting surfaces in the surrounding fluid medium, is Rayleigh's celebrated formula:

$$p = -\frac{i\omega\rho'}{2\pi} \int_{S} \frac{\dot{w}(r, \theta)}{r'} e^{-i\omega(t-r'/c)} \, dS \tag{2.14}$$

where $dS$ is the area of the element at point $Y$, $S$ the total plate area, $\dot{w}(r, \theta)e^{i\omega t}$ the plate velocity at point $Y$, and $r'$ is the distance from $Y$ to $X$. Using the fact that $r' = \sqrt{r_0^2 + r^2 - 2rr_0 \cos(\theta - \theta_0)}$, this can be rewritten in the form

$$p(r, \theta) = \frac{\omega\rho'}{2\pi} \int_{S} \frac{dw(r_0, \theta_0)}{dt} \left(\sin\frac{\omega r'}{c}\right)\frac{r_0}{r'} \, dr_0 \, d\theta_0$$

$$+ \frac{\rho'}{2\pi} \int_{S} \frac{d^2w(r_0, \theta_0)}{dt^2} \left(\cos\frac{\omega r'}{c}\right)\frac{r_0}{r'} \, dr_0 \, d\theta_0 \tag{2.15}$$

This is a very complicated formula to evaluate in general. The second term represents an effective inertia of the acoustic medium, since $i\omega\dot{w}$ is equal to $\ddot{w}$, the local acceleration. The first term represents a damping effect. The full equation of motion of a plate with an acoustic medium on one side is easily written down using equation (2.7):

$$D\nabla^4 w + \rho Hb \frac{d^2w}{dt^2} = F(r, \theta)e^{i\omega t} - p(r, \theta) \tag{2.16}$$

If the inertial and damping terms are separated, this becomes

$$D\nabla^4 w + \rho bH \frac{\partial^2 w}{\partial t^2} + \frac{\omega\rho'}{2\pi} \int_{S} \frac{dw}{dt}(r_0, \theta_0) \sin\left(\frac{\omega r'}{c}\right)\frac{r_0}{r'} \, dr_0 \, d\theta_0$$

$$+ \frac{\rho'}{2\pi} \int_{S} \frac{d^2w}{dt^2}(r_0, \theta_0) \cos\left(\frac{\omega r'}{c}\right)\frac{r_0}{r'} \, dr_0 \, d\theta_0 = F(r, \theta)e^{i\omega t} \tag{2.17}$$

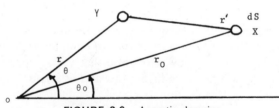

**FIGURE 2.6.**   Acoustic damping.

This is an integral-differential equation for $w(r, \theta, t)$ which may be solved, for example, by modal analysis.

**Acoustic damping effects in aircraft and machine elements.**  As the foregoing discussion shows, acoustic damping can sometimes be a very important factor in controlling structural response, but its order of magnitude is often too small to be useful. This can happen if the density of the fluid medium is too low in relation to the massiveness of the structure or if acoustic pressures from some parts of the vibrating structure cancel out those from the other parts, as would happen for modes of vibration in which adjacent areas vibrate in antiphase with each other. For spacecraft, acoustic damping does not exist. For heavy machines, the air is much too thin to exert significant pressures on the surface. For some thin, lightweight, stiffened structures such as aircraft panels, the acoustic damping can on occasion be important though it is rarely the main source of damping. What is important is that the acoustic damping can, in principle, be estimated.

### 2.2.2.  Linear Air Pumping

The fluid in which a structure is immersed can provide other damping mechanisms besides the radiation of energy away as sound waves. Two examples are illustrated in Figure 2.7 [2.3, 2.4]. If the vibrating structure is backed by a nearly airtight volume, as often happens in the construction of complex structures, the entrapped air is alternately compressed and rarefied by the motion of the panel, leading to a pressure increment $\Delta p e^{i\omega t}$ proportional to the panel motion $W(x, y)e^{i\omega t}$, as in Figure 2.7a. If the enclosed air is totally encapsulated, no energy dissipation can occur. If, however, there are any small leaks, the pressure increment will be $\Delta p e^{i(\omega t + \varepsilon)}$, where $\varepsilon$ is a phase angle resulting from the losses through the leak. The flow through the leak may be laminar or turbulent, depending on the magnitude of $W$, the enclosed volume $V_0$, the size of the leak, and the type of mode in which the panel is responding. For a high-order mode, for example, for which some parts of the surface are out of phase with others, the flow may be quite small, and hence the damping effect will also be small. The effect can be estimated by assuming adiabatic compression and rarefaction to take place so that $pV^\gamma$ is a constant, where $p$ is the instantaneous pressure and $V$ the instantaneous volume. If laminar flow occurs in the leak, we can assume that the flow rate through the leak of area $S_L$, $v$, is proportional to $\Delta p$ at each instant:

$$v = \alpha \Delta p e^{i\omega t} \tag{2.18}$$

We can therefore write

$$[p_0 + \Delta p e^{i\omega t}]\left[ V_0 - \int_s W e^{i\omega t}\, dS + \int_t v e^{i\omega t}\, dt \right]^\gamma = p_0 V_0^\gamma \tag{2.19}$$

$$\therefore \quad \Delta p = \frac{p_0 \gamma}{V_0}\left( 1 + \frac{i p_0 \gamma \alpha}{V_0 \omega} \right) \int_s W\, dS \tag{2.20}$$

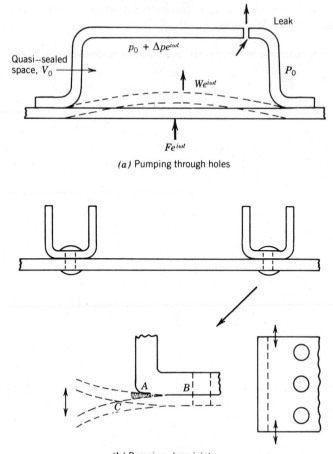

*(a)* Pumping through holes

*(b)* Pumping along joints

**FIGURE 2.7.**  Air pumping mechanism.

where $\omega$ is frequency of the vibration, $V_0$ the equilibrium volume, $p_0$ the atmospheric pressure, $W$ the amplitude of the panel response at each point of the surface, and $S$ the surface area. Note that $\Delta p$ is complex, having one component in phase with $W$ and another 90 degrees out of phase with $W$, that is a damping term. The equation of motion now becomes

$$DV^4W + \rho bH \frac{d^2w}{dt^2} - \frac{p_0\gamma}{V_0}\left[1 + i\frac{\gamma\alpha p_0}{V_0\omega}\right]\int_s W \, dS = F(x, y)e^{i\omega t} \qquad (2.21)$$

This is again an integral-differential equation that may be solved for $W$. Note that the damping term depends on $\int_s (iW/\omega) \, dS$, which can be interpreted as $\int_s (\dot{w}/\omega^2) \, dS$, so that the damping depends on the average velocity of the plate surface. Note also that the damping depends on $1/\omega^2$, so that this mechanism becomes less efficient at high frequencies. Note also that the damping term depends on $(\gamma p_0/V_0)$ and hence rapidly decreases with decreasing $p_0$ and increasing $V_0$. Equation (2.21), though approximate at best, does illustrate the main parameters involved and their effects. Finally, note that the constant of proportionality $\alpha$ between $v$ and $\Delta p$ depends on the flow path(s) within the leak(s) and cannot readily be calculated.

## 2.2.3.  Coulomb Friction Damping

Frictional forces arising from the relative motion of two contacting surfaces are generally modeled by a constant force proportional to the normal load between the surfaces and directed against the velocity vector at each instant. The motion of a system incorporating friction must therefore be analyzed taking account of this "piecewise linear" behavior. Figure 2.8 illustrates some of the ways in which frictional damping might arise in a simple structure, whether naturally or by intent. For the damped joint, if a force $Fe^{i\omega t}$ is applied to the beam, it responds linearly according to the classical theory until the stretching of the middle surface is sufficiently great to introduce significant in-plane stretching loads which modify the equation of motion. If this in-plane load, which oscillates at twice the frequency $\omega$, is greater than $\mu P$, where $\mu$ is the coefficient of friction and $P$ is the applied static load at the joint, then slip will occur at the joint, and this gives rise to energy dissipation at the joint. Similarly for the two-layer beam, the response will become nonlinear as soon as the mid-plane shear stress exceeds $\mu N$, where $N$ is the applied static pressure. For the damped-rivetted joint, the rivet will inhibit motion of the ends of the beam but will not prevent all motion within the joint. For the link to ground, the point $X_0$ will not move until the local shear force exceeds $\mu N$. In each of these cases analysis is quite difficult and tedious because of the nonlinearity itself and because of the great uncertainty concerning distribution of static loads within actual joints, but a review of the literature will show that many efforts have been made to understand these problems [2.5–2.13]. For the purposes of this book it is instructive to examine a few simple problems to illustrate some of the possible approaches.

Let us examine the simplest of all possible structures, namely a single degree of freedom system as illustrated in Figure 2.9. For this system, if $F < \mu N$, the mass $m$ will not move. If $F \geq \mu N$, motion will take place without stopping, the sign of the frictional force varying with the sign of the velocity $\dot{w}$, so that the equation of motion becomes

$$m\ddot{w} + kw = F(t) - \mu N \operatorname{sgn}(\dot{w}) \qquad (2.22)$$

using $\mu N \operatorname{sgn}(\dot{w})$ as an approximation to the true (usually unknown) force-velocity law of the system. If we seek an exact solution of this equation, we

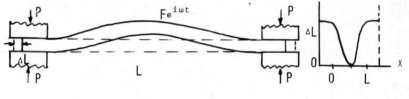

(A) CLAMPED JOINT

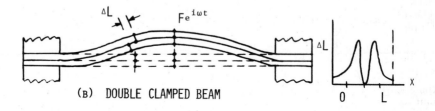

(B) DOUBLE CLAMPED BEAM

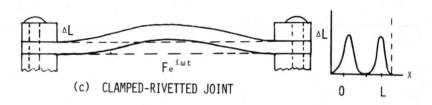

(c) CLAMPED-RIVETTED JOINT

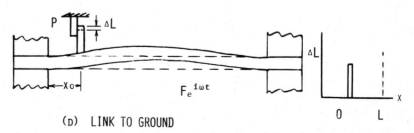

(D) LINK TO GROUND

**FIGURE 2.8.** Friction damping.

must consider each half period, for which $\dot{w}$ is alternately positive and negative, separately. Suppose that, at instants $t = 2\pi n/\omega (n = 0, 1, 2, ...)$, $w = +A$, and that at instants $t = (2n - 1)\pi/\omega$, $w = -A$, where $A$ is the amplitude (presently unknown) of the motion. Then during the first half period, for which $w > 0$, we seek a solution of the form

$$w(t) = C_1 \cos pt + C_2 \sin pt + a + \frac{A_0 \cos (\omega t + \gamma)}{1 - \omega^2/\omega_0^2} \qquad (2.23)$$

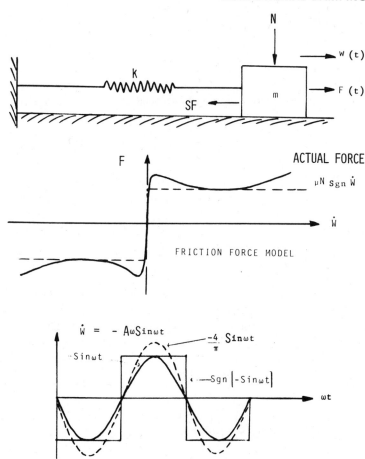

**FIGURE 2.9.**   One degree of freedom system with friction damping.

where $\omega_0^2 = k/m$, $a = \mu N/k$, $A_0 = F/k$, and $C_1$ and $C_2$ are constants of integration. From the initial conditions we have

$$w(0) = -w\left(\frac{\pi}{\omega}\right) = A$$

$$\dot{w}(0) = \dot{w}\left(\frac{\pi}{\omega}\right) = 0$$

so we have four equations from which to determine $C_1$, $C_2$, $A_0$, and $\gamma$. After some manipulation we get

$$w(t) = a\left[1 - \frac{\cos \omega_0(t - \pi/2\omega)}{\cos (\omega_0 \pi/2\omega)}\right] + \frac{A_0 \cos (\omega t + \gamma)}{1 - \omega^2/\omega_0^2} \tag{2.24}$$

with

$$\tan \gamma = -\left[\left(\frac{A_0}{a\omega_0 \tan{(\pi\omega_0/2\omega)}(1 - \omega^2/\omega_0^2)}\right)^2 - 1\right]^{-1/2} \tag{2.25}$$

valid for $0 \le t \le \pi\omega$. If no stopping is to occur, then $w > 0$, and so

$$\frac{ap \sin \omega_0(t - \pi/2\omega)}{\cos{(\omega_0\pi/2\omega)}} - \frac{A_0 \sin{(\omega t + \gamma)}}{1 - \omega^2/\omega_0^2} < 0$$

is the condition for motion without stops. Using a similar approach, an identical solution is obtained for the second half of the cycle, for which $w < 0$. Note that $w \to \infty$ as $\omega \to \omega_0$, so that friction does not prevent infinite amplitudes from occurring at resonance in this simple case. However, this conclusion is not always true for more complex systems.

**Approximate solution.**   We have been able to find an exact solution to this simple problem, but that is not always easily done in more complex problems and an approximate solution may be of value. The basis of this approximate method, known as the method of harmonic balance, is to replace the nonlinear term $\mu N \operatorname{sgn}(\dot{w})$ by a series of harmonic components. For this problem, if $w$ is assumed to be of the form $A \cos \omega t$, then $\dot{w} = -A\omega \sin \omega t$, and hence

$$\operatorname{sgn}(\dot{w}) = \operatorname{sgn}(-A\omega \sin \omega t) = -\frac{4}{\pi} \frac{A}{|A|} \sin \omega t + \cdots \tag{2.26}$$

Neglecting higher-order terms, equation (2.22) therefore becomes

$$m\ddot{w} + kw - \frac{4}{\pi} \mu N \frac{A}{|A|} \sin \omega t = F \cos{(\omega t + \gamma)} \tag{2.27}$$

Since we have assumed $w = A \cos \omega t$, substitution into equation (2.27) gives two equations, corresponding to the coefficients of $\sin \omega t$ and $\cos \omega t$, from which $A$ and $\gamma$ can be determined. After some manipulations we get

$$|A| = \frac{\sqrt{A_0^2 - (4a/\pi)^2}}{|1 - \omega^2/\omega_0^2|} \tag{2.28}$$

and

$$\tan \gamma = \frac{4\mu N/\pi}{|A|(k - m\omega^2)} \tag{2.29}$$

This approximate solution differs somewhat from the exact solution (2.24) but retains the essential qualitative features, and the two solutions approach full agreement when $A_0 \gg 4a/\pi$.

## 2.3.  DAMPING IN MATERIALS

### 2.3.1.  Introductory Remarks

After all external sources of damping have been accounted for, there still remain a very large number of mechanisms whereby vibrational energy can be dissipated within the volume of a material element as it is cyclically deformed. We shall not in any way endeavor to review all of these mechanisms, only a few of which are dominant at any one time. Table 2.1 [2.14] summarizes many of the most important mechanisms and the frequency and temperature ranges over which they are generally most effective. All these mechanisms are associated with internal reconstructions of the micro and/or macro structure, ranging from crystal lattice to molecular scale effects. Included are magnetic effects (magnetoelastic and magnetomechanical hysteresis), thermal effects (thermoelastic phenomena, thermal conduction, thermal diffusion, and thermal flow), and atomic reconstruction (dislocations, concentrated defects of crystal lattices, phonoelectronic effects, stress relaxation at grain boundaries, phase processes in solid solutions, blocks in polycrystalline materials, etc.) [2.15–2.18].

Regardless of the precise physical mechanisms involved, all real materials dissipate some energy, no matter how little, during cyclic deformation. Such effects are often highly nonlinear, so detailed analysis of response with such damping mechanisms is usually very difficult. However, experimental measurements of behavior of samples of specific materials can be qualitatively, and sometimes quantitatively, understood in terms of the measured specific damping energy $D$ (the energy dissipated per unit volume per cycle) for various strain levels, as illustrated in Figure 2.10. This figure illustrates that $D$ is very small for most conventional structural materials, somewhat higher for certain very unique high damping alloys, for which one damping mechanism or another has been enhanced, and highest of all for many polymeric rubberlike materials, not usually used as basic constructional elements, for which again one particular damping mechanism is dominant.

Another approach toward quantifying the internal damping behavior of materials is through the hysteresis loop, again presumed to be obtained from experimental measurements on a material sample [2.15, 2.18]. Figure 2.11 illustrates a type of hysteresis loop which is believed to be representative of typical constructional metal alloys and even of some high damping alloys. The loop is extremely thin, unless the metal is strained into its plastic range, and is not easily observed directly. However, the cusped shape seems to be quite a common feature. Many nonlinear analyses of damped response of structures

**TABLE 2.1   Mechanisms of Damping[a]**

| Type of damping / Type of internal structure reorganization | LINEAR | | NONLINEAR | |
|---|---|---|---|---|
| | Relaxation $f(\omega,T)$ | Resonant $f(\omega)$ | Hysteretic $f(\varepsilon)$ | Viscous $f(\omega,\varepsilon,T)$ |
| Magnetic | 1 Magnetoelastic relaxation<br>2 Eddy currents | 2 Magneto-elastic resonance | 3 Ferromagnetic hysteresis | |
| Thermal | 5 Thermoelastic damping (thermal diffusion, thermal currents) | | 6 Thermal hysteresis | |
| ATOMIC — Damping from dislocations | 9 Relaxation of dislocations-low temperature range<br>10 Relaxation of dislocations-high temperature range | 11 Dislocation resonance | 12 Dislocation hysteresis<br><br>14 Deformation Hysteresis | 13 Damping depending on history of deformation |
| Damping depending not only on dislocations | 15 Relaxation on grain boundaries | | | 16 Viscoelastic delay micro-creep |
| Point defects | 17 Relaxation on point defects | | | |
| Damping depending on solvent atoms | 18 SNOECK'S damping<br>19 Ordering in solid solutions<br>20 KOESTER'S damping<br>21 Damping caused by phase processed in solid solutions | | | |
| Electronic Mechanisms | 7 Electronic absorption of ultrasounds<br>8 Phonon and phono-electronic mechanisms | | | |

The right-hand columns (Point defects downward) contain two charts:

Frequency chart (ω, [Hz]):
```
                         6
5 14 15          13             2  7
18 19 20                      
16                 1           
        5 10 17 21     9  12        11        ω
0   1   10    10^3      10^6       10^9  [Hz]
        Frequency - Hz
```

Temperature chart (T °C):
```
                5
              1  3
            10  11
        13  13  13  20
            11
    9         8          12
2  7      17              6 15 16
                         10
-270      0 20 100   200 230 300  T °C
    Temperature °C    Tp.ch   T CURIE
```

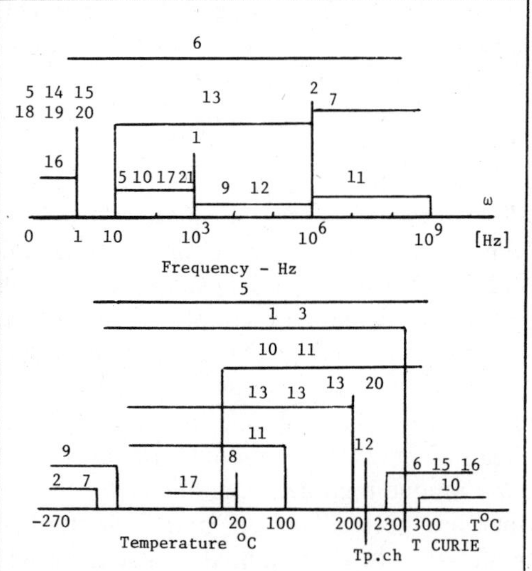

[a] This table is reproduced from reference 2.14, by permission of the author.

have been carried out using analytical representations of such a hysteresis loop, each half of the loop having a different functional form. One representation is

$$\vec{\sigma} = E\left\{\varepsilon - \frac{v}{n}\left[(\varepsilon_0 + \varepsilon)^n - 2^{n-1}\varepsilon_0^n\right]\right\}$$

$$\bar{\sigma} = E\left\{\varepsilon + \frac{v}{n}\left[(\varepsilon_0 - \varepsilon)^n - 2^{n-1}\varepsilon_0^n\right]\right\}$$

(2.30)

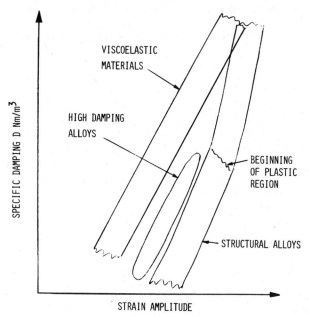

FIGURE 2.10. Effect of strain amplitude on energy dissipation for various types of materials.

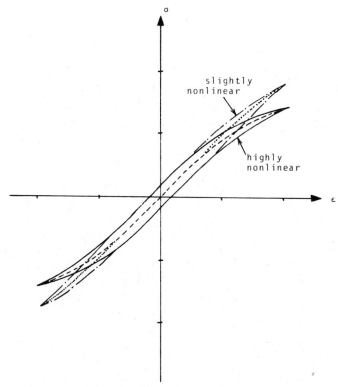

FIGURE 2.11. Nonlinear hysteresis loops.

where $\vec{\sigma}$ is the stress during the loading part of the cycle and $\overleftarrow{\sigma}$ is that during the unloading part. An alternate form, somewhat simpler, is

$$\vec{\overleftarrow{\sigma}} = E(\varepsilon)\left\{\varepsilon \pm \eta(\varepsilon)\varepsilon_0\left|1 - \frac{\varepsilon^2}{\varepsilon_0^2}\right|^{|n|}\right\} \tag{2.31}$$

with $E(\varepsilon) = E/(1 + \alpha|\varepsilon|^\beta)$. Hysteresis loops corresponding to these equations are illustrated in Figure 2.11. The identification of the parameters in these equations is not a simple task, requiring as it does a very accurate set of measured hysteresis loops at various strain levels, and at various frequencies and temperatures. For most structural metals, the deviation of the hysteresis loop from a single line is extremely small, and hence the material damping is insignificant in comparison with that from most other commonly operating damping mechanisms.

The question of how well one can identify the parameters is therefore rather academic except in very unusual situations, such as when the structure is in a vacuum where most extraneous damping mechanisms disappear, or when the material is a special high damping alloy. Thus, although the aforementioned equations represent a framework for prediction of dynamic response, they will not be taken further in this book since the materials used to obtain very high levels of damping in structures behave quite differently from most structural metals or high damping alloys, as will be seen presently.

## 2.3.2. High Damping Alloys

Certain alloy systems have been developed that possess very specific atomic structures conducive to high damping [2.19–2.22]. These alloys are not usually the best adapted to practical construction purposes, since the gain in damping is often at the expense of stiffness, strength, durability, corrosion resistance, cost, machinability, or long-term stability. However, special situations arise where such materials can be used with considerable advantage. For this reason we shall examine the damping properties of one such alloy in a little detail. Because of the highly nonlinear nature of these materials, only experimentally observed modal damping levels and natural frequencies will be presented, and interpretation in terms of hysteresis loop parameters will be omitted, since it is not at all simple to perform such calculations.

One such alloy is known as Sonoston, a commercially available alloy marketed for a variety of properties, only one of which is the high damping level. The modal damping levels were measured by both the bandwidth method and from the ratio of peak modal response to input acceleration at the root of the beam specimen. The resonant frequencies decrease very slightly with increasing stress, indicating a "softening" nonlinearity, and the modal damping is very weakly dependent on mode number and frequency but does depend on peak stress level and on temperature. Figures 2.12 and 2.13 show some plots of modal damping versus peak dynamic stress for various mode numbers and

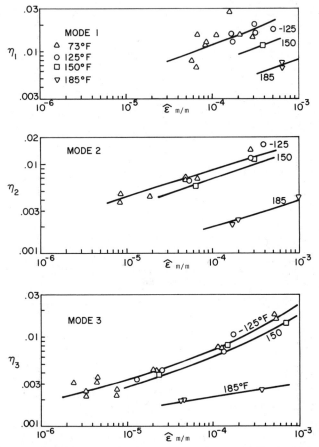

**FIGURE 2.12.**  Damping of Sonoston beams as function of strain amplitude, mode number, and temperature.

temperatures. Note the reversible loss of damping as the temperature increases beyond about 85°C (185°F).

### 2.3.3.  Composite Materials

In common usage a composite material is defined as a combination of two or more materials on a macroscopically homogeneous level (e.g., fibers of one stiff material embedded uniformly, but with directionality, in a matrix of another). Examples of such composite systems include boron fibers in aluminum or titanium matrix (metal matrix composites) or carbon fibers in epoxy matrix (nonmetallic matrix composites), as illustrated in Figure 2.14. Another type of composite structure, not classically fitting the definition, is the brazed or welded honeycomb core, which in isolation is stiff only in compression, sandwiched between two metal sheets and the assembly brazed together at all

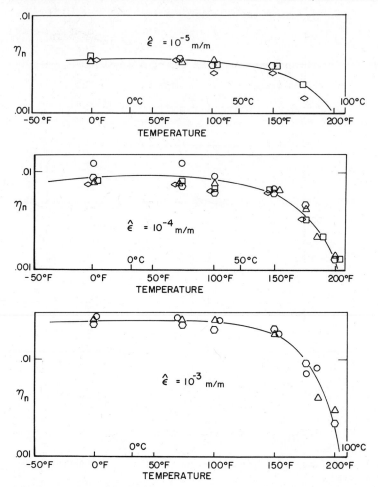

**FIGURE 2.13.** Effect of strain and temperature on damping of Sonoston beam.

touching points. The aim in each case is to increase the stiffness and reduce the weight of the structure, and for this reason composites are being used in structures where such features are highly advantageous, especially in aerospace structures. The advantages unfortunately are accompanied by some disadvantages such as low natural resistance to erosion and impact damage in composites, very high costs for honeycomb, and difficulties in repairing damaged structures [2.23–2.26]. As for conventional metals composites have not generally been developed with high damping as a primary objective. For example, boron-aluminum or carbon-epoxy composites are very low damped but are also highly nonlinear in behavior, so it is often difficult to distinguish in tests between the effects of damping and the effects of the nonlinearity.

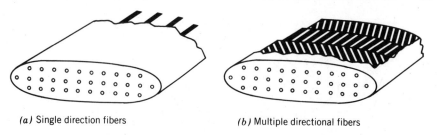

*(a)* Single direction fibers         *(b)* Multiple directional fibers

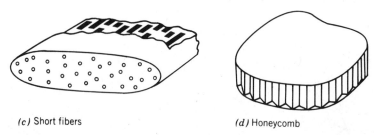

*(c)* Short fibers                    *(d)* Honeycomb

**FIGURE 2.14.** Composite structures. (*a*) Single direction fibers. (*b*) Multiple directional fibers. (*c*) Short fibers. (*d*) Honeycomb.

### 2.3.4. Viscoelastic Materials

Viscoelastic damping is exhibited strongly in many polymeric and glassy materials, and this mechanism of internal damping has many possibilities for industrial application and is consequently the main topic of this book. Polymeric materials are made up of long molecular chains, such as the organic chain illustrated in Figure 2.15. The carbon atoms join strongly together and can be branched so that the long chains can be strongly or weakly linked, according to the composition and processing of the polymer. The damping arises from relaxation and recovery of the polymer network after it has been deformed, and a strong dependence exists between frequency effects and temperature effects because of the direct relationship between material temperature and molecular motion [2.27–2.32]. The chemistry and processing of polymeric materials will not be discussed further, but the specific material dynamic behavior resulting from that chemistry is of very great importance for noise and vibration control and will be discussed in considerable depth.

Glasses are characterized not by long networks, as in polymers, but by short-term order and long-term disorder, as illustrated in Figure 2.15. The inorganic oxides of which the glass is composed form different lattice geometries, depending on the elements involved and their proportions. Damping again arises from relaxation processes after deformation of the glass, recovery

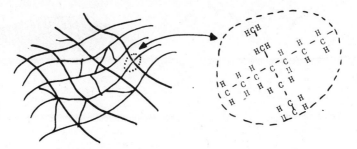

TYPICAL POLYMERIC STRUCTURE NETWORK

- $Si^{4+}$ – SILICON

- $O^{2-}$ – OXYGEN

$\bigcirc Na^{+}$ – SODIUM

TYPICAL GLASS STRUCTURE (SODIUM-SILICATE GLASS)

**FIGURE 2.15.**  Polymer and glass structures.

being due not to the original distribution of short networks but to other conditions of thermodynamic equilibrium [2.33–2.38]. Since the glass is not cross-linked, as can also happen in a polymer, creep can occur (i.e., deformation under continuous load application is also continuous, if usually slow). For a cross-linked polymer, however, the static stiffness can be quite high, and creep will not occur. By proper tailoring, polymeric materials can be manufactured to possess a wide variety of damping, strength, durability, creep resistance, thermal stability, and other desirable properties, over selected temperature and frequency ranges. Similar tailoring can be accomplished for glasses at higher temperatures. In each case of course distinct natural limitations exist, which should be respected, such as upper temperature limits of each material before irreversible damage occurs.

### 2.3.5.  Characterization of Damping Materials' Behavior

The basis of the mathematical approach to modeling damping phenomena is rheology, the science of deformation and flow of matter. The first direction in which rheology has developed, called the microscopic theory, is based on the discrete models of modern physics and uses the results regarding the internal structure of matter to describe processes performed inside the medium in terms of atomic and molecular interactions. The second direction, which is usually that of most utility in engineering, is called the macroscopic approach and encompasses theories based on phenomenological aspects of physics. The macroscopic approach of rheology works in terms of state equations based on the laws of thermodynamics of irreversible processes, which can be written in the very general form

$$f[D_1(\sigma), D_2(\varepsilon), t, T, \cdot, \cdot] = 0 \qquad (2.32)$$

where $f$ represents a vector function of variables, $\sigma$ is the stress tensor, $\varepsilon$ is the strain tensor, $t$ is the time, $T$ the temperature, other variables represent the physiochemical properties of the medium and external environmental conditions, and $D_1$ and $D_2$ represent differential, integral, or combined operators, generally nonlinear. The state equations are generally models of material behavior and, depending on the effect of external excitation (external forces, temperature fields, magnetic fields, chemical reactions, radiation, etc.), describe the materials to some degree of approximation. Generally, at the present level of the state of the art, experimental data are used to build a mathematical model for each specific material being evaluated. The remainder of this chapter will be directed toward an understanding of some aspects of this process, with primary emphasis on the behavior of specific classes of high damping materials, such as polymers or vitreous enamels, for which linear behavior (within limits) presents the prospect of some particularly elegant and practically useful approximations.

**Standard linear model.**  One of the best known representations of the state equation is known as the standard linear model [2.39–2.44], and it gives the following relationship between $\sigma$ and $\varepsilon$:

$$\sigma + \alpha \frac{d\sigma}{dt} = E\varepsilon + \beta E \frac{d\varepsilon}{dt} \qquad (2.33)$$

This particular equation represents a more complex relationship between stress and strain than either Hooke's law ($\sigma = E\varepsilon$) or the simple dashpot-spring combination, for which $\sigma = E\varepsilon + E\beta(d\varepsilon/dt)$, though it still represents real material behavior only with difficulty. This will become clearer as we discuss other methods of representation. The equations do serve, however, to illustrate some aspects of rheological behavior.

Consider first the case where a constant stress $\sigma_0$ is applied to an initially unstressed material specimen at time $t = 0$. Then $d\sigma/dt = 0$ and (2.33) gives

$$\frac{d\varepsilon}{dt} + \frac{\varepsilon}{\beta} = \frac{\sigma_0}{\beta E} \qquad (2.34)$$

which has the solution

$$\varepsilon = \frac{\sigma_0}{E} + Ae^{-t/\beta} \tag{2.35}$$

with the initial condition $\varepsilon = 0$ at $t = 0$, that is,

$$A + \frac{\sigma_0}{E} = 0$$

$$\therefore \quad \varepsilon = \left(\frac{\sigma_0}{E}\right)(1 - e^{-t/\beta})$$

Alternatively, if we apply an initial strain $\varepsilon = \varepsilon_0$ suddenly at $t = 0$ and follow the variation of stress, then we put $d\varepsilon/dt = 0$ in equation (2.33) to get

$$\sigma + \alpha \frac{d\sigma}{dt} = E\varepsilon_0$$

The corresponding solution is

$$\sigma = E\varepsilon_0 + \beta e^{-t/\alpha} \tag{2.36}$$

with the initial condition $\sigma = 0$ at $t = 0$, so that

$$\sigma = E\varepsilon_0(1 - e^{-t/\alpha}) \tag{2.37}$$

where $\alpha$ is known as the constant of stress relaxation.

Third, if we apply harmonically varying stress and strain, of the type $\sigma = \sigma_0 e^{i\omega t}$ and $\varepsilon = \varepsilon_0 e^{i\omega t}$, then equation (2.33) gives

$$\sigma_0 = E\varepsilon_0 \left(\frac{1 + i\omega\beta}{1 + i\omega\alpha}\right) \tag{2.38}$$

which can also be written

$$\sigma_0 = (E' + iE'')\varepsilon_0 \tag{2.39}$$

where

$$E' = \left(\frac{1 + \omega^2\alpha\beta}{1 + \omega^2\alpha^2}\right)E \tag{2.40}$$

$$E'' = E\frac{\omega(\beta - \alpha)}{1 + \omega^2\alpha^2} \tag{2.41}$$

Equations (2.40) and (2.41) reflect some aspects of real material behavior. However, the variation of $E'$ and $E''$ with frequency is much more rapid than is usually observed in real polymeric materials.

**Generalized standard model.**    The limitations of the simple form of the standard model can be reduced [2.41] by introducing additional derivatives of $\sigma$ and $\varepsilon$ in equation (2.33) to give

$$\sigma + \sum_{n=1}^{\infty} \alpha_n \frac{d^n \sigma}{dt^n} = E\varepsilon + E \sum_{n=1}^{\infty} \beta_n \frac{d^n \varepsilon}{dt^n} \tag{2.42}$$

For harmonic response, of the form $\sigma = \sigma_0 e^{i\omega t}$, $\varepsilon = \varepsilon_0 e^{i\omega t}$, this now gives

$$\sigma_0 = \frac{E\varepsilon_0 \left[ 1 + \sum_{n=1}^{\infty} \beta_n (i\omega)^n \right]}{\left[ 1 + \sum_{n=1}^{\infty} \alpha_n (i\omega)^n \right]} \tag{2.43}$$

which can again be written in the form

$$\sigma_0 = (E' + iE'')\varepsilon_0 \tag{2.44}$$

where $E'$ and $E''$ are now much more complicated functions of $\omega$, and by proper choices of $\beta_n$ and $\alpha_n$ can be tailored to fit measured data for $E'$ and $E''$ as a function of frequency. The disadvantage of the model is the fact that a substantial number of terms (i.e., values of $\alpha_n$ and $\beta_n$) are needed to model adequately a real material over a very wide frequency range. This is inconvenient but not especially difficult to deal with, since the expressions for $E'$ and $E''$ now become

$$\frac{E'}{E} = \frac{AC + BD}{C^2 + D^2} \tag{2.45}$$

$$\frac{E''}{E} = \frac{BC + AD}{C^2 + D^2} \tag{2.46}$$

where

$$A = 1 - \beta_2 \omega^2 + \beta_4 \omega^4 \cdots i^{2n-2} \beta_{2n-2} \omega^{2n-2} \tag{2.47}$$

$$B = \beta_1 \omega - \beta_3 \omega^3 + \beta_5 \omega^5 \cdots i^{2n-1} \beta_{2n-1} \omega^{2n-1} \tag{2.48}$$

$$C = 1 - \alpha_2 \omega^2 + \alpha_4 \omega^4 \cdots i^{2n-2} \alpha_{2n-2} \omega^{2n-2} \tag{2.49}$$

$$D = \alpha_1 \omega - \alpha_3 \omega^3 + \alpha_5 \omega^5 \cdots i^{2n-1} \alpha_{2n-1} \omega^{2n-1} \tag{2.50}$$

**Generalized derivatives.** In order to reduce the number of terms required by the generalized standard model to take adequate account of the slower rate of change of properties with frequency seen in real materials, the integral derivatives used hitherto can be replaced by fractional derivatives [2.39–2.41], that is,

$$\sigma(t) + \sum_{n=1}^{\infty} a_n D^{\alpha_n} \sigma(t) = E\varepsilon(t) + E \sum_{n=1}^{\infty} b_n D^{\beta_n} \varepsilon(t) \tag{2.51}$$

where the generalized derives $D^{\alpha_n}$ and $D^{\beta_n}$ are defined as

$$D^{\alpha_n}[x(t)] = \frac{1}{\Gamma(1 - \alpha_n)} \frac{d}{dt} \int_0^t \frac{x(\tau)}{(t - \tau)^{\alpha_n}} \, dt \tag{2.52}$$

with $0 < \alpha_n < 1$ and $\Gamma$ is the gamma function. Note that as with the generalized standard model this definition allows one to obtain solutions in the time domain by use of the Laplace transform. What is of interest is the form to which equation (2.51) reduces when $\sigma(t) = \sigma_0 e^{i\omega t}$ and $\varepsilon(t) = \varepsilon_0 e^{i\omega t}$:

$$\sigma_0 \left[ 1 + \sum_{n=1}^{\infty} a_n (i\omega)^{\alpha_n} \right] = E\varepsilon_0 \left[ 1 + \sum_{n=1}^{\infty} b_n (i\omega)^{\beta_n} \right]$$

This can be expressed in the complex form $\sigma_0 = E^* \varepsilon_0 = (E' + iE'')\varepsilon_0$ with

$$\frac{E'}{E} = \text{Re} \left\{ \frac{1 + \sum\limits_{n=1}^{\infty} b_n (i\omega)^{\beta_n}}{1 + \sum\limits_{n=1}^{\infty} a_n (i\omega)^{\alpha_n}} \right\} \tag{2.53}$$

$$\frac{E''}{E} = \text{Im} \left\{ \frac{1 + \sum\limits_{n=1}^{\infty} b_n (i\omega)^{\beta_n}}{1 + \sum\limits_{n=1}^{\infty} a_n (i\omega)^{\alpha_n}} \right\} \tag{2.54}$$

These equations simplify considerably if the parameters are well chosen, in which case no more than one value of $b_n$ and one value of $a_n$ is often needed. If we let $i = e^{i\pi/2}$, then $i^{\beta_n} = e^{i\pi\beta_n/2} = \cos(\pi\beta_n/2) + i\sin(\pi\beta_n/2)$;

$$\therefore \quad \frac{E^*}{E} = \frac{1 + b_1(i\omega)^{\beta_1}}{1 + a_1(i\omega)^{\alpha_1}} = \frac{A + iB}{C + iD} \tag{2.55}$$

where

$$A = 1 + b_1 \cos\left(\frac{\beta_1 \pi}{2}\right)\omega^{\beta_1} \tag{2.56}$$

$$B = b_1 \sin\left(\frac{\beta_1 \pi}{2}\right)\omega^{\beta_1} \tag{2.57}$$

$$C = 1 + a_1 \cos\left(\frac{\alpha_1 \pi}{2}\right)\omega^{\alpha_1} \tag{2.58}$$

$$D = a_1 \sin\left(\frac{\alpha_1 \pi}{2}\right)\omega^{\alpha_1} \tag{2.59}$$

$$\therefore \quad \frac{E'}{E} = \frac{AC + BD}{C^2 + D^2} \tag{2.60}$$

and

$$\frac{E''}{E} = \frac{AD - BC}{C^2 + D^2} \tag{2.61}$$

**Complex modulus.** So far we have seen that, for the generalized standard model and the generalized derivative model of material behavior, we have started with a set of hypothetical relationships in the time domain, then converted to the frequency domain by focusing attention to stress-strain-time relationships of the form $e^{i\omega t}$ in order to obtain the stress-strain relationship in the frequency domain:

$$\sigma_0 = E^*(\omega)\varepsilon_0 = [E'(\omega) + iE''(\omega)]\varepsilon_0 \tag{2.62}$$

and then identifying parameters in the series expansion so as to make the analytical expressions for $E'(\omega)$ and $E''(\omega)$ fit whatever experimental data are available. It would be at least as logical to start with the experimental measurements, which must be made in all events, to define $E'(\omega)$ and $E''(\omega)$ directly as functions of frequency and use the experimentally measured data directly in further predictions of response [2.18, 2.27, 2.30]. In this case we define the equation of motion in the frequency domain, not the time domain, and we have to return to the time domain via the Fourier transform to study, for example, transient phenomena. Once one has thought about the matter for a while, this approach is just as satisfactory from a philosophical point of view as starting in the time domain and transforming to the frequency domain. The advantage is that experimental data, usually obtained in the frequency domain, are used directly, whereas the disadvantage is that time domain problems must be solved by transform theory; note, however, that for the generalized derivative representation, we have to use transform theory too!

## 2.4. THE COMPLEX MODULUS APPROACH

### 2.4.1. Introductory Remarks

Now that we have seen that the complex modulus approach represents a valid method of describing viscoelastic material behavior, more convenient in some ways than the generalized standard model or the generalized derivative model, and less convenient in others, we can relate it to other observable phenomena. First, let us recall that when we use the complex notation $e^{i\omega t}$, we are using a convenient mathematical fiction to combine two functions, namely $\cos \omega t$ and $\sin \omega t$, each of which individually can equally well represent harmonic motion in the time domain. For strain or stress variations of the form $\varepsilon = \varepsilon_0 \sin \omega t$ or $\varepsilon = \varepsilon_0 \cos \omega t$, equation (2.62) must be rearranged:

$$\sigma = E'\varepsilon + \frac{E''}{|\omega|}\frac{d\varepsilon}{dt} \tag{2.63}$$

since $i|\omega|\varepsilon = d\varepsilon/dt$ for harmonic motion. Note the use of $|\omega|$ since negative values of $\omega$ would otherwise not lead to negative values of $E''$ for negative frequencies. If a strain $\varepsilon_0 \sin \omega t$, which is an observable quantity in a properly constructed test system, is applied to a material specimen, then the corresponding stress $\sigma(t)$ is given by

$$\sigma(t) = E'\varepsilon_0 \sin \omega t + \frac{E''}{|\omega|}\omega\varepsilon_0 \cos \omega t \tag{2.64}$$

$$= E'\varepsilon_0 \sin (\omega t + \varepsilon')\sqrt{1 + \eta^2} \tag{2.65}$$

where $\tan \varepsilon = \eta = E''/E'$; that is, the stress leads the strain (or the strain lags the stress) by an angle $\varepsilon'$ depending on the loss factor $\eta$. This result can equally well be predicted using the complex notation directly of course. What is important is that this phase lag (or lead) can be observed and can be used as a direct measure of $\eta$ in a suitable experiment.

### 2.4.2. Hysteresis Loops

Another well-established type of relationship is obtained by plotting $\sigma$ against $\varepsilon$ for all instants of a cycle of oscillation. Such a graph is illustrated in Figure 2.16 and takes the form of an elliptical or near-elliptical loop. This can be seen from equation (2.64) by recognizing that if $\varepsilon = \varepsilon_0 \sin \omega t$, then $\varepsilon_0 \cos \omega t = \pm\varepsilon_0\sqrt{1 - \sin^2 \omega t} = \pm\varepsilon_0\sqrt{1 - \varepsilon^2/\varepsilon_0^2}$. Therefore

$$\sigma = E'\varepsilon \pm E''\sqrt{\varepsilon_0^2 - \varepsilon^2} \tag{2.66}$$

where the $+$ or $-$ sign depends on whether $\cos \omega t$ is positive or negative. This is the equation of an elliptical-type figure, which can be observed on an

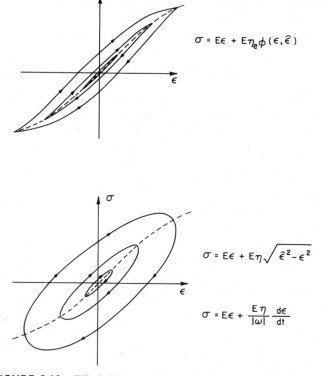

$$\sigma = E\epsilon + E\eta_e\phi(\epsilon,\hat{\epsilon})$$

$$\sigma = E\epsilon + E\eta\sqrt{\hat{\epsilon}^2 - \epsilon^2}$$

$$\sigma = E\epsilon + \frac{E\eta}{|\omega|}\frac{d\epsilon}{dt}$$

**FIGURE 2.16.** Elliptical hysteresis loops for linear viscoelastic material.

oscilloscope screen in a suitable test system. Note that the shape of the ellipse does not change as the maximum strain $\varepsilon_0$ changes. The shape does change, however, as the loss factor $\eta = E''/E'$ changes.

### 2.4.3. Energy Dissipation

The energy dissipation during a cycle of deformation of unit volume of the specimen material is given by

$$D = \oint \sigma \, d\varepsilon = \int_0^{2\pi/\omega} \sigma\left(\frac{d\varepsilon}{dt}\right) dt$$

$$= \int_0^{2\pi/\omega} (E'\varepsilon_0 \sin \omega t + \eta E'\varepsilon_0 \cos \omega t)\omega\varepsilon_0 \cos \omega t \, dt$$

$$= \eta E'\omega\varepsilon_0^2 \int_0^{2\pi/\omega} \cos^2 \omega t \, dt$$

$$\therefore \quad D = \pi\eta E'\varepsilon_0^2 \tag{2.67}$$

Since the maximum energy $U$ stored is $\frac{1}{2}E'\varepsilon_0^2$, it follows that $\eta = D/2\pi U$ is an important measure of the damping capability of the material; the larger the value of $D$, the larger is $\eta$ and the thicker is the hysteresis loop.

### 2.4.4.  Relationship between Various Moduli

The relationships between the moduli of classical elasticity are carried over into the realm of viscoelasticity in the frequency domain, simply by replacing the real moduli by the corresponding complex moduli. For homogeneous, isotropic materials, only three moduli are needed to describe all states of stress, namely Young's modulus, the shear modulus, and the bulk modulus. These three moduli, or their equivalents in other mathematical descriptions, represent, respectively, purely extensional deformation, shear deformation (no volume change), and compressive (or purely volumetric) deformation, as illustrated in Figure 2.17.

**Extensional modulus.**  The complex Young's modulus $E^* = E' + iE'' = E'(1 + i\eta_e)$ is the viscoelastic counterpart of the classical elastic Young's modulus. For a specimen of initial length $L$, with initial cross-sectional area $S$ subjected to loads along its length, the extensional stress $\sigma_e$ is defined as load/area, the strain $\varepsilon_e$ as $\Delta L/L$, and

$$\sigma_e = E^*(\omega)\varepsilon_e = E'(1 + i\eta_e)\varepsilon_e \qquad (2.68)$$

The corresponding reduction in width of the specimen, $\Delta H = v\varepsilon_e H$, is related to Poisson's ratio, $v$, which can be complex. The corresponding change in volume is

$$\Delta V = (1 - 2v)V\varepsilon_e \qquad (2.69)$$

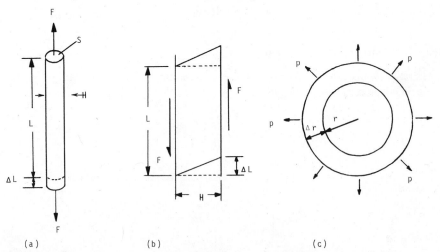

(a)          (b)          (c)

**FIGURE 2.17.**   Types of strain and stress elements. (a) Tension and compression. (b) Shear. (c) Bulk.

**Shear modulus.**   For a prismatic specimen of load-carrying area $S$ and thickness $H$, the shear stress $\sigma_s$ is defined as load/area and the shear strain as $\varepsilon_s = \Delta L/L$, and

$$\sigma_s = G^* \varepsilon_s = G'(1 + i\eta_s)\varepsilon_s \qquad (2.70)$$

**Poisson's ratio.**   The relationship between $G'$ and $E'$ is, from the classical elasticity theory,

$$E' = 2(1 + v)G' \qquad (2.71)$$

and $\eta_e \simeq \eta_s$, so that the imaginary part of $v$ is usually very small. From equation (2.71)

$$v = \left(\frac{E'}{2G'}\right) - 1 \qquad (2.72)$$

and the variation of $v$ with frequency and temperature can be determined by independent measurements of $E'$ and $G'$. Typically for elastomeric materials in their rubbery region, $v \simeq 0.5$, that is, $E = 3G$. In the lower temperature range, where the elastomer or polymer is very stiff, $v \simeq 0.33$, and so $E \simeq 2.67G$. Very little data exist to define $v$ well as a function of frequency and temperature. However, Figure 2.18 illustrates the variation of $E'$ and $G'$ with reduced frequency (see Chapter 3) for 3M-467 adhesive, obtained from independent tests,

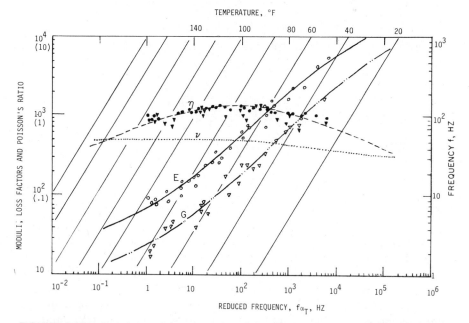

**FIGURE 2.18.**   Complex modulus properties of viscoelastic polymer as function of reduced temperature, and frequency ($E$, $G$ in lb/in.$^2$).

and the variation of $v = (E'/2G') - 1$ shows the transition from 0.5 in the rubbery region to about 0.3 in the glassy region (see Data Sheet 034).

### 2.4.5. Relationship between Harmonic and Transient Responses

Using the complex modulus approach, we can solve any physical problem by replacing $E$ with a complex number $E' + iE''$, $E'(1 + i\eta)$, or $k(1 + i\eta)$, where it is recognized that $E'$, $E''$, $\eta$, and $k$ are all functions of frequency. For a single degree of freedom system in the frequency domain, for example, the response $W(\omega)$ to a force $F(\omega)$ is

$$W(\omega) = \frac{F(\omega)}{k(1 + i\eta) - m\omega^2} \tag{2.73}$$

If $F(\omega)$ is considered to be the Fourier transform of a force $f(t)$ in the time domain, then

$$f(t) = \frac{1}{2\pi} \int_{\infty}^{\infty} F(\omega)e^{i\omega t}\, d\omega \tag{2.74}$$

and

$$F(\omega) = \int_{-\infty}^{\infty} f(t)e^{-i\omega t}\, dt$$

are the corresponding Fourier transform pair. If, for example, $f(t) = F\delta(t)$, then

$$F(\omega) = \int_{-\infty}^{\infty} F\delta(t)e^{-i\omega t}\, dt = F$$

Using the inverse transform, therefore

$$w(t) = \frac{F}{2\pi} \int_{-\infty}^{\infty} \frac{e^{i\omega t}\, d\omega}{k(1 + i\eta) - m\omega^2} \tag{2.75}$$

This integral has caused many problems of considerable concern to dynamicists [2.42–2.44]. For example, if (erroneously) we let $k$ and $\eta$ be constants, then $w(t)$ will exist for $t < 0$ and an apparent noncausality arises. However, if $k$ and $\eta$ correspond to actual measured stiffness and damping behavior of a real material, and (2.75) is evaluated numerically (as it must), this apparent paradox disappears! It is important to settle this issue, which has been a matter of controversy for many years. Equation (2.75) can evaluated numerically using the fact that $k(\omega) = k(-\omega)$ and $\eta(\omega) = -\eta(-\omega)$ so that

$$\frac{w(t)}{F} = \frac{1}{\pi} \int_0^\infty \frac{[k(\omega) - m\omega^2] \cos \omega t + k(\omega)\eta(\omega) \sin \omega t}{[-m\omega^2 + k(\omega)]^2 + k^2(\omega)\eta^2(\omega)} \, d\omega \qquad (2.76)$$

Some illustrative examples will be discussed in Section 2.5.2.

## 2.5.  EXAMPLES AND ILLUSTRATIONS

### 2.5.1.  Two Degree of Freedom System with Friction Damping

It is more difficult to obtain exact solutions for systems with more than one degree of freedom subject to frictional damping at a point, but approximate solutions are not difficult to achieve using the method of harmonic balance. Consider the system illustrated in Figure 2.19a. The receptances at two points of interest, namely 1 and 2, are obtained from experimental measurements or from finite element analysis. The discrete two degree of freedom model illustrated can account for the first two modes, and the corresponding receptances fairly accurately provided that the parameters $m_1$, $m_2$, $k_1$, $k_2$, and $k_3$ are properly chosen, as was discussed in Chapter 1. With these parameters known, it is possible to use the model illustrated in Figure 2.19b, for which the equations of motion are for $k_4 = \infty$

$$m_1 \ddot{w}_1 + k_1(w_1 - w_2) + k_3 w_1 = F_1$$
$$m_2 \ddot{w}_2 + k_1(w_2 - w_1) + k_2 w_2 + \mu N \, \text{sgn} \, \dot{w}_2 = 0 \qquad (2.77)$$

These equations are solved by assuming that $w_2$ is a harmonic function of time of the form $w_2(t) = W_2 \cos(\omega t)$, so that $\dot{w}_2 = -\omega W_2 \sin \omega t$ and $\text{sgn}(\dot{w}_2) = \text{sgn}(-\omega W_2 \sin \omega t)$ must be expanded as a Fourier series again, as in equation (2.26).

$$\therefore \quad \mu N \, \text{sgn} \, \dot{W}_2 = -\frac{4\mu N}{\pi} \frac{W_2}{|W_2|} \sin \omega t$$

Let

$$W_1 = W_{11} \cos \omega t + W_{12} \sin \omega t$$

$$F_1 = F_{11} \cos \omega t + F_{12} \sin \omega t$$

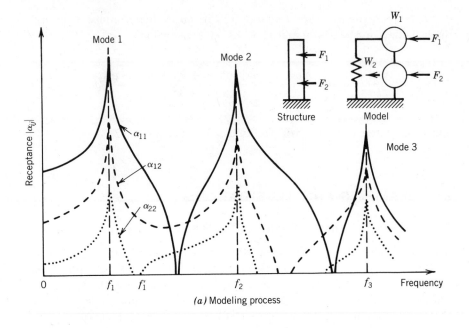

*(a)* Modeling process

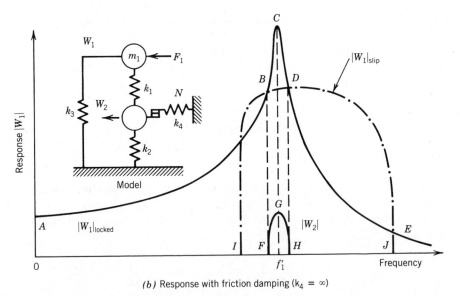

*(b)* Response with friction damping ($k_4 = \infty$)

**FIGURE 2.19.** Response of two degree of freedom system with friction damping. (*a*) Modeling process. (*b*) Response with friction damping ($k_4 = \infty$).

Substituting into Equation (2.77) and comparing terms in sin $\omega t$ and cos $\omega t$ gives the four equations:

$$
\left.\begin{aligned}
(-m_1\omega^2 + k_1 + k_3)W_{11} - k_1 W_2 &= F_{11} \\
(-m_1\omega^2 + k_1 + k_3)W_{12} &= F_{12} \\
(-m_2\omega^2 + k_1 + k_2)W_2 + W_{11}k_1 &= 0 \\
W_{12}k_1 - \frac{4\mu N}{\pi}\frac{W_2}{|W_2|} &= 0
\end{aligned}\right\}
\tag{2.78}
$$

From the second and fourth of these equations, we obtain $W_{12}$ and $F_{12}$. From the first and third equations we obtain $W_{11}$ and $W_2$. First we examine $F_1$:

$$
F_1 = F_{11} \cos \omega t + F_{12} \sin \omega t
$$

$$
= F_{11} \cos \omega t + \frac{4\mu N}{\pi k_1}\frac{W_2}{|W_2|}(k_1 + k_3 - m_1\omega^2) \sin \omega t
$$

$$
\therefore \quad |F_1|^2 = F_{11}^2 + \left(\frac{4\mu N}{\pi}\right)^2\left(1 + \frac{k_3}{k_1} - \frac{m_1\omega^2}{k_1}\right)^2
$$

From the first and third equations, we have

$$
|W_2| = \frac{F_{11}}{\Delta k_1}
$$

$$
\therefore \quad \frac{|W_2|}{|F_1|} = \frac{1}{k_1\Delta}\sqrt{1 - \left(\frac{4\mu N}{\pi F_1}\right)^2\left(1 + \frac{k_3}{k_1} - \frac{m_1\omega^2}{k_1}\right)^2}
\tag{2.79}
$$

with

$$
\Delta = \left(1 + \frac{k_3}{k_1} - \frac{m_1\omega^2}{k_1}\right)\left(1 + \frac{k_2}{k_1} - \frac{m_2\omega^2}{k_1}\right) - 1
\tag{2.80}
$$

Similarly it is found that

$$
\frac{|W_1|}{|F_1|} = \frac{1}{k_1\Delta}\sqrt{\left(1 + \frac{k_2}{k_1} - \frac{m_2\omega^2}{k_1}\right)^2\left\{1 - \beta^2\left(1 + \frac{k_3}{k_1} - \frac{m_1\omega^2}{k_1}\right)^2\right\} + \beta^2\Delta^2}
\tag{2.81}
$$

with $\beta = 4\mu N/\pi F_1$. Figure 2.19b illustrates the type of response expected from these equations.
If

$$
\beta^2\left(1 + \frac{k_3}{k_1} - \frac{m_1\omega^3}{k_1}\right)^2 > 1
$$

$|W_2/F_1|$ does not exist, and slip does not occur (i.e., mass $m_2$ is locked to ground). In that case the response is given by

$$\frac{|W_1|}{|F_1|} = \frac{1}{|k_1 + k_3 - m_1\omega^2|} \tag{2.82}$$

On the other hand, if

$$\beta^2\left(1 + \frac{k_3}{k_1} - \frac{m_1\omega^2}{k_1}\right)^2 < 1$$

slip does occur and $|W_1/F_1|$ and $|W_2/F_1|$ both exist. Note that the solution $|W_2/F_1|$ exists only over the frequency range $FH$, whose width is equal to the intersection of the points $BD$ of the solution for $(W_1/F_1)_{\text{slip}}$ with $|W_1/F_1|$ for $m_2$ locked. This means that certain frequency ranges exist over which the mass $m_2$ does not move relative to the damper and that, when resonance is approached, the loads on the damper increase until finally they are large enough for the maximum sustainable friction load $\mu N$ to be reached, after which only the solution with slip is allowable. Depending on the value of $\mu N$, the response will vary from that shown in Figure 2.19a, for which $m_2$ is free and $\mu N = 0$, to the curve $ABCDE$ in Figure 2.19b, for which $\mu N = \infty$. Another important feature is that, if the second natural frequency $f_2$ lies outside the bounds $IJ$ of the solution for $|W_1/F_1|$ with slip, at a particular value of $\mu N$, then the second mode will not be controlled by friction even though the first mode is. Similarly, if $f_1$ lies inside the bounds $IJ$, as it will for $\mu N$ sufficiently small, then also no control of response in mode 1 takes place.

### 2.5.2.  Calculations for Transient Response

Two cases will be considered, namely a low damping single degree of freedom system having an elastomeric spring element and a high damping system with a viscoelastic material. Figures 2.20 and 2.21 illustrate the variation of the stiffness $k$ and loss factor $\eta$ with frequency at room temperature for two materials BTR (a silicone elastomer) and 3M-467 (a viscoelastic adhesive). In each case, for the sake of illustration, the stiffness $k$ was chosen to be 2910 lb/in. ($5.11 \times 10^5$ N/m) at 100 Hz, and the mass $m$ was 2.849 lb (0.007382 mass units or 1.295 kg). In order to evaluate equation (2.76) numerically, we represent $k$ and $\eta$ for each material by simple empirical functions of the frequency $f$(Hz):

For BTR

$$k = 1158(1 + 100f)^{0.10} \tag{2.83}$$

$$\eta = 0.15 \tag{2.84}$$

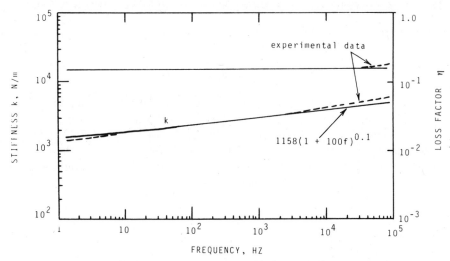

**FIGURE 2.20.** Variation of $k$ and $\eta$ with frequency for BTR (20°F).

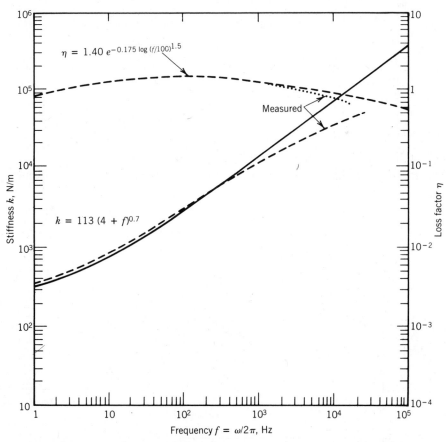

**FIGURE 2.21.** Variation of $k$ and $\eta$ with frequency for 3M-467 at 70°F.

For 3M-467

$$k = 113(4 + f)^{0.70} \qquad (2.85)$$

$$\eta = 1.4e^{-0.175|\log(f/100)|^{1.5}} \qquad (2.86)$$

These expressions have no special significance beyond the fact that they adequately represent the variation of $k$ and $\eta$ in the frequency range $0 \le f \le 10^4$ Hz. Other expressions could equally well be used. Figures 2.22 and 2.23 show the response $w(t)$ for fixed $k$ and $\eta$, viscous (fixed $k$, $\eta$ replaced by $\eta f/100$) and variable $k$ and $\eta$. Only the case of fixed $k$ and $\eta$ leads to noncausality. These cases will be discussed further in Chapter 4.

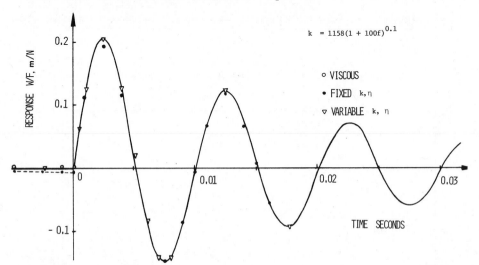

**FIGURE 2.22.** Response of low damping one degree of freedom system to impulse excitation.

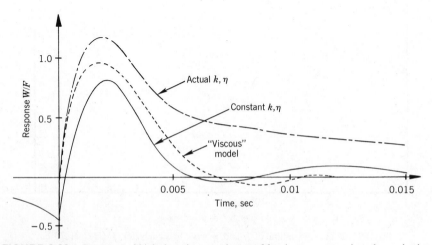

**FIGURE 2.23.** Response of high damping one degree of freedom system to impulse excitation.

# REFERENCES

2.1. E. J. Richards and D. J. Mead (eds.), *Noise and Acoustic Fatigue in Aeronautics*, Wiley, New York, 1968.

2.2. W. C. Strutt (Lord Rayleigh), *The Theory of Sound*, Vols. 1 and 2, 2nd ed., Dover, New York, 1965.

2.3. E. E. Ungar, "Vibration energy losses at joints in metal structures," *Shock Vib. Assoc. Environ., Bull.*, **33**, Pt. IV, 189–199 (1964).

2.4. E. E. Ungar, "Energy dissipation at structural joints; mechanisms and magnitudes," Flight Dynamics Laboratory Report FDL-TR-14-98, 1964.

2.5. R. Plunkett, "Friction damping," *Damping Applications for Vibration Control*, ed. P. J. Torvik, ASME Publication AMD-Vol. 38, pp. 65–74, 1981.

2.6. R. S. H. Richardson and H. Nolle, "Energy dissipation in rotary structural joints," *J. Sound Vib.*, **54**(4), 577–588 (1977).

2.7. A. Schlesinger, "Vibration isolation in the presence of Coulomb friction," *J. Sound Vib.*, **63**(2), 213–224 (1979).

2.8. C. F. Beards and J. L. Williams, "The damping of structural vibration by rotational slip in joints," *J. Sound Vib.*, **53**(3), 333–340 (1977).

2.9. S. W. E. Earles, "Theoretical estimation of the frictional energy dissipation in a simple lap joint," *J. Mech. Eng. Sci.*, **8**(2), 207–214 (1966).

2.10. D. I. G. Jones and A. Muszynska, "Design of turbine blades for effective slip damping at high rotational speeds," *Shock Vib. Bull.*, **69**, Pt. 2, 87–90 (1979).

2.11. A. Muszynska, D. I. G. Jones, T. Lagnese, and L. Whitford, "On nonlinear response of multiple blade systems," *Shock Vib. Bull.*, **51**, Pt. 3, 89–110 (1981).

2.12. J. H. Griffin, "Friction damping of resonant stresses in gas turbine engine airfoils," *J. Eng. Power*, **102**(2), 329–333 (1980).

2.13. A. V. Srinivasan and D. G. Cutts, "Dry friction damping mechanisms in engine blades," ASME paper 82-GT-162, 1982.

2.14. A. Muszynska, "Tlumienie wewnetrzne w ukladach mechanicznych (Internal damping in mechanical systems)," *Dynamika Maszyn*, Polish Acad. Sci., Ossolineum, Warsaw, pp. 164–212, 1974 (in Polish).

2.15. G. S. Pisarenko, "Vibration of elastic systems taking account of energy dissipation in the material," Wright Air Development Center Report WADD-TR-60-582, 1962 (translated from Russian).

2.16. A. J. Nowick, "Internal friction in metals," *Prog. Metal Phys.*, **4** (1953).

2.17. C. M. Zener, *Elasticity and Anelasticity of Metals*, Univ. of Chicago Press, Chicago, 1948.

2.18. B. J. Lazan, *Damping of Materials and Members in Structural Mechanics*, Pergamon, New York, 1968.

2.19. G. E. Bowie, J. F. Nachman, and A. N. Hammer, "Exploitation of Cu-Rich Damping Alloys: Part 1—The search for alloys with high damping at low stress," ASME paper 71-VIBR-106, 1971.

2.20. G. F. Weissmann and W. Babington, "A high damping magnesium alloy for missile application," *J. Environ. Sci.* (October 1966).

2.21. D. Birchon, "Hidamets, metals to reduce noise and vibration," *The Engineer*, London, 1966.

2.22. D. I. G. Jones and W. J. Trapp, "Influence of additive damping on resonance fatigue of structures," *J. Sound Vib.*, **17**(2), 157–185 (1971).

2.23. C. W. Bert, "Fundamental frequencies of orthotropic plates with various planforms and edge conditions," Shock Vib. Bull., **47**(2), 89–94 (1977).

2.24. R. D. Adams and D. G. C. Bacon, "The dynamic properties of unidirectional fibre-reinforced composites in flexure and tension," *J. Composite Materials*, **7**, 53–67 (1973).

2.25.   R. F. Gibson and R. Plunkett, "Dynamic mechanical behavior of fiber-reinforced composites: measurement and analysis," *J. Composite Materials*, **10**, 325–341 (1976).

2.26.   R. L. Sierakowski and C. T. Sun, "Experimental investigation of the dynamic response of cantilever plates," *Shock Vib. Bull.*, **44**, Pt. 5, 89–98 (1974).

2.27.   J. C. Snowdon, *Vibration and Shock in Damped Mechanical Systems*, Wiley, New York, 1968.

2.28.   P. Grootenhuis, "Vibration control with viscoelastic materials," *Environmental Engineering*, Proc. SEE, No. 38, 1969.

2.29.   H. Oberst, "Reduction of noise by the use of damping materials," *Trans. Roy. Soc.*, **A263**, 441 (1968).

2.30.   J. D. Ferry, *Viscoelastic Properties of Polymers*, 2nd ed., Wiley, 1970.

2.31.   H. Oberst, "Uber die Dampfung Biegeschwingunge Dunner Blech Durch fest Haftende Belage," *Acustica*, **4**, 181–194 (1952).

2.32.   L. C. Rogers, "On modelling viscoelastic behavior," *Shock Vib. Bull.*, **51** (1981).

2.33.   D. I. G. Jones, "High temperature damping of dynamic systems," *Shock Vib. Digest*, 13–18 (October 1979 and May 1982).

2.34.   G. A. Graves, C. M. Cannon, and B. Kumar, "A study to determine the effect of glass compositional variation on vibration damping properties," AFWAL-TR-80-4061, Air Force Wright Aeronautical Labs, 1980.

2.35.   A. D. Nashif, "Materials for vibration control in engineering," *Shock Vib. Bull.*, **43**, 145–151 (1973).

2.36.   A. D. Nashif, W. D. Brentnall, and D. I. G. Jones, "A vibration damping treatment for high temperature gas turbine applications, *Shock Vib. Bull.*, **53**(4) 29–40 (1983).

2.37.   P. Srindharan and R. Plunkett, "Damping in porcelain enamel coatings," AFML-TR-71-193, Air Force Materials Lab., 1971.

2.38.   A. D. Nashif, "Enamel coatings for high temperature damping materials," *Ceramic Bull.*, **53**(12), 846–849 (1974).

2.39.   R. L. Bagley and P. J. Torvik, "A generalized derivative model for an elastomeric damper," *Shock Vib. Bull.*, **49** (1979).

2.40.   M. Caputo, "Vibrations of an infinite plate with a frequency independent Q," *JASA*, **60**(3), 634–639 (1976).

2.41.   L. C. Rogers, "Operators and fractional derivatives for viscoelastic constitutive equations," *J. Rheology*, **27**(4), 351–372 (1983).

2.42.   S. H. Crandall, "The role of damping in vibration theory," *J. Sound Vib.*, **11**(1), 3–18 (1970).

2.43.   R. H. Scanlan, "Linear damping models and causality in vibrations," *J. Sound Vib.*, **13**(4) (1970).

2.44.   T. K. Caughey, "Vibration of dynamic systems with linear hysteretic damping," Proc. 4th U.S. National Congress on Applied Mechanics, Vol. 1, p. 87, 1962.

# 3

# BEHAVIOR AND
# TYPICAL PROPERTIES
# OF DAMPING
# MATERIALS

## ADDITIONAL SYMBOLS

| | |
|---|---|
| $E_0$ | static (equilibrium) Young's modulus |
| $E, E(\omega)$ | Young's modulus |
| $\breve{E}$ | minimum value of $E$ |
| $\hat{E}$ | maximum value of $E$ |
| $f$ | frequency (Hz) |
| $G, G(\omega)$ | shear modulus |
| $n$ | nondimensional parameter in curve fit of complex modulus data |
| $T, T_0, T_\infty$ | temperatures |
| $\alpha_T, \alpha_\varepsilon$ | shift factors |
| $\varepsilon, \varepsilon_0$ | strains |
| $\eta, \eta(\omega)$ | loss factors |
| $\lambda$ | extensional strain ($\lambda = 1 + \varepsilon$) |
| $\rho, \rho_0$ | density of damping material at temperatures $T$, $T_0$, respectively |

## 3.1. INTRODUCTION

In the previous chapter, it was shown that the dynamic properties of linear rubberlike materials can be represented by any two of the following three parameters, namely the storage modulus, loss modulus, and loss factor. For the purposes of this chapter only the storage modulus and the loss factor will be used to represent the damping properties. The damping properties of rubberlike materials vary with different environments. For example, the variation with temperature of the frequency response of a typical metal cantilever beam coated with a damping layer is illustrated in Figure 3.1. The temperature range shown straddles the region where the material has good damping capabilities, whereas the frequency range shown includes the first four modes of vibration of the beam. It is evident, in this figure, that the modal behavior of the damped beam, and therefore the degree to which the damping treatment is capable of dissipating the unwanted vibrational energies, is strongly influenced by both temperature and frequency. Therefore a good understanding of such effects, both separately and collectively, on the variation of the damping properties is necessary before specific treatments can be designed. The most important environmental factors are temperature, frequency, dynamic load, and

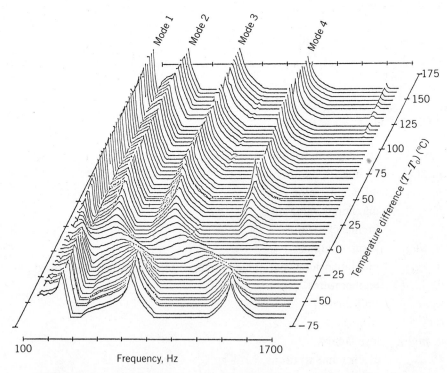

**FIGURE 3.1.** Typical temperature spectrum map for a composite cantilever beam.

static preload. To a lesser extent the damping properties also depend on a number of other environmental factors, such as aging, vacuum, radiation, and oil. In this chapter the effects of the most important environmental parameters will be discussed, along with some typical analytical representations that characterize the properties of the materials [3.1–3.7].

## 3.2.  EFFECTS OF ENVIRONMENTAL FACTORS

An understanding of the variation of damping properties with environment is essential for effective noise and vibration control treatments to be designed. The fact that such properties vary with environment should be treated as an advantage rather than as a detriment, because the variation in the damping properties of materials permits some design flexibility. The following sections will describe how each environment affects the properties of typical rubberlike damping materials.

### 3.2.1.  Effects of Temperature

Temperature is usually considered to be the single most important environmental factor affecting the properties of damping materials [3.3]. This effect is illustrated in Figure 3.2, where four distinct regions can be observed. The first is the so-called glassy region, where the material takes on its maximum value for the storage modulus while having extremely low values for the loss factor. The modulus in the glassy region changes slowly with temperature, while the loss factor increases significantly with increasing temperature. The second region is characterized by having a modulus that decreases rapidly with increasing temperature, while the loss factor takes on its maximum value. The third is the rubbery region where both the modulus and the loss factor usually take on somewhat low values and vary slowly with temperature. The fourth region is typical of a few damping materials such as vitreous enamels and thermoplastics. In this region the material continues to soften with increasing temperature, as it melts, while the loss factor takes on a very high value. Although the fourth region is important for complete characterization of the damping properties, it is usually of little use in the design of damped systems because of instability and other unwanted physical properties. It should also be noted that, for most rubberlike materials, such as cross-linked polymers, this fourth region does not exist. For the remainder of this chapter, the damping properties will be discussed with regard to the first three regions: glassy, transition, and rubbery.

Although the behavior illustrated in Figure 3.2 is typical of all rubberlike materials, different materials have different specific properties, characterized mainly by the various levels of the storage modulus and loss factor within each temperature region and the location of each region with respect to temperature, as illustrated in Figure 3.2. Typical values of the storage modulus

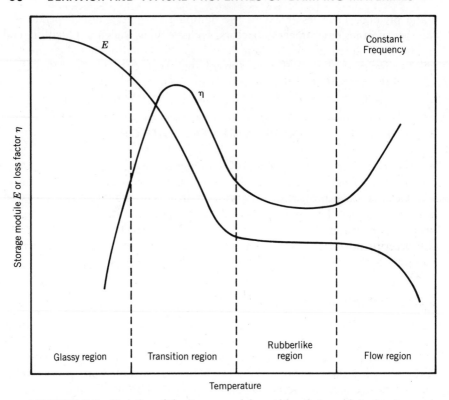

**FIGURE 3.2.** Variation of the storage modulus and loss factor with temperature.

could be as high as $10^8$ KPa in the glassy region and as low as 10 KPa in the rubbery region. The width of the transition region could vary anywhere from 20° C for an unfilled viscoelastic material to as high as 200° C or 300° C for a vitreous enamel. The widths of the rubbery region could vary anywhere from 50° C to 300° C, as for many silicone materials. The loss factor in the glassy region is usually below $10^{-2}$ or $10^{-3}$, whereas it can reach values as high as 1 or 2 in the transition region. Typical loss factor values in the rubbery region are usually between 0.1 to 0.3 for many materials, depending on their composition.

### 3.2.2. Effects of Frequency

The effect of frequency on the damping properties of a typical rubberlike material is illustrated in Figure 3.3 for a number of different temperatures. These temperatures are selected to illustrate the effects of frequency on the damping properties in the glassy, transition, and rubbery regions. The most important effect of frequency is that the modulus always increases with increasing frequency. However, as illustrated in Figure 3.3, this increase is rather small in both the glassy and rubbery regions, whereas it takes on its greatest

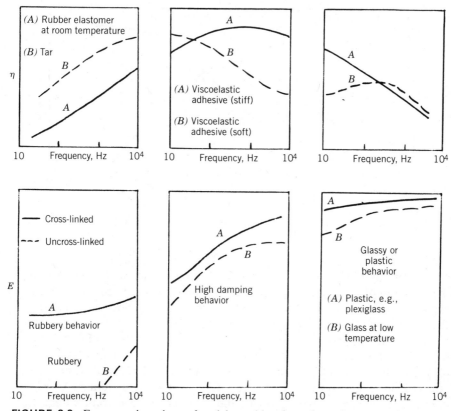

**FIGURE 3.3**  Frequency dependence of modulus and loss factor for various types of materials.

rate of change in the transition region. As far as the loss factor is concerned, it can be seen in Figure 3.3 that it increases with increasing frequency in the rubbery region, takes on its maximum value in the transition region, and then decreases with increasing frequency in the glassy region. The variation of the material properties illustrated in Figure 3.4 is typical for materials with high damping capabilities over a frequency range of about two or three decades. If

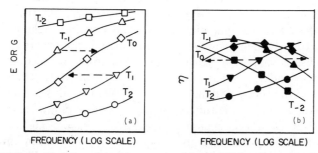

**FIGURE 3.4.**  Variation of the (a) storage modulus and (b) loss factor with frequency.

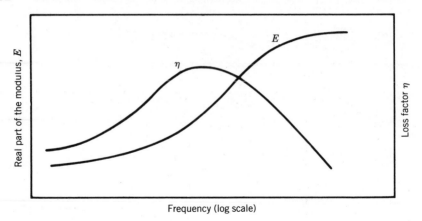

**FIGURE 3.5.** Variation of the real part of the modulus and loss factor with frequency.

a considerably larger frequency range is of interest, such as 10 decades, then the variation of the damping properties with frequency at fixed temperature is expected to take the shape illustrated in Figure 3.5. A close examination of this figure reveals that it is qualitatively the inverse of the temperature behavior, but to a lesser degree; that is, it takes several decades of frequency to reflect the same change of behavior as a few degrees of temperature. This phenomenon is one of the most important aspects of viscoelasticity theory, especially in regard to the characterization of viscoelastic materials. As will be discussed later, this behavior provides the basis for the temperature-frequency superposition principle which is used to transform material properties from the frequency domain to the temperature domain, and vice versa.

### 3.2.3. Effects of Cyclic Dynamic Strain

The effects of dynamic strain amplitude on the damping properties of materials are very difficult to measure [3.7]. This is because high strain amplitudes usually result in high energy dissipation in the material, which causes the temperature of the material to rise rapidly, so that the two effects of temperature and strain amplitude become combined. This is especially true if one tries to measure the effects of dynamic strain in the transition region where the damping is extremely high. However, in the rubbery region where the modulus and the loss factor vary more slowly with temperature, the effect of temperature becomes secondary to that of dynamic strain amplitude. Therefore most investigations of the effects of dynamic strain amplitude on the damping properties have been limited to the rubbery region. A typical plot of such behavior, for a material in its rubbery region, is illustrated in Figure 3.6.

The variation of the damping properties with dynamic strain amplitude is similar to that of temperature. However, the effect is much smaller than that of

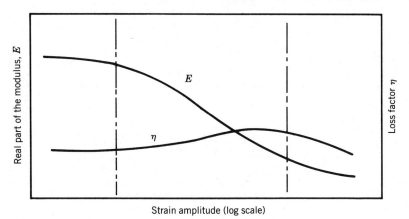

**FIGURE 3.6.**   Variation of the real part of the modulus and loss factor with strain amplitude.

temperature. Here again there exist three regions, namely the linear, the transition, and the equilibrium. The variation of the modulus with dynamic strain amplitude depends on the composition of the material. For instance, if the material has a significant amount of fillers such as carbon black, then its nonlinearity is greater, as illustrated in Figure 3.7.

### 3.2.4.  Effects of Static Preload

The effects of static preload on the dynamic properties of materials are usually most important in the rubbery region, as illustrated in Figure 3.8. It is seen that the modulus increases with increasing preload, whereas the loss factor decreases.

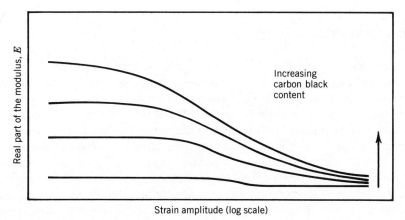

**FIGURE 3.7.**   Variation of the real part of the modulus of natural rubber with strain amplitude for different contents of carbon black.

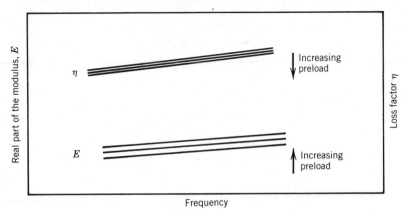

**FIGURE 3.8.**   Variation of the real part of the modulus and loss factor with increasing preload.

### 3.2.5.  Effects of Other Environmental Factors

The damping properties of rubberlike materials are often affected by other environmental factors such as aging, oil exposure, high temperature exposure, vacuum, and pressure. The changes could be very significant, if extremes are encountered, such as high temperature exposure. It is therefore usually advisable to study each possible application, for each damping material, and perform the tests necessary to evaluate such effects. Generally speaking, if a material is continually exposed to a specific environment, such as high temperature, then its properties are likely to display increased modulus values and decreased loss factor values, similar to the effects of static preload.

### 3.3.  ANALYTICAL MODELING

Based on the foregoing discussions, it is evident that the properties of damping materials change considerably with different environments. At first sight this seems to be a rather complex phenomenon to grasp and describe analytically. However, it turns out that there are a number of principles that can be used to overcome such difficulties. One such principle is the fact that the effect of frequency is the inverse of that of temperature. Similar approaches can also be taken for the effects of static and dynamic strain amplitudes. The analytical representation of frequency effects on the dynamic properties will be examined first. The effects of temperature will then be combined with those of frequency. The effects of both the static and dynamic strain amplitudes will subsequently be incorporated with the preceding two effects, so that it will be possible to describe the behavior of damping materials under combined environments.

### 3.3.1.  Representation of Frequency Effects

Using the complex modulus representation, which has been discussed in previous chapters, Young's modulus of a rubberlike material can be written as

$$E^*(\omega) = E(\omega)[1 + i\eta(\omega)] \tag{3.1}$$

where the two functions of frequency, $E(\omega)$ and $\eta(\omega)$, can be represented analytically in a number of different ways. One of the approaches that has been suggested [3.5, 3.6], and found to fit measured data on a variety of rubberlike materials, is

$$E(\omega) = \check{E} + \hat{E}(1 - \Phi) \tag{3.2}$$

where $\check{E}$ and $\hat{E}$ are the minimum and maximum values of the modulus of the material with respect to frequency and $\Phi$ is a function of frequency satisfying the following conditions:

$$\lim_{\omega \to 0} \Phi = 1 \quad \text{and} \quad \lim_{\omega \to \infty} \Phi = 0 \tag{3.3}$$

A function satisfying these two limits is

$$\Phi = \frac{1}{1 + (\beta\omega)^n} \tag{3.4}$$

where $\beta$ and $n$ are constants depending on the material. Concerning the loss factor $\eta(\omega)$, it has been shown [3.2, 3.5] to be approximately equal to

$$\eta(\omega) = \frac{\pi}{2} \frac{d[\ln E(\omega)]}{d[\ln (\omega)]} \tag{3.5}$$

Thus the two expressions for the modulus and loss factor of the material can be written in terms of frequency in the form

$$E(\omega) = \check{E} + \hat{E}\left[ 1 - \frac{1}{1 + (\beta\omega)^n} \right] \tag{3.6}$$

$$\eta(\omega) = \frac{n\pi\hat{E}(\beta\omega)^n}{2E(\omega)[1 + (\beta\omega)^n]^2} \tag{3.7}$$

Figure 3.9 shows the effects of different values of $\beta$, $n$, and $\check{E}$ on the computed values of $E(\omega)$ and $\eta(\omega)$ from equations (3.6) and (3.7). It is interesting to note that the two functions $E(\omega)$ and $\eta(\omega)$ can be completely described

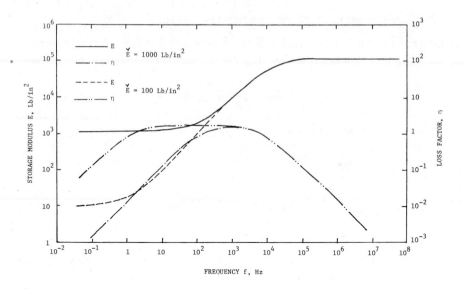

**FIGURE 3.9.** (*a*) Effects of different $\breve{E}$ on damping properties ($\hat{E} = 10^5$ lb/in$^2$).

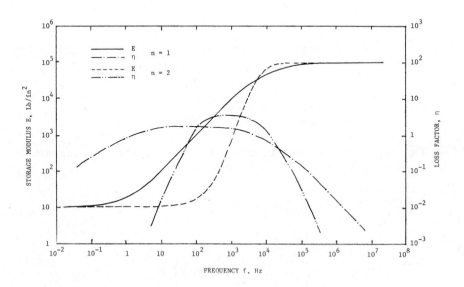

**FIGURE 3.9.** (*b*) Effects of different $\eta$ on the damping properties ($\hat{E} = 10^5$ lb/in$^2$).

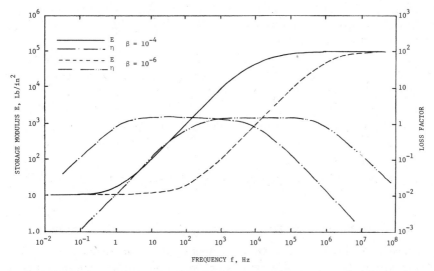

**FIGURE 3.9.** (c) Effects of different $\beta$ on the damping properties ($\hat{E} = 10^5$ lb/in$^2$).

by the four constants $\hat{E}$, $\check{E}$, $\beta$, and $n$, which can be determined from experimentally measured data. To determine these constants from measured data, so that analytical expressions of the material properties can be generated, it is useful to note the expression for the loss modulus of the material, which is the product of the storage modulus and the loss factor:

$$E''(\omega) = E(\omega)\eta(\omega) = \frac{\pi d[E(\omega)]}{2d[\ln{(\omega)}]} = \frac{n\pi\hat{E}(\beta\omega)^n}{2[1 + (\beta\omega)^n]^2} \qquad (3.8)$$

The loss modulus attains its maximum value, with respect to frequency, when its derivative with respect to $\omega$ is zero:

$$\therefore \quad \frac{d(E\eta)}{d\omega} = 0 \qquad (3.9)$$

substituting equation (3.8) into (3.9) yields:

$$(\beta\omega)^n = 1 \qquad (3.10)$$

Equation (3.10) can then be used to generate the expression for the maximum value of the loss modulus:

$$[E(\omega)\eta(\omega)]_{\text{max}} = \frac{n\pi\check{E}}{8} \qquad (3.11)$$

Thus the constant $n$ can easily be determined from the maximum values of the storage and loss moduli. It can be seen from the foregoing discussion that the analytical expressions for the variation of the dynamic properties in terms of frequency can be determined by first plotting the experimentally measured storage and loss moduli against $\omega$ and then determining the four constants as follows:

1.  Note the maximum values of the storage and loss moduli, the minimum value of the storage modulus, and the location of $\omega$ where the loss modulus is at its maximum.
2.  Compute $n$ and $\beta$ from equations (3.11) and (3.10), respectively.

### 3.3.2.   Representation of Frequency-Temperature Effects

The inverse relationship between the temperature and frequency effects is very useful in establishing the temperature-frequency superposition principle [3.2, 3.3]. By this principle measurements of the damping properties as a function of frequency made at different temperatures can be collapsed on one master graph, if the appropriate temperature shift factor is used. Considerable work has been done to establish such shift factors, along with the necessary graphical and analytical representations.

**Temperature-frequency superposition principle.**  If the effects of both frequency and temperature on the damping behavior of materials are to be taken into account, one of the most useful techniques for presenting the experimental data is the temperature-frequency equivalence (reduced frequency) principle for linear viscoelastic materials [3.2, 3.3]. In this approach $(T_0\rho_0/T\rho)E$ and $\eta$ are plotted against the so-called reduced frequency parameter $\omega\alpha_T$, where $\omega$ is the actual frequency, $\alpha_T$ is a function of the absolute temperature $T$, and $T_0$ is a reference temperature, again on the absolute scale. Usually $T_0/T$ and the density ratio $\rho_0/\rho$ are regarded as being 1.0 over a wide temperature range and are ignored. The preparation of "master curves" of $E$ and $\eta$ against $\omega\alpha_T$ is extremely useful for extrapolating test results obtained under a wide set of conditions. For example, in a test series one might have data over the frequency range 100 to 1000 Hz and the temperature range 0° C to 100° C and wish to estimate the properties at 50° C and 2 Hz. In order to achieve this goal, one first uses the available data to produce a best fitting set of master curves. The process is most satisfactorily accomplished empirically by judging the factor $\alpha_T$ on the basis of the shift needed to make the curve of log $E$ versus log (frequency), from Figure 3.4, at temperature $T_i$ $(i = \pm 1, 2, \ldots)$ match as closely as possible the curve $E$ versus frequency at temperature $T_0$, while matching the curves of $\eta$ versus frequency at temperatures $T$ and $T_0$, to produce curves like those shown in Figure 3.10. In this way the limitations of

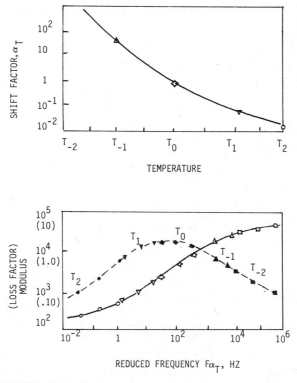

**FIGURE 3.10.** Typical reduced frequency and shift factor plots.

the measuring techniques can be at least partly compensated for. Typical graphs of $\alpha_T$ versus temperature are shown in Figure 3.11.

**A reduced temperature nomogram.** The graph of $E$ and $\eta$ versus reduced frequency $\omega\alpha_T$ represents a fundamental relationship between the various parameters and variables. However, its use to directly read off the modulus and loss factor for any given frequency and temperature is impeded by the inconvenience of reading $\alpha_T$, calculating $\omega\alpha_T$, and then reading off $E$ and $\eta$ for each data point. However, if we relabel the scales on this graph, we can readily create an extremely simple nomogram to do this tedious (if simple) calculation for us [3.8–3.10]. We merely use the right-hand scale for frequency, as indicated in Figure 3.12. Then for $f = 1$, $f\alpha_T = \alpha_T$, so that the points corresponding to the values of $\alpha_T(= f\alpha_T)$ for specific selected temperatures $T_0$, $T_1$, $T_2$, ..., can be drawn in. Similarly for $f = 10$ Hz, $f\alpha_T = 10\alpha_T$, and again the points can be drawn in. If the points corresponding to each $T_i$ ($i = \pm 1$, $\pm 2, \ldots$) are then filled in, they will form oblique lines as indicated. To use the resulting nomogram, for each specific $f$ and $T_i$ one then moves up the oblique

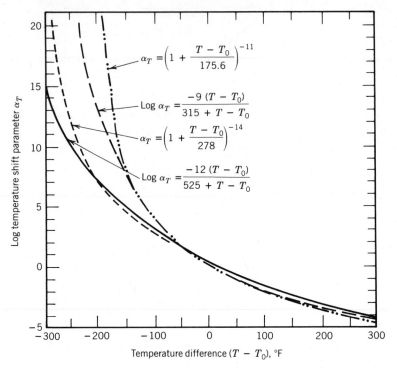

FIGURE 3.11. Typical variation of the shift factor with temperature.

$T_i$ line until it crosses the horizontal $f$ line. The intersection point $X$ corresponds to the proper value of $f\alpha_T$. One then moves vertically to read off the values of $E$ and $\eta$.

Another possible use for such a nomogram is to reduce the data in the first place. For if one selects the proper position for $T_0$ and the interval $(\Delta T)$ between $T_0$ and $T_1$, $T_2$, and so on, then the grid lines can be used to place the test data in position (in effect, by calculating $f\alpha_T$ according to the assumed $T_0$ and $\Delta T$). Only one combination of values of $T_0$ and $\Delta T$ will give an adequate reduction of the data—just as in the original reduced-frequency approach the values of $\alpha_T$ are dictated by the test data!

**Analytical modeling for single transition material.** The analytical expression for the variation of the damping properties with frequency can be extended to include the effects of temperature if the shift factor $\alpha_T$ is known as a function of temperature, so that equations (3.6) and (3.7) can be rewritten in the form

$$E(\omega, T) = \breve{E} + \hat{E}\left[1 - \frac{1}{1 + (\beta\omega\alpha_T)^n}\right] \qquad (3.12)$$

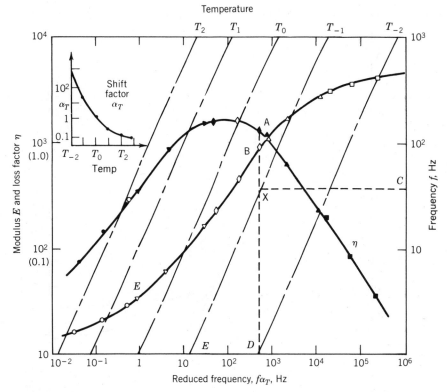

**FIGURE 3.12.**  Variation of $E$ and $\eta$ with reduced frequency $f\alpha_T$. Also construction of reduced temperature nomogram.

and

$$\eta(\omega, T) = \frac{n\pi\hat{E}(\beta\omega\alpha_T)^n}{2E[1 + (\beta\omega\alpha_T)^n]^2} \tag{3.13}$$

where $\alpha_T$ is still to be determined as a function of temperature. Of the various forms that have been suggested for the shift factor $\alpha_T$ [3.2], the following relation is selected for modeling purposes:

$$\log \alpha_T = \frac{-C_1(T - T_0)}{(T - T_\infty)} \tag{3.14}$$

where $C_1$, $T_0$, and $T_\infty$ are material constants to be determined experimentally. $T_0$ is usually selected to be approximately where the material takes on its maximum loss modulus, whereas $T_\infty$ is selected to be somewhere in the glassy region. The main advantage of this representation, for the shift factor, is the

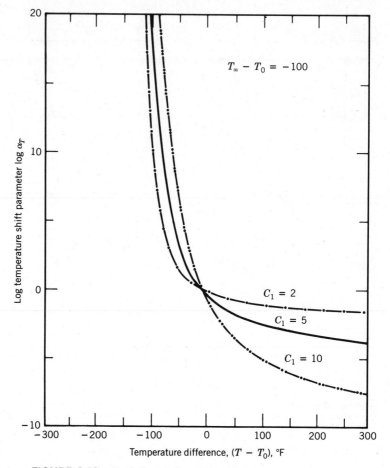

**FIGURE 3.13.** Typical variation of the shift factor with temperature.

ability to select easily the width of the transition region with temperature. Figures 3.13 and 3.14 illustrate the effects of different values selected for the material constants on $\log \alpha_T$.

### 3.3.3. Representation of Frequency-Dynamic Strain Effects

The frequency-dynamic strain equivalence has been found [3.7] to be similar to that of the frequency-temperature one. For nonlinear materials in their rubbery region, high dynamic strain amplitude effects are equivalent to low frequency, whereas low dynamic strain amplitude effects are equivalent to high frequency. Thus the frequency-dynamic strain superposition principle can be used in much the same manner as for the case of temperature and frequency.

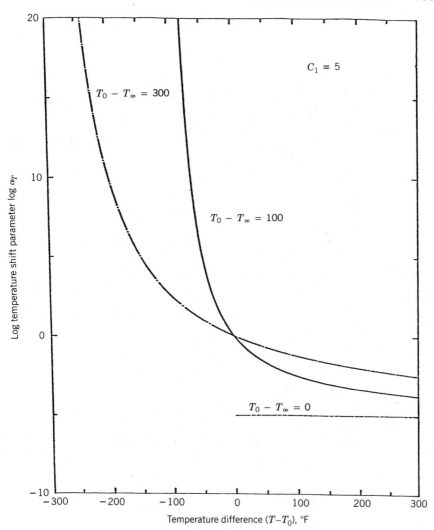

**FIGURE 3.14.**   Typical variation of the shift factor with temperature.

Because of the similarities between the temperature and dynamic strain amplitude effects, as seen in Figures 3.2 and 3.6, it is possible to use the approach illustrated in Figures 3.4 and 3.10 to collapse the data on one master graph. Figure 3.15 represents the variation of the modulus and loss factor with frequency for five different strain amplitude levels, for a typical filled rubber at a temperature within its rubbery region. A reference strain level $\varepsilon_0$ is chosen to be between the low strain values of $\varepsilon_1$ and $\varepsilon_2$ and the high ones of $\varepsilon_3$ and $\varepsilon_4$. Using the same approach as discussed in Section 3.3.2. the data of Figure 3.15 can be collapsed on one master graph as shown in Figure 3.16 by using a shift factor $\alpha_\varepsilon$. Equations can now be considered for the frequency-dynamic strain

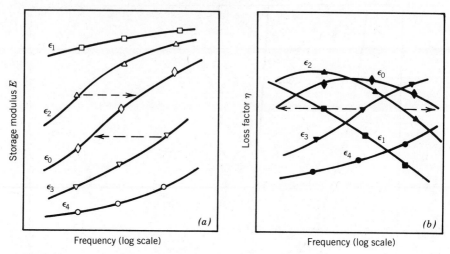

**FIGURE 3.15.** Variation of the complex storage modulus and loss factor with frequency and strain.

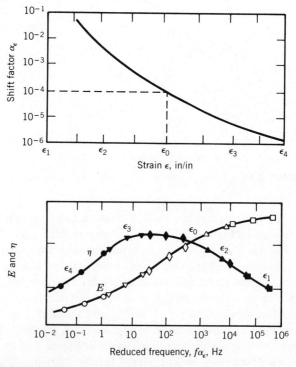

**FIGURE 3.16.** Typical reduced frequency and shift factor plots.

**104**

equivalence, similar to those of the frequency-temperature one. As an example, the shift factor $\alpha_\varepsilon$ has been shown [3.7] to take on the form

$$\log \alpha_\varepsilon = \frac{-C_1(\varepsilon - \varepsilon_0)}{C_2 + \varepsilon - \varepsilon_0} \tag{3.15}$$

where $C_1$ and $C_2$ are material constants that must be determined experimentally from measured data.

### 3.3.4.  Representation of Preload Effects

The effect of static preload on the dynamic properties of materials has been illustrated in Figure 3.8. For many applications, such as in engine mounts, this effect is rather important, especially when good isolation characteristics are required at high frequencies. In such applications the effects of temperature are also present because of the engine environment. Thus, to design rubber materials in various isolation or damping configurations, it is necessary to evaluate their properties as functions of the combined static and dynamic strains. However, because of the large number of combinations of parameters, the testing of materials becomes difficult to perform. An alternative is to use an approach by which the effects of the various environments can be separated so that it would be sufficient to test a given material for its static and dynamic properties, independently of each other, and then use the analysis to predict their combined effects. A general theory has been developed [3.11] to predict the combined linear dynamic and nonlinear static behavior of viscoelastic materials. A similar approach which yields simpler results will be discussed here and is based on the use of the Mooney–Rivlin equation [3.12, 3.13]. The nonlinear static representation by the Mooney–Rivlin equation will first be discussed and will then be extended to include the dynamic effects.

**Representation of nonlinear static strain effects.**  The Mooney–Rivlin equation is normally used to describe the nonlinear static stress-strain relationship for extensional strains $\lambda$ of magnitude up to $\lambda = 2$ or 3. For simple extension this relationship is

$$\sigma = 2(C_1\lambda + C_2)\left(\lambda - \frac{1}{\lambda^2}\right) \tag{3.16}$$

where $\lambda$ is the extensional strain, which relates to the engineering strain, $\varepsilon$, by

$$\lambda = \varepsilon + 1 \tag{3.17}$$

Equation (3.16) therefore becomes

$$\sigma = 2[C_1(\varepsilon + 1) + C_2]\left[\varepsilon + 1 - \frac{1}{(\varepsilon + 1)^2}\right] \tag{3.18}$$

To simplify equation (3.18), as $\varepsilon$ becomes small, we rewrite it in the form

$$\sigma = 2[C_1(\varepsilon + 1) + C_2]\left[\frac{\varepsilon(\varepsilon^2 + 3\varepsilon + 3)}{(\varepsilon + 1)^2}\right] \tag{3.19}$$

$$\therefore \quad \frac{\sigma}{\varepsilon} = 2[C_1(\varepsilon + 1) + C_2]\left[\frac{\varepsilon^2 + 3\varepsilon + 3}{(\varepsilon + 1)^2}\right] \tag{3.20}$$

For small strains equation (3.20) gives

$$\frac{\sigma}{\varepsilon} = 6(C_1 + C_2) \tag{3.21}$$

Thus the static Young's modulus $E_0$ can be expressed as

$$E_0 = 6(C_1 + C_2) \tag{3.22}$$

**Representation of combined static and dynamic nonlinear effects.**   For rubberlike materials subjected to a linear dynamic load superimposed on a nonlinear static load, it has been suggested [3.2] that the stress can be factored into a function of frequency, $\omega$, and a function of strain, $\lambda$, or

$$\sigma(\omega, \lambda) = F(\lambda)G(\omega) \tag{3.23}$$

The function $F(\lambda)$ could take on several forms, one of which is given by equation (3.16). Thus equation (3.23) could be rewritten as

$$\sigma(\omega, \lambda) = 2(C_1\lambda + C_2)\left(\lambda - \frac{1}{\lambda^2}\right)G(\omega) \tag{3.24}$$

Equation (3.24) can now be used to describe the combined static and dynamic modulus properties of rubberlike materials by substituting it into the following expression:

$$E(\omega, \lambda) = \lambda\frac{d\sigma}{d\lambda} \tag{3.25}$$

$$\therefore \quad E(\omega, \lambda) = [C_1 F_1(\lambda) + C_2 F_2(\lambda)]G(\omega) \tag{3.26}$$

where

$$F_1(\lambda) = 2\left[2\lambda^2 + \frac{1}{\lambda}\right] \tag{3.27}$$

$$F_2(\lambda) = 2\left[\lambda + \frac{2}{\lambda^2}\right] \tag{3.28}$$

Using the complex modulus approach to describe the damping behavior of rubberlike materials, equation (3.26) can be rewritten as

$$E(\omega, \lambda)[1 + i\eta(\omega, \lambda)] = \{C_1[1 + i\eta_1(\omega)]F_1(\lambda) + C_2[1 + i\eta_2(\omega)]F_2(\lambda)\}G(\omega)$$
(3.29)

The real and imaginary parts of equation (3.29) can be used to describe the real part of the modulus, and the loss factor, of the material:

$$E(\omega, \lambda) = [C_1F_1(\lambda) + C_2F_2(\lambda)]G(\omega)$$
(3.30)

$$\eta(\omega, \lambda) = [C_1F_1(\lambda)\eta_1 + C_2F_2(\lambda)\eta_2]G(\omega)$$
(3.31)

To determine the constants $\eta_1$, $\eta_2$, and $G$, it is necessary to note that [3.2] that $\eta_1$ can be assumed to be zero. Also as $\lambda \to 1$, the following expressions can be written:

$$\lim_{\lambda \to 1} E(\omega, \lambda) = E$$

$$\lim_{\lambda \to 1} \eta(\omega, \lambda) = \eta$$

$$\lim_{\lambda \to 1} F_1(\lambda) = 6$$

$$\lim_{\lambda \to 1} F_2(\lambda) = 6$$

Hence equations (3.30) and (3.31) become

$$E = 6(C_1 + C_2)G$$
(3.32)

and

$$\eta = 6C_2\eta_2 G$$
(3.33)

Also

$$6(C_1 + C_2) = E_0$$
(3.34)

where $E_0$ is the static equilibrium modulus. Thus equations (3.30) and (3.31) become

$$E(\omega, \lambda) = [C_1F_1(\lambda) + C_2F_2(\lambda)]\frac{E(\omega)}{6(C_1 + C_2)}$$
(3.35)

$$\eta(\omega, \lambda) = \frac{(C_1 + C_2)F_2(\lambda)}{C_1F_1(\lambda) + C_2F_2(\lambda)}\eta(\omega)$$
(3.36)

**TABLE 3.1. Static Tension and Compression Data**

| $\lambda$ | 0.81 | 0.86 | 0.91 | 1.1 | 1.19 | 1.37 | 1.49 | 1.76 | 2.05 |
|---|---|---|---|---|---|---|---|---|---|
| $\sigma$(psi) | $-26$ | $-19$ | $-14$ | 12 | 28 | 49 | 64 | 97 | 133 |

**TABLE 3.2. Combined Static and Dynamic Tension Data**

| Frequency (Hz) | $\lambda$ | $E(\omega, \lambda)$ (psi) | $\eta(\omega, \lambda)$ |
|---|---|---|---|
| 0.05 | 1.1 | 160 | 0.036 |
| | 1.19 | 170 | 0.027 |
| | 1.37 | 203 | 0.027 |
| | 1.76 | 254 | 0.017 |
| | 2.05 | 300 | |
| 0.1 | 1.1 | 161 | 0.047 |
| | 1.19 | 173 | 0.027 |
| | 1.37 | 206 | 0.026 |
| | 1.76 | 255 | 0.022 |
| | 2.05 | 305 | |
| 0.50 | 1.1 | 161 | 0.050 |
| | 1.19 | 176 | 0.032 |
| | 1.37 | 206 | 0.032 |
| | 1.76 | 258 | 0.025 |
| | 2.05 | 309 | |
| 1.0 | 1.1 | 164 | 0.058 |
| | 1.19 | 187 | 0.029 |
| | 1.37 | 212 | 0.036 |
| | 1.76 | 266 | 0.028 |
| | 2.05 | 313 | |
| 2.0 | 1.1 | 167 | 0.063 |
| | 1.19 | 184 | 0.031 |
| | 1.37 | 215 | 0.044 |
| | 1.76 | 274 | 0.034 |
| 5.0 | 1.1 | 170 | 0.056 |
| | 1.19 | 191 | 0.046 |
| | 1.37 | 223 | 0.047 |
| 10.0 | 1.1 | 180 | |
| | 1.19 | | |
| | 1.37 | 231 | 0.064 |

Equations (3.35) and (3.36) provide a representation of the modulus and loss factor of rubberlike materials subjected to combined static and dynamic loadings, based on static and dynamic properties which can be measured independently of each other. To investigate the validity of the representation used in equations (3.35) and (3.36), measurements reported in [3.11] for rubberlike materials at different static and dynamic loading conditions will be used. These results are given in Table 3.1 for the static measurements and in Table 3.2 for the combined static and dynamic measurements.

The constants $C_1$ and $C_2$ are determined by plotting the data of Table 3.1, in terms of $\bar{\sigma}$ against $\lambda$, based on equation (3.16), where

$$\bar{\sigma} = \frac{\sigma}{2(\lambda - 1/\lambda^2)}$$

as shown in Figure 3.17. It can be seen from the line drawn between the points that $C_1 + C_2 = 24$ and $C_1 = 14$. Therefore the values to be used for this material are $C_1 = 14$ and $C_2 = 10$, which gives a static modulus value at $\lambda = 1$ of $E(\lambda)|_{\lambda \to 1} = 144 \text{ lb/in}^2$. Equations (3.35) and (3.36) can now be used assuming that the material properties are measured statically to obtain $C_1$ and $C_2$ and measured dynamically without preload to determine $E(\omega)$. Since the dynamic properties of the material (Table 3.2) were measured with different combined static preload conditions, an extrapolation of the data was made to determine the values of the modulus and loss factor at different frequencies but without

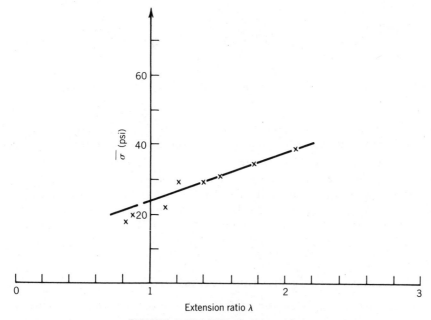

**FIGURE 3.17.** Variation of $\bar{\sigma}$ with $\lambda$.

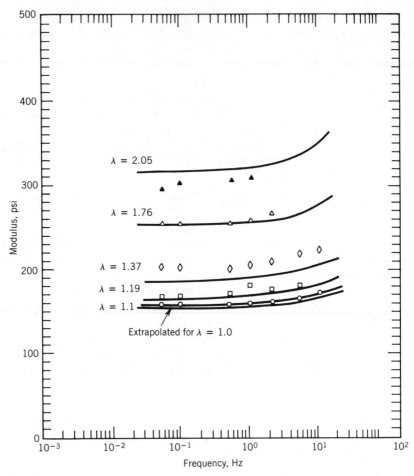

**FIGURE 3.18.**   Variation of the real part of the modulus with frequency and preload.

preload effects. Those extrapolated values were then used to predict the properties at all the other preload values. The results are given in Figures 3.18 and 3.19. It can be seen in these figures that equations (3.35) and (3.36) do accurately represent the combined static and dynamic behavior of the material. In other words, it is often sufficient to obtain the material properties statically, to determine $C_1$ and $C_2$, and then dynamically without preload, to determine $E(\omega)$ and $\eta(\omega)$, and then use equations (3.35) and (3.36) to predict the combined behavior in terms of the modulus $E(\omega, \lambda)$ and loss factor $\eta(\omega, \lambda)$.

### 3.3.5.   General Analytical Representation

A more general representation of the nonlinear behavior of rubberlike materials under combined static and dynamic loading can be developed if the

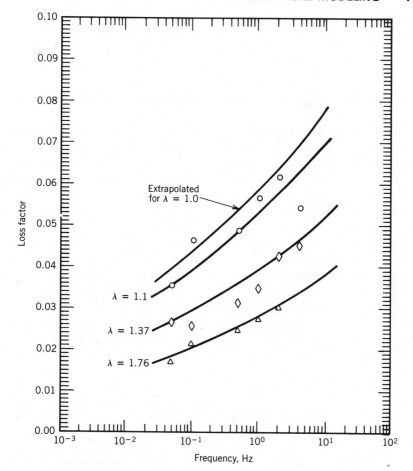

**FIGURE 3.19.**   Variation of the loss factor with frequency and preload.

combined stress is factored into two components, one containing the nonlinear static behavior and the other containing the nonlinear (or linear) dynamic behavior. Such a representation could be of the form

$$E(\lambda, \omega, T, \varepsilon) = [C_1 F_1(\lambda) + C_2 F_2(\lambda)] \frac{E(\omega, T, \varepsilon)}{6(C_1 + C_2)} \qquad (3.37)$$

$$\eta(\lambda, \omega, T, \varepsilon) = \frac{(C_1 + C_2) F_2(\lambda)}{C_1 F_1(\lambda) + C_2 F_2(\lambda)} \, \eta(\omega, T, \varepsilon) \qquad (3.38)$$

where $E(\lambda, \omega, T, \varepsilon)$ and $\eta(\lambda, \omega, T, \varepsilon)$ are the modulus and loss factor as a function of the static extension ratio $\lambda$, frequency $\omega$, temperature $T$, and dynamic engineering strain amplitude $\varepsilon$. The constants $C_1$ and $C_2$ are determined from static measurements only, whereas the modulus $E(\omega, T, \varepsilon)$ and $\eta(\omega, T, \varepsilon)$ are determined from dynamic measurements only on the material.

## 3.4.  PROPERTIES OF TYPICAL DAMPING MATERIALS

Materials are usually evaluated for their dynamic properties by a number of different measurement techniques, depending on the type of environment of interest. For example, the vibrating beam technique [3.3, 3.14–3.16] is often used to characterize the linear dynamic properties in terms of temperature and frequency for either shear or extensional deformation. For investigating the effects of static and dynamic loadings, impedance [3.17, 3.18] and resonance [3.3, 3.19–3.20] techniques are often used. The approximate analytical or graphical representation is then used to describe the material properties. From such a representation the material properties for a desired condition can be extrapolated, but extrapolations made far outside the range of the material measurements might be questionable. This is because the various reducing principles have not been fully verified over wide ranges of environment. In this section a general discussion will be given regarding measurements of the damping properties of a material and how the measured data can be curve fitted by using the analytical representation discussed previously.

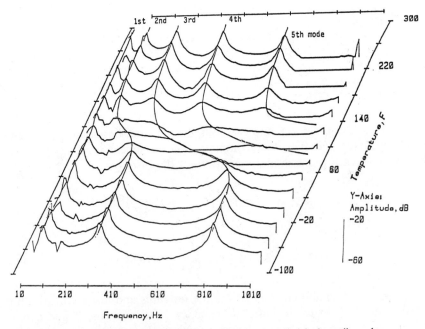

**FIGURE 3.20.**  Temperature spectrum map for a sandwiched cantilever beam.

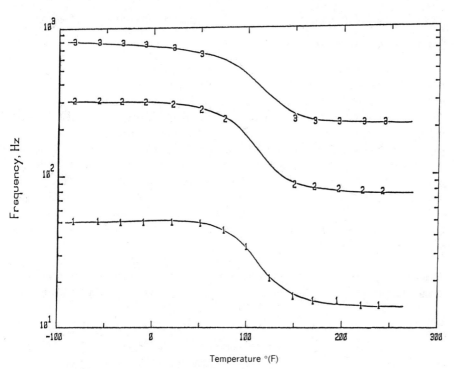

FIGURE 3.21. Variation of the resonant frequency with temperature.

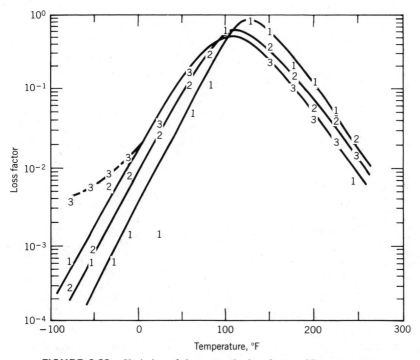

FIGURE 3.22. Variation of the composite loss factor with temperature.

**113**

## 3.4.1. Measurement and Analysis

The first consideration in making measurements of the damping properties of materials is to select the appropriate measurement technique. Since the effects of both temperature and frequency are to be demonstrated, the vibrating beam technique was selected. Also, since the material is often used in a constrained layer type of application, its properties under shear loading must generated. Thus the sandwich beam is considered. The frequency response of a typical damped sandwich beam is illustrated in Figure 3.20 for a number of different temperatures, ranging from the glassy region to the rubbery region of the material. The frequency and damping for each mode of vibration, in terms of temperature, are shown in Figures 3.21 and 3.22. Either the actual data from these two figures or the data read off the smoothed graphs can be used to compute the material properties. However, the use of smoothed data is recommended if the scatter is excessive. Such computations are usually made with geometry of the beam, and the undamped frequencies, assumed known. The damping properties for each resonance peak, and each temperature, are

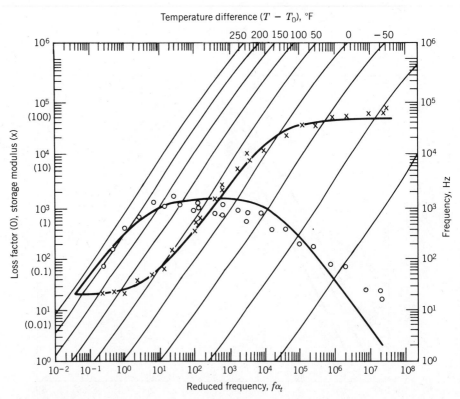

**FIGURE 3.23.** Variation of the storage modulus and loss factor with reduced frequency (typical material).

then calculated [3.14] and plotted in terms of reduced variables, using the temperature frequency nomogram as shown in Figures 3.23 and 3.24. Superimposed on these figures are the analytical representations to describe the storage modulus, loss modulus, and loss factor. For this material (3.12), (3.13), and (3.14) were used. For these equations the following constants were found to give a good fit:

$$\hat{G} = 3.4 \times 10^4 \text{ psi}$$
$$\check{G} = 20 \text{ psi}$$
$$(\eta\hat{G}) = 1.2 \times 10^4 \text{ psi}$$
$$(\eta\check{G}): \quad \text{location with reduced frequency} = 1.5 \times 10^4 \text{ Hz}$$
$$T_0 = 125° \text{ F}$$
$$T_\infty = 300° \text{ F}$$
$$C_1 = 8$$

Using these constants in equations (3.12), (3.13), and (3.14), the damping properties of the material can be described continuously in terms of both temperature and frequency. The material properties can now be easily plotted or used in design calculations.

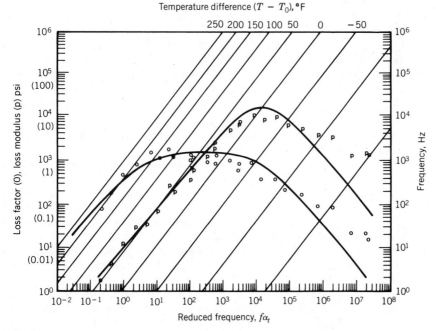

**FIGURE 3.24.**   Variation of the loss modulus and loss factor with reduced frequency (typical material).

## 3.4.2.  Properties of Typical Materials

An approach similar to the preceding one has been used to summarize the damping properties of many commercially available materials, as exemplified in Chapter 7.

# REFERENCES

3.1.   J. C. Snowdon, *Vibration and Shock in Damped Mechanical Systems*, Wiley, New York, 1968.

3.2.   J. D. Ferry, *Viscoelastic Properties of Polymers*, 2nd ed., Wiley, 1970.

3.3.   D. I. G. Jones, "Temperature-frequency dependence of dynamic properties of damping materials," *J. Sound Vib.*, **33**(4), 451–470 (1974).

3.4.   P. Grootenhuis, "Vibration control with viscoelastic materials," *Environmental Engineering*, Proc. SEE, No. 28, May 1969.

3.5.   F. Schwartzl, *Physica* **17**, 830–923 (1951).

3.6.   J. Heyboer, P. Dekking, and A. J. Staverman, Proc. 2nd International Congress on Rheology, p. 123, Oxford, 1953.

3.7.   G. E. Warnaka and H. T. Miller, "Strain-frequency temperature relationships in polymers," *J. Eng. Industry* **B90**(3), 491–498 (1968).

3.8.   D. I. G. Jones, "A reduced-temperature nomogram for characterization of damping material behavior," *Shock Vib. Bull.*, **48**, Pt. 2, 13–22 (1978).

3.9.   D. I. G. Jones and J. P. Henderson, "Specification of damping material performance," *Shock Vib. Bull.*, **48**, Pt. 2, 1–11 (1978).

3.10.   D. I. G. Jones, "An attractive method for displaying material damping data," *AIAA J.*, **18**(8), 644–649 (1981).

3.11.   J. L. Sullivan, K. N. Morman, and R. A. Pett, "A non-linear viscoelastic characterization of a natural rubber gum vulcanizate," *Rubber Chem. Tech.*, **53**(5) (1980).

3.12.   M. Mooney, "A theory of large elastic deformations," *J. Appl. Phys.*, **11**, 582–592 (1940).

3.13.   R. S. Rivlin, "Torsion of a rubber cylinder," *J. Appl. Phys.*, **18**, 444–449 (1947).

3.14.   "Standard for measuring vibration damping properties of materials," American Society for Testing and Materials, E756-80, 1980.

3.15.   A. D. Nashif, "A new method for determining the damping properties of viscoelastic materials," *Shock Vib. Bull.*, **36**, 37–47 (1967).

3.16.   F. S. Owens, "Elastomers for damping over wide temperature ranges," *Shock Vib. Bull.*, **36**, Pt. 4, 25–35 (1967).

3.17.   J. L. Edwards and D. R. Hicks, "Useful range of a mechanical impedance technique for measurement of dynamic properties of materials," *J. ASA*, **52**, 1053–1056 (1972).

3.18.   D. I. G. Jones and A. Muszynska, "On the modal identification of multiple degree of freedom systems from experimental data," *Shock Vib. Bull.*, **53**(2), 91–110 (1983).

3.19.   C. M. Cannon, A. D. Nashif, and D. I. G. Jones, "Damping measurements on soft viscoelastic materials using a tuned damper technique," *Shock Vib. Bull.*, **38**, Pt. 3, 151–163 (1968).

3.20.   D. L. Hunston, W. D. Bascom, E. E. Wells, F. D. Fahey, and J. L. Bitner, "Viscoelastic characterization of structural adhesive via forced oscillation experiments," *Adhesion and Absorption of Polymers* (ed. Lieng-Huang Lee), Plenum, New York, 1980.

# 4

# MODELING OF STRUCTURAL RESPONSE OF DAMPED SYSTEMS

## ADDITIONAL SYMBOLS

| | |
|---|---|
| $a$ | $C/2m$, damping parameter |
| $A_1, B_1$ | amplitudes of forced response |
| $A$ | amplification factor at resonance |
| $C_0, C_1, \ldots, C_3$ | nondimensional parameters |
| $d_{jk}$ | parameters ($j, k = 0, 1, 2, \ldots, 8$) |
| $D_s$ | energy dissipated in cycle |
| $F_s(t)$ | stiffness force |
| $F_D(t)$ | damping force |
| $F(t)$ | exciting force |
| $\bar{F}(\omega)$ | Fourier transform of $F(t)$ |
| Im | imaginary part |
| $k^*$ | complex modulus $[= k' + ik'' = k'(1 + i\eta)]$ |
| $l_1, l_2$ | partial lengths of beam |
| $\ln(x)$ | natural logarithm of argument $x$ |
| $L$ | length of suspension material |
| $m, n$ | integers |

| | |
|---|---|
| $n$ | fraction of peak amplitude |
| Re | real part |
| $T$ | torsional spring constant |
| $U, U_s$ | maximum energy stored in cycle |
| $_j[U]_i$ | transfer matrix |
| $\bar{W}(\omega)$ | Fourier transform of $w(t)$ |
| $w_c$ | complementary function |
| $w_0(t)$ | displacement of base |
| $W_0$ | peak amplitude of base movement |
| $\bar{W}_0(\omega)$ | Fourier transform of $w_0(t)$ |
| $w_p$ | particular integral |
| $W_r$ | amplitude of relative displacement between mass and base |
| $W_{rms}$ | root mean square response amplitude |
| $W_{st}$ | static displacement amplitude |
| $\{Z\}_j$ | state vector at point $j$ |
| $\alpha$ | receptance of 1 DOF system |
| $\alpha_D, \alpha_Q$ | direct and quadrature receptances |
| $\beta$ | $(\rho b H / EI)^{1/4}$ |
| $\delta(t)$ | Dirac delta function of argument $t$ |
| $\Delta \omega$ | frequency bandwidth |
| $\varepsilon, \alpha, \sigma$ | phase angles |
| $\lambda$ | dimensional parameter |
| $\zeta$ | damping ratio |
| $\omega_D$ | damped natural frequency (free vibration) |
| $\omega_0$ | undamped natural frequency |
| $\omega_r$ | damped resonant frequency (forced vibration) |

## 4.1  INTRODUCTION

When damping is deliberately introduced into a structure, specific changes are made in the manufacture of selected parts so that as the structure vibrates, and these parts deform, they act in turn cyclically to strain the added viscoelastic elements and dissipate energy. If damping materials are to be used to solve successfully structural vibration problems, it is necessary to understand not only the behavior of the damping materials but also the structural dynamics associated with the problem. To facilitate this understanding, it is often cost-effective to study a simplified mathematical model representing, in some respects the dynamic characteristics of the structure. This mathematical model can range in complexity from a single degree of freedom system, represented by a single lumped mass on a spring, to a sophisticated analytical description of a continuous system with distributed mass, stiffness, and damping properties combined with a distributed forcing function. The sophistication of the model used in a problem-solving situation will depend not only on the complexity of the structure but also on the time and other resources available for the solution to the problem.

In Chapter 1 some of the techniques used to analyze response of structures were discussed. It is the aim of this chapter to continue and expand that discussion. We shall first examine in considerable detail the very simplest damped structure, namely a single degree of freedom system, with various types of damping and subject to various types of excitation. Since damping can rarely, if ever, be measured directly but must be inferred from the response parameters measured in actual tests, such as displacements or accelerations, it follows that we can learn a great deal from the response of a single degree of freedom system as it is modified by damping. This knowledge can then be applied, with care, to far more complex systems. Also an understanding of steady state simple harmonic response is important not only because many of the problems that occur in vibrating structures exhibit single frequency steady state excitation but also because such an understanding forms a conceptual basis for frequency and time domain solutions for other kinds of excitations.

## 4.2 STEADY STATE RESPONSE OF A SINGLE DEGREE OF FREEDOM SYSTEM

### 4.2.1. Force Excitation—Viscous Damping

The simplest system used to illustrate steady state response is the linear single degree of freedom oscillator illustrated in Figure 4.1. Although this system does not accurately represent most structures in real life, it does display some of the essential features of complex real structures. The system consists of a mass $m$ attached to a spring $k$, with damping being represented classically by a

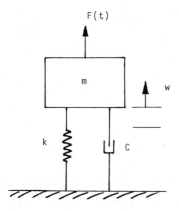

**FIGURE 4.1.** Single degree of freedom system with viscous damping.

viscous dashpot, so that damping is proportional to velocity. With an excitation force $F(t)$ applied to the mass, the system will respond with a displacement $w(t)$, considered positive in the upward direction.

The force on the mass, resulting from elongation of the spring, is

$$F_s(t) = -kw(t)$$

and the force due to movement of the damper is

$$F_d(t) = -C\dot{w}(t)$$

with forces considered positive in the direction of positive displacement. Newton's second law gives

$$F(t) + F_s(t) + F_d(t) = m\ddot{w} \tag{4.1}$$

and yields the differential equation

$$m\ddot{w}(t) + C\dot{w}(t) + kw(t) = F(t) \tag{4.2}$$

Letting the excitation force be a steady state harmonic function $F(t) = F \cos \omega t$, equation (4.2) becomes the second-order linear differential equation

$$m\ddot{w}(t) + C\dot{w}(t) + kw(t) = F \cos \omega t \tag{4.3}$$

The complete solution of this equation is the algebraic sum of the complementary (or transient) solution, obtained by solving the homogeneous equation, for which the right-hand side of equation (4.3) is set equal to zero, and a particular (or steady state) solution of the equation including the forcing function on the right-hand side. The complementary solution of this equation can be shown to be of the form:

$$w_c = e^{-at}(C_1 \sin \omega_D t + C_2 \cos \omega_D t) \tag{4.4}$$

where

$$a = \frac{C}{2m}$$

$$\omega_D = \sqrt{\frac{k}{m} - \left(\frac{C}{2m}\right)^2}$$

This is a harmonic response at the damped natural frequency of the system, with amplitude that decays with time. The arbitrary constants, $C_1$ and $C_2$, are obtained by substituting the initial conditions into the complete solution of the equation.

The particular solution $w_p$ is any function $w(t)$ that satisfies the differential equation. It is readily shown that a particular solution of equation (4.3) is

$$w_p = \frac{F \cos(\omega t - \varepsilon)}{\sqrt{(k - m\omega^2)^2 + \omega^2 C^2}} \tag{4.5}$$

with

$$\varepsilon = \tan^{-1}\left[\frac{C\omega}{(k - m\omega^2)}\right] \tag{4.6}$$

Thus the complete solution of equation (4.3) is

$$w = w_c + w_p = e^{-at}(C_1 \sin \omega_D t + C_2 \cos \omega_D t) + A_1 \cos(\omega t - \varepsilon) \tag{4.7}$$

with

$$A_1 = \frac{F}{\sqrt{(k - m\omega^2)^2 + \omega^2 C^2}} \tag{4.8}$$

which shows that the response of this single degree of freedom system is the sum of a transient oscillation, at the natural frequency $\omega_D$, whose amplitude depends on initial conditions and decays with time, and a steady state oscillation at the frequency $\omega$ of the exciting force and lagging the excitation by a phase angle $\varepsilon$. The transient response due to the initial conditions quickly dies away, but the steady state response remains as long as the excitation is present, as illustrated in Figure 4.2. The steady state, or particular solution, behavior of this single degree of freedom system illustrates what happens when a harmonically varying force is applied to a resonant structure. As the frequency increases, the inertia term $-m\omega^2 w$ steadily increases until it is eventually equal to the stiffness $kw$. This is the state known as resonance. At this very important frequency the "stiffness" and "inertia" forces cancel each other out, and the amplitude of vibration is limited only by the damping. At frequencies

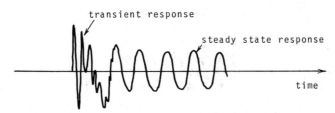

FIGURE 4.2.   Start-up transients for steady state response.

well above resonance the inertia term completely dominates, and the response amplitude becomes very small and is out of phase with the excitation (i.e., $\varepsilon \simeq 180°$). When the frequency of excitation $\omega$ is very low, the motion of the system is dominated by the spring stiffness $k$ and the response of the mass is in phase with the excitation (i.e., $\varepsilon \simeq 0$). At low frequencies the amplitude of the steady state dynamic displacement is approximately equal to the static displacement that would be caused by a constant force $F$. Three important frequencies can be identified for this viscous case:

1.  The undamped natural frequency, $\omega_0 = \sqrt{k/m}$, derived from equation (4.4) by letting $C = 0$.
2.  The damped natural frequency from equation (4.4)

$$\omega_D = \sqrt{\frac{k}{m} - \left(\frac{C}{2m}\right)^2} \tag{4.9}$$

3.  The resonant frequency for which $|w_p/F|$ is a maximum. To find this latter frequency, note that, by equation (4.5),

$$\left|\frac{w_p}{F}\right| = \frac{1}{\sqrt{(k - m\omega^2)^2 + \omega^2 C^2}} \tag{4.10}$$

For this function to be a maximum, $(k - m\omega^2)^2 + \omega^2 C^2$ must be a minimum. Therefore

$$d\frac{|F/w_p|^2}{d\omega} = 2(k - m\omega^2)(-2m\omega) - 2C^2\omega = 0$$

$$\therefore \quad \omega = \omega_r = \sqrt{\frac{k}{m}\left(1 - \frac{C^2}{2km}\right)} \tag{4.11}$$

Two more resonant frequencies can be calculated for the single degree of freedom system with viscous damping. These resonances are the frequencies at which maximum velocity, or maximum acceleration, occurs for a given excitation force, that is,

$$\omega_{r,\text{vel}} \equiv \text{the frequency at which } \left|\frac{\dot{w}_p}{F}\right| \text{ is a maximum}$$

$$\therefore \quad \omega_{r,\text{vel}} = \sqrt{\frac{k}{m}} = \omega_0 \tag{4.12}$$

and

$$\omega_{r,\,accel} \equiv \text{the frequency at which } \left| \frac{\ddot{w}_p}{F} \right| \text{ is a maximum}$$

$$\therefore \quad \omega_{r,\,accel} = \sqrt{\frac{k}{m}\left(1 - \frac{C^2}{2km}\right)^{-1}} \tag{4.13}$$

In many practical problems the differences between these resonant frequencies are small, and are often neglected.

### 4.2.2.   Force Excitation–Hysteretic Damping

The viscous damping assumption used in this analysis was chosen primarily for mathematical convenience, since it yields the only known simple solution for transient response, modeled by the complimentary function. However, for a large number of engineering problems, the effect of damping on the steady state response, represented by the particular solution, is of primary concern. Hysteretic damping, based on the concept of a complex modulus, can often be effectively utilized in the calculation of steady state response. Hysteretic damping was used by Bishop [4.1]. However, Lazan [4.2] considered damping to be associated with hysteresis loop effects, so that "rate-independent linear damping" would be more descriptive. One of the major advantages of assuming hysteretic damping is the possibility of utilizing the correspondence principle in complicated elastic analyses, where a complex number can be substituted for the real value of the modulus to account for damping. The principal difference between the viscous damped system and the hysteretic system is that for the viscous system the energy dissipated per cycle depends linearly on the frequency of oscillation, whereas for the hysteretic case it is independent of the frequency. An example will illustrate the point (Figure 4.3). Let the viscous

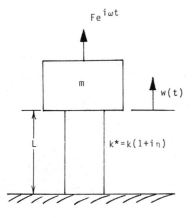

**FIGURE 4.3.**   Single degree of freedom system with hysteretic damping.

damping coefficient in equation (4.5) be $C = k\eta/\omega$. Then the particular solution shown in equation (4.5) becomes

$$w = w_p = B_1 \cos(\omega t - \varepsilon) \tag{4.14}$$

where

$$B_1 = \frac{F}{\sqrt{(k - m\omega^2)^2 + k^2\eta^2}}$$

$$\varepsilon = \tan^{-1}\frac{k\eta}{k - m\omega^2}$$

considering the steady state solution of equation (4.3), with $w_c$ assumed to be zero.

Recognizing that $e^{i(\omega t - \varepsilon)} = \cos(\omega t - \varepsilon) + i\sin(\omega t - \varepsilon)$, then $w = \text{Re}[B_1 e^{i(\omega t - \varepsilon)}]$, where $i^2 = -1$ and Re denotes the real part of the complex number, and hence

$$\dot{w} = \text{Re } B_1 i\omega e^{i(\omega t - \varepsilon)} = i\omega w \tag{4.15}$$

Therefore the differential equation (4.3) can be written in the form

$$m\ddot{w} + k(1 + i\eta)w = \text{Re}(Fe^{i\omega t}) \tag{4.16}$$

or

$$m\ddot{w} + k^*w = Fe^{i\omega t} \tag{4.17}$$

which is a valid form for the solution only for steady state response, with Re implied, and $k^* = k(1 + i\eta)$ is the complex stiffness of the suspension, accounting for both stiffness and damping. Thus our single degree of freedom system can be represented as shown in Figure 4.3, where

$$k^* = \frac{E^*S}{L} \tag{4.18}$$

where $E^* = E(1 + i\eta) = E' + iE''$ is the complex Young's modulus of the suspension system, $\eta$ is the loss factor of the suspension material, $S$ is the cross-sectional area, and $L$ the undeformed length of the suspension. In real materials $E$ and $\eta$ will depend on frequency and temperature, and these variations must be allowed for when using these equations. However, the hysteretic assumption, for which $E$, or $k$, and $\eta$ are assumed constant over a limited frequency range at a specific temperature can be very useful. $E$, $k$, and $\eta$

cannot be constant over the entire frequency domain, however, since this would imply a finite energy dissipation rate at zero frequency, among other things.

### 4.2.3. Comparison of Viscous and Hysteretic Damping

Displacement response functions for the single degree of freedom system with viscous damping and with hysteretic damping are shown in Figure 4.4$a$ and $b$, respectively. It can be seen the peak response $|W/F|$ occurs at a frequency lower than the undamped natural frequency for the viscous case, but in the case of the hysteretic damping the resonant peak always occurs at the undamped natural frequency. Energy dissipated per cycle, due to viscous damping, can be calculated by integrating the product of the force and the displacement over one cycle of vibration.

$$D_S = \oint F \, dw = \int_0^{2\pi/\omega} F\dot{w} \, dt$$

$$= F \int_0^{2\pi/\omega} \cos \omega t[-A_1\omega \sin (\omega t - \varepsilon)] \, dt$$

$$\therefore \quad D_S = \pi C\omega A_1^2 \tag{4.19}$$

Likewise in the case of hysteretic damping, using the relationship for $\dot{w}$ in equation (4.15)

$$D_S = F \int_0^{2\pi/\omega} \cos \omega t \, \text{Re}[B_1 i\omega e^{i(\omega t - \varepsilon)}] \, dt$$

$$\therefore \quad D_S = \pi k\eta B_1^2 \tag{4.20}$$

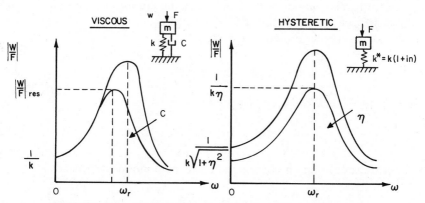

**FIGURE 4.4.**  Dynamic response of viscous and hysteretic systems.

**TABLE 4.1. Force Excitation of a Single Degree of Freedom System with Viscous or Hysteretic Damping**

|  | Viscous Damping | Hysteretic Damping |
|---|---|---|
| Differential equation | $m\ddot{w} + C\dot{w} + kw = F\cos\omega t$ | $m\ddot{w} + k(1 + i\eta)w = \text{Re}[Fe^{i\omega t}]$ |
| Particular solution (steady state forced vibration) | $W = A_1 \cos(\omega t - \varepsilon)$ $$A_1 = \frac{F}{\sqrt{(k - m\omega^2)^2 + \omega^2 C^2}}$$ | $W = B_1 \cos(\omega t - \varepsilon)$ $$B_1 = \frac{F}{\sqrt{(k - m\omega^2)^2 + k^2\eta^2}}$$ |
| Energy dissipated per cycle | $D = \pi C\omega A_1^2$ | $D = \pi k\eta B_1^2$ |
| Resonant frequency | Decreases with increasing value of $C$ | Independent of the value of $\eta$ |
| Static displacement at $\omega = 0$ | $\dfrac{F}{k}$ | Depends on the value of $\eta$* |
| Resonant amplitude | Depends on all equation parameters | Independent of mass |

* Since static displacement cannot depend on damping for a noncreeping solid, $F/K$ is commonly used as a reference static displacement for hysteretic damping.

The differences between viscous and hysteretic damping cases for a single degree of freedom system with force excitation are summarized in Table 4.1.

### 4.2.4. Base Excitation of a Single Degree of Freedom System

For many problems, such as a tuned damper attached to the surface of a structure, the one degree of freedom system is excited not by an external force applied to the mass $m$ but instead by oscillating motions of the base to which it is attached, as illustrated in Figure 4.5. For viscous damping the equation of motion for harmonic motion of the base is

$$m\ddot{w} + C(\dot{w} - \dot{w}_0) + k(w - w_0) = 0 \tag{4.21}$$

If $w_0 = W_0 e^{i\omega t}$, then equation (4.21) becomes

$$m\ddot{w} + C\dot{w} + kw = (k + iC\omega)W_0 e^{i\omega t} \tag{4.22}$$

The steady state solution corresponding to base motion $W_0 \cos\omega t$ is

$$w = W_0 \sqrt{\frac{k + (C\omega)^2}{(k - m\omega^2)^2 + (C\omega)^2}} \cos(\omega t - \alpha - \varepsilon_1) \tag{4.23}$$

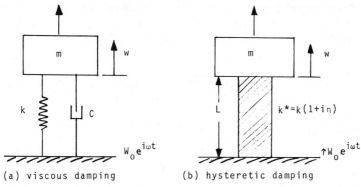

**FIGURE 4.5.**  Base excitation of single degree of freedom system. (*a*) Viscous damping. (*b*) Hysteretic damping.

where

$$\alpha = \tan^{-1}\left(\frac{C\omega}{k}\right)$$

$$\varepsilon_1 = -\tan^{-1}\left[\frac{C\omega}{(k - m\omega^2)}\right]$$

For hysteretic damping, $k + iC\omega$ is effectively replaced by $k(1 + i\eta)$, and the solution becomes

$$w = W_0\sqrt{\frac{1 + \eta^2}{(1 - \xi^2)^2 + \eta^2}}\cos(\omega t - \delta - \varepsilon_2) \tag{4.24}$$

where

$$\xi^2 = \frac{m\omega^2}{k} \tag{4.25}$$

$$\left.\begin{array}{l}\delta = \tan^{-1}\eta \\[2mm] \varepsilon_2 = \tan^{-1}\left(\dfrac{\eta}{1 - \xi^2}\right) = \tan^{-1}\left(\dfrac{k\eta}{k - m\omega^2}\right)\end{array}\right\} \tag{4.26}$$

Energy dissipated in a single degree of freedom system with base excitation can be determined from equation (4.19) for viscous damping:

$$D_S = \pi C\omega W_r^2 \tag{4.27}$$

or from equation (4.20) for hysteretic damping:

$$D_S = \pi k\eta W_r^2 \tag{4.28}$$

where $W_r$ is the amplitude of the relative displacement between the mass and the base. To determine $W_r$ in terms of $W_0$ for hysteretic damping, consider the forces acting on the mass and apply Newton's second law to obtain

$$k(1 + i\eta)W_r\, e^{i(\omega t + \varepsilon)} = -m\omega^2 \ddot{w} \tag{4.29}$$

which reduces, given $w = W_r\, e^{i(\omega t + \varepsilon)} + w_0 e^{i\omega t}$, to:

$$W_r = \frac{m\omega^2 W_0}{[(k - m\omega^2)^2 + (\eta k)^2]^{1/2}} \tag{4.30}$$

which when substituted into (4.28) yields

$$D_S = \frac{\pi\eta k\xi^4 W_0^2}{(1 - \xi^2)^2 + \eta^2} \tag{4.31}$$

where $\xi$ is defined in (4.25). Likewise for the case of a single degree of freedom system with viscous damping and base excitation, energy dissipation per cycle can be shown to be

$$D_S = \frac{\pi C\omega\xi^4 W_0^2}{(1 - \xi^2)^2 + (C\omega/k)^2} \tag{4.32}$$

### 4.2.5.  Effects of Real Material Behavior

As discussed in Chapter 3, the dynamic behavior of linear rubberlike (or viscoelastic) materials can be described in terms of a complex modulus $k(1 + i\eta)$ where the stiffness $k$ and loss factor $\eta$ depend on both frequency and temperature. Thus neither the viscous damping assumption nor the hysteretic damping assumption can truly describe the behavior of a single degree of freedom system consisting of a mass on a viscoelastic spring. Fortunately, however, the properties of most materials change relatively slowly with frequency, so that the variation of properties under isothermal conditions can be modeled by using values of complex moduli determined from a relatively few discrete tests over a wide range of frequencies. Figure 4.6 illustrates nondimensional energy dissipation in a low damped single degree of freedom system with base excitation for three different cases: hysteretic damping, viscous damping, and a real material for which the variations of $k$

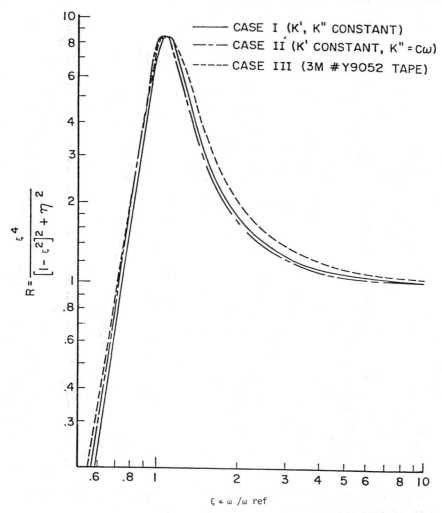

**FIGURE 4.6.** Performance of damper with moderate loss factor $\eta$.

and $\eta$ with frequency were determined experimentally. The nondimensional energy dissipation ratio $R$ is the ratio of energy dissipated in the system to the energy which would be dissipated in a similar system with the mass fixed in space. Figure 4.7 shows the corresponding behavior of the single degree of freedom system with a high damping adhesive as the spring. The temperature was constant at 71.6° F, and the physical dimensions of the damper were assumed to be such that the resonant frequency was 100 Hz. The hysteretic and viscous damping properties were chosen to coincide with measured material properties at the frequency where $R$ was a maximum for each case [4.5].

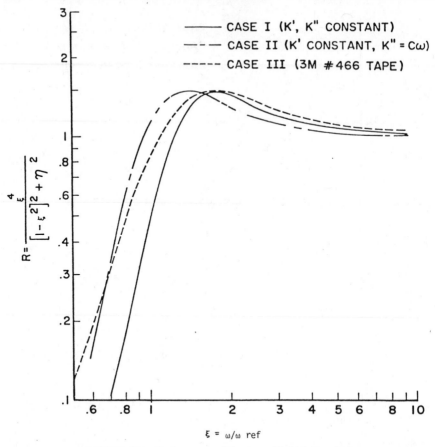

**FIGURE 4.7.**   Performance of damper with high loss factor.

## 4.3.   DETERMINATION OF DAMPING FROM STEADY STATE HARMONIC RESPONSE

As stated previously, damping cannot be measured directly but instead is deduced from the response characteristics of selected vibrating systems. The steady state response of a single degree of freedom system, excited by a harmonic force of constant amplitude, can be used to determine damping through the observation of several characteristics, including the bandwidth of the frequency response, the amplitude of response at resonance, Nyquist plots, hysteresis loops, and dynamic stiffness.

### 4.3.1.   Damping from Half-Power Bandwidth

One common method of determining damping is to measure the frequency bandwidth, between points on the response curve, for which the response is

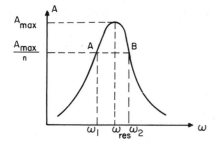

**FIGURE 4.8.**   Fractional-power bandwidth.

some fraction of the resonant response of the system, such as the points $A$ and $B$ illustrated in Figure 4.8. The usual convention is to consider points $A$ and $B$ to be located at frequencies where the amplitude of response is $1/\sqrt{2}$ times the maximum response. The bandwidth at these points is frequently referred to as the "half-power bandwidth," a term borrowed from the analysis of electrical systems in which amplitude is a measure of voltage and electrical power is proportional to the square of voltage. This amplitude ratio of $(1/\sqrt{2})$ corresponds to reduction of amplitude measured in decibels of

$$20 \log_{10}\left(\frac{1}{\sqrt{2}}\right) = -3.01 \text{ dB}$$

Thus a measurement associated with an amplitude ratio of $(1/\sqrt{2})$ is frequently referred to as a "3 dB bandwidth." More generally, the damping of a system can be determined from the bandwidth between any two points $A$ and $B$ associated with any amplitude ratio $1/n$, where $n$ is greater than 1. Although there are practical limits of measurement, values of $n$ greater than $\sqrt{2}$ or less than $\sqrt{2}$ can occasionally be of use.

Consider the viscous damping system shown in Figure 4.5. Substituting the frequency of maximum amplitude, $\omega_{res}$, from equation (4.11) into the particular solution of forced response, equation (4.7), the amplitude at resonance can be shown to be

$$(W_p)_{res} = \frac{F}{k}\left[\frac{1}{2(C/2\sqrt{km})(1 - C^2/4km)^{1/2}}\right] \tag{4.33}$$

To find the frequencies of points $A$ and $B$ where the amplitude is $1/n$ times $(W_p)_{res}$, the response from equation (4.10) is equated to $(1/n)$ times the response from equation (4.33) to yield

$$\frac{1}{[1 - (m\omega^2/k)]^2 + C^2\omega^2/k^2} = \frac{1}{(Cn/\sqrt{km})\sqrt{1 - C^2/4km}}$$

The two solutions $\omega_i$, $i = 1, 2$, are given by the quadratic equation:

$$\therefore \quad \left(\frac{m\omega_i^2}{k}\right)^2 - 2\left(1 - \frac{2C^2}{4km}\right)\left(\frac{m\omega_i^2}{k}\right) + 1 - 4n^2\left(\frac{C^2}{4km}\right)\left(1 - \frac{C^2}{4km}\right) = 0$$

This quadratic equation has two solutions:

$$\omega_{1,2}^2 = \frac{k}{m}\left[1 - 2\left(\frac{C^2}{4km}\right) \pm 2\sqrt{n^2 - 1}\left(\frac{C}{2\sqrt{km}}\right)\sqrt{1 - \frac{C^2}{4km}}\right] \qquad (4.34)$$

For $C^2/4km \ll 1$ this gives

$$\sqrt{m/k}(\omega_{1,2}) = 1 \pm \sqrt{n^2 - 1}\left(\frac{C}{2\sqrt{km}}\right) \qquad (4.35)$$

$$\therefore \quad \frac{\Delta\omega}{\omega_{res}} = \frac{\omega_2 - \omega_1}{\omega_{res}} = 2\sqrt{n^2 - 1}\left(\frac{C}{2\sqrt{km}}\right) \qquad (4.36)$$

Therefore for $n = \sqrt{2}$ we have

$$\frac{\Delta\omega}{\omega_{res}} = 2\left(\frac{C}{2\sqrt{km}}\right) = 2\zeta \qquad (4.37)$$

where the term $\zeta = C/2\sqrt{km} = C/C_c$ is known as the damping ratio, and $C_c$ is the critical damping of the system ($C_c = 2\sqrt{km}$). Similar calculations can be made for the hysteretic damping system shown in Figure 4.5. In this case the amplitude at resonance is

$$(W_p)_{res} = \frac{F}{k\eta}$$

where $\omega_{res} = \sqrt{k/m}$. The frequencies at points $A$ and $B$ of Figure 4.8, where the response is $1/n$ times $(W_p)_{res}$, are given by

$$\omega_{1,2} = \sqrt{\left(\frac{k}{m}\right)[1 \pm \eta\sqrt{n^2 - 1}]}$$

Hence for $n = \sqrt{2}$

$$\frac{\Delta\omega}{\omega_{res}} = \sqrt{1 + \eta} - \sqrt{1 - \eta} \qquad (4.38)$$

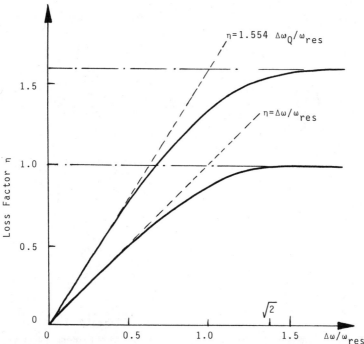

**FIGURE 4.9.**   Graphs of $\eta$ versus $\Delta\omega/\omega_{res}$ for one degree of freedom system.

and for $\eta \ll 1$

$$\frac{\Delta\omega}{\omega_{res}} \simeq \left(1 + \frac{\eta}{2}\right) - \left(1 - \frac{\eta}{2}\right) = \eta \qquad (4.39)$$

The relationship between $\Delta\omega/\omega_{res}$ and $\eta$ is linear only for small $\eta$, as shown in Figure 4.9. Note that for values of $\eta > 1$ no frequency $\omega_1$ exists, with the hysteretic damping assumption, for which the amplitude of response is $|W_p|/\sqrt{2}$. In fact for $n > 1$ the "peak" amplitude is less than the static displacement $(F/k)$. This is evident not only for the hysteretic damping case but also for the case where $\eta(\omega)$ and $k(\omega)$ are determined from measurements of real materials, as illustrated in Figure 4.10.

### 4.3.2.   Resonant Response Amplitude

The resonant amplification factor $A$, which is defined as the ratio of the amplitude of response at resonance to the displacement if the force is applied statically, is a measure of damping in a single degree of freedom system under force excitation.

$$\therefore \quad A = \frac{(W_p)_{res}}{F/k} \qquad (4.40)$$

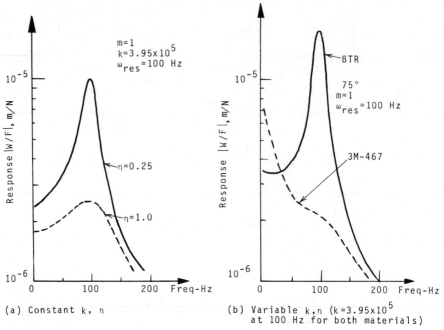

**FIGURE 4.10.** Effects of material properties on single degree of freedom system response. (*a*) Constant $k, \eta$ (*b*) Variable $k, \eta$ ($k = 3.96 \times 10^5$ at 100 Hz for both materials).

From equations (4.10) and (4.11) it can be seen that in the case of viscous damping

$$A = \frac{1}{2\zeta\sqrt{1 - \zeta^2}} \tag{4.41}$$

where $\zeta = C/C_c = C/2\sqrt{km}$. For $\zeta \ll 1$ this reduces to the familiar relationship:

$$A = \frac{1}{2\zeta} \tag{4.42}$$

Likewise for hysteretic damping, as in equation (4.14), with $\omega = \sqrt{k/m}$,

$$A = \sqrt{\frac{1 + \eta^2}{\eta}} \tag{4.43}$$

that is, for $\eta \ll 1$

$$A = \frac{1}{\eta} \tag{4.44}$$

$A$ can also be expressed in terms of the ratio of the maximum energy stored in a system to the energy dissipated per cycle [4.3], that is, the $Q$ of the system:

$$A \simeq Q = \frac{2\pi U_s}{D_s} \qquad (4.45)$$

with

$$U_s = \int_{w=0}^{W_p} F \, dw = \tfrac{1}{2}kW_p^2 \qquad (4.46)$$

where $W_p$ is the amplitude of the sinusoidal response. In the case of the base-excited single degree of freedom system, the energy stored is

$$U_s = \tfrac{1}{2}kW_r^2 \qquad (4.47)$$

where $W_r$ is the amplitude of the relative displacement between the base and the mass and $D_s$ is as defined in equations (4.27) and (4.28). It can also be shown from equation (4.30) that

$$A = \left. \left| \frac{W_r}{W_0} \right| \right|_{\text{max}} \qquad (4.48)$$

in a base-excited system.

### 4.3.3.  Nyquist Diagram

Another measure of the damping in a system is the relationship between the component of response that is in-phase with the force (real) and the component 90 degrees out of phase (imaginary), as plotted in the complex plane as a Nyquist diagram. Consider a single degree of freedom system represented by the differential equation

$$m\ddot{w} + k(1 + i\eta)w = Fe^{i\omega t}$$

A particular solution describing the steady state response can be written

$$w_p = W_p^* e^{i\omega t}$$

where $W_p^*$ is the complex valued function

$$W_p^* = \frac{1 - \xi^2 - i\eta}{(1 - \xi^2)^2 + \eta^2} W_{st} \qquad (4.49)$$

with $\xi^2 = \omega^2 m/k$ and $W_{st} = F/k$. It is further seen that

$$W_p^* = \text{Re } W_p^*(\omega) + i \text{ Im } W_p^*(\omega) = |W_p^*|e^{-i\varepsilon(\omega)}$$

where

$$\left.\begin{array}{l} \varepsilon = \tan^{-1}\dfrac{\eta}{(1 - \xi^2)} \\[4mm] \therefore \quad \text{Re } W_p^* = \dfrac{(1 - \xi^2)W_{st}}{(1 - \xi^2)^2 + \eta^2} \\[4mm] \text{Im } W_p^* = \dfrac{-\eta W_{st}}{(1 - \xi^2)^2 + \eta^2} \end{array}\right\} \qquad (4.50)$$

Therefore the amplitude of vibration response is

$$B = |W_p^*| = \sqrt{[\text{Re } W_p^*]^2 + [\text{Im } W_p^*]^2} = \frac{1}{\sqrt{(1 - \xi^2)^2 + \eta^2}} \qquad (4.51)$$

Taking the real part of the excitation, $Fe^{i\omega t}$, and the real part of the response, equation (4.51), we obtain the previous results shown in equation (4.14). A plot of Im $W_p^*(\omega)$ versus Re $W_p^*(\omega)$ in the complex plane is shown in Figure 4.11 and forms a circle with the equation

$$[\text{Re } W_p^*(\omega)]^2 + \left[\text{Im } W_p^*(\omega) + \frac{W_{st}}{2\eta}\right]^2 = \frac{W_{st}}{2\eta}$$

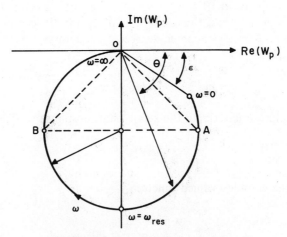

**FIGURE 4.11.**   Nyquist diagram.

and having a radius $W_{st}/2\eta$. When $\omega = 0$, $\xi = 0$, equation (4.50) gives

$$\text{Re } W_p^*(0) = \frac{W_{st}}{(1 + \eta^2)}$$

$$\text{Im } W_p^*(0) = \frac{W_{st}\eta}{(1 + \eta^2)}$$

$$\tan \varepsilon|_{\omega=0} = \eta$$

When $\omega = \omega_1 = \omega_{\text{res}}\sqrt{1 - \eta}$, equation (4.50) gives

$$\text{Re } W_p^*(\omega_1) = \frac{W_{st}}{2\eta} \quad \text{and} \quad \text{Im } W_p^*(\omega_1) = \frac{-W_{st}}{2\eta} \tag{4.52}$$

which corresponds to point $A$ in Figure 4.8. Likewise when $\omega = \omega_2 = \omega_{\text{res}}\sqrt{1 + \eta}$,

$$\text{Re } W_p^*(\omega_2) = \frac{-W_{st}}{2\eta}$$

$$\text{Im } W_p^*(\omega_2) = \frac{-W_{st}}{2\eta} \tag{4.53}$$

which corresponds to point $B$ in Figure 4.8.

$$\therefore \quad |W_p^*|_{\omega^i} = \frac{W_{st}}{\eta\sqrt{2}} = \frac{(W_p)_{\text{max}}}{\sqrt{2}} \tag{4.54}$$

showing that points $A$ and $B$ in Figure 4.11 are the same as points $A$ and $B$ in Figure 4.8, being the "half-power bandwidth" points discussed in the previous section. Nyquist diagrams for a force-excited single degree of freedom system with hysteretic damping and two different loss factors are illustrated in Figure 4.12. These diagrams were derived from the plots of the real and imaginary components of $W_p^*/W_{st}$ shown in Figure 4.13. Of course, as previously discussed, the values of $k$ and $\eta$ in real viscoelastic materials are functions of both temperature and frequency. Figure 4.14 illustrates two Nyquist diagrams calculated for single degree of freedom systems having viscoelastic springs with properties based on measured data for the different materials. One material (3M-467) is a pressure sensitive adhesive with high loss factor and material properties that rapidly change with frequency and temperature. The other material (BTR) is a filled silicone rubber with a much lower loss factor and with slower changes in material properties as a function of frequency and temperature. Figure 4.15 shows the variation with frequency of Re $W_p^*(\omega)$ and Im $W_p^*(\omega)$.

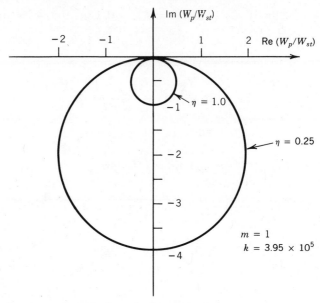

**FIGURE 4.12.** Nyquist diagram for single degree of freedom system with hysteretic damping.

### 4.3.4. Hysteresis Loops

A plot of the amplitude of instantaneous force versus instantaneous displacement (or stress versus strain) in a material for all values of time during steady state forced vibration is referred to as a hysteresis loop. For linear damping, including viscous damping, hysteretic damping, and linear rate-dependent damping, where $k$ and $\eta$ are functions of frequency, it has been shown [4.2] that hysteresis loops are elliptical in shape. To visualize the hysteresis loop for a force-excited single degree of freedom system with viscous damping, consider the force and displacement time histories illustrated in Figure 4.16, which are

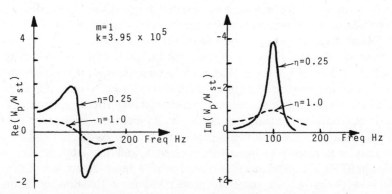

**FIGURE 4.13.** Graphs of $\mathrm{Re}(W_p/W_{st})$ and $\mathrm{Im}(W_p/W_{st})$ versus frequency for single degree of freedom system with hysteretic damping ($m = 1$, $k = 3.95 \times 10^5$).

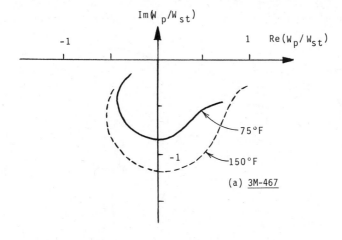

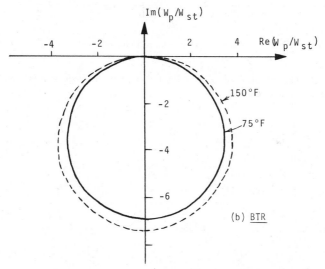

**FIGURE 4.14.**   Nyquist diagram for variable $k$, $\eta$ material. ($a$) 3M-467. ($b$) BTR; $k = 3.95 \times 10^5$ at 100 Hz, $m = 1$.

simply the plots of $W_p$ from equation (4.23). The locus of simultaneous values of force $F$ and displacement $W_p$ form the elliptical hysteresis loop shown in Figure 4.17. Similarly a hysteresis loop can be generated for a force-excited single degree of freedom system with hysteretic damping as shown in Figure 4.18. Figure 4.19 shows an experimentally measured hysteresis loop for a base-excited single degree of freedom system with a polyurethane foam spring [4.3]. In this experiment data were obtained that showed smooth elliptical hysteresis loops. The measured area of each ellipse agreed well with the area calculated from measured values of the major and minor axes.

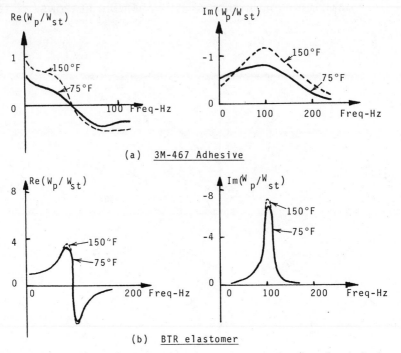

(a)   3M-467 Adhesive

(b)   BTR elastomer

**FIGURE 4.15.**   Graphs of $\mathrm{Re}(W_p/W_{st})$ and $\mathrm{Im}(W_p/W_{st})$ versus frequency for single degree of freedom system with variable $k$, $\eta$.

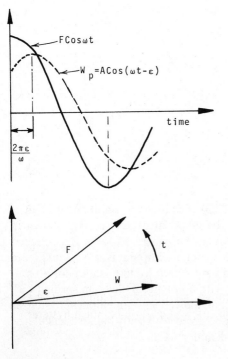

**FIGURE 4.16.**   Time histories of force and response.

**140**

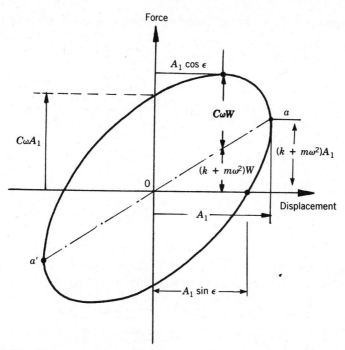

**FIGURE 4.17.** Hysteresis loop for force-excited single degree of freedom system with viscous damping.

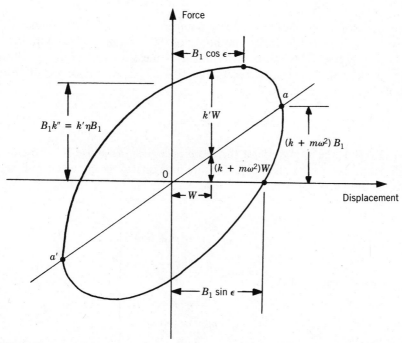

**FIGURE 4.18.** Hysteresis loop for force-excited single degree of freedom system with hysteretic damping.

**141**

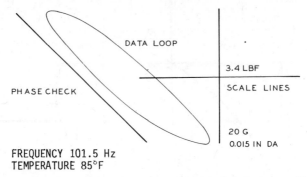

DATA LOOP

3.4 LBF

SCALE LINES

PHASE CHECK

20 G
0.015 IN DA

FREQUENCY 101.5 Hz
TEMPERATURE 85°F

**FIGURE 4.19.**   Measured hysteresis loop from a base-excited single degree of freedom system.

The area of a hysteresis loop is equal to the energy dissipated in the system, $D_s$, as calculated from equations (4.19), (4.20), (4.27), and (4.28). If there is no damping, the hysteresis loop collapses to the $a$–$0$–$a$ line, which represents force displacement behavior in a perfectly elastic system. If the mass changes, the slope of the $0$–$a$ line changes, as does the shape of the ellipse, but the area remains constant.

### 4.3.5.   Damping from Quadrature Bandwidth

In some cases, particularly for highly damped systems, it is advantageous to measure the bandwidth of the amplitude of the imaginary or quadrature response, Im $W_p^*(\omega)$. This can be determined by measurement of the bandwidth of the peak in the quadrature receptance [4.4]. The receptance $\alpha = W_p/F$ is a complex quantity, given by

$$\alpha = \alpha_D + i\alpha_Q = |\alpha|e^{-i\varepsilon} \tag{4.55}$$

$|\alpha| = |W_p/F|$ is the amplitude of the receptance, $\varepsilon$ is the phase, $\alpha_D$ is the direct (real) receptance, and $\alpha_Q$ is the quadrature (imaginary) receptance. From equations (4.50) and (4.51)

$$\left.\begin{aligned}
|\alpha| &= \frac{1}{\sqrt{(k - m\omega^2)^2 + (k\eta)^2}} \\[2mm]
\varepsilon &= \tan^{-1}\!\left(\frac{k\eta}{k - m\omega^2}\right) \\[2mm]
\alpha_D &= \frac{k - m\omega^2}{(k - m\omega^2)^2 + (k\eta)^2} \\[2mm]
\alpha_Q &= \frac{k\eta}{(k - m\omega^2)^2 + (k\eta)^2}
\end{aligned}\right\} \tag{4.56}$$

As discussed in Section 4.3.1, it is impossible to obtain the "half-power bandwidth" from the response amplitude of a single degree of freedom system having $\eta \geq 1.0$. However, as illustrated in Figure 4.15, for some high damping systems it may be more useful to consider instead the bandwidth of the imaginary part of the response, $|\mathrm{Im}\, W_p^*|$ or $\alpha_Q$, versus frequency. When the amplitude of the imaginary part of the response is considered, the frequencies of points $A$ and $B$ for which the value of $\alpha_Q$ is $1/n$ times the peak value of $\alpha_Q$, are given by

$$\omega_{1,2Q} = \sqrt{\left(\frac{k}{m}\right)[1 \pm \eta\sqrt{n-1}]} \qquad i = 1, 2 \tag{4.57}$$

Let $\Delta\omega_Q = \omega_{2Q} - \omega_{1Q}$;

$$\frac{\Delta\omega_Q}{\omega_{\mathrm{res}}} = \sqrt{1 + \eta\sqrt{n-1}} - \sqrt{1 - \eta\sqrt{n-1}} \tag{4.58}$$

and for $n = \sqrt{2}$

$$\frac{\Delta\omega_Q}{\omega_{\mathrm{res}}} = \sqrt{1 + 0.6436\eta} - \sqrt{1 - 0.6436\eta} \tag{4.59}$$

Therefore for $\eta \ll 1$

$$\frac{\Delta\omega_Q}{\omega_{\mathrm{res}}} = (1 + 0.3218\eta) - (1 - 0.3218\eta) = 0.6436\eta \tag{4.60}$$

$$\therefore \quad \eta \simeq 1.554 \frac{\Delta\omega_Q}{\omega_{\mathrm{res}}} \tag{4.61}$$

It will now be noted that, if $n = \sqrt{2}$, $\omega_{1Q}$ exists for values of $\eta \leq 1.554$; that is, the half-power bandwidth of the quadrature (imaginary) response can theoretically be determined for $\eta < 1.554$, whereas the half-power bandwidth for the total response amplitude curve exists only for $\eta < 1$. Figure 4.9 shows the relationship between quadrature half-power bandwidth $\Delta\omega_Q/\omega_{\mathrm{res}}$ and loss factor $\eta$.

### 4.3.6. Dynamic Stiffness

Equation (4.61) gives the complex ratio of displacement to excitation in the form:

$$\alpha = \frac{W_p^*}{F} = \frac{1}{(k - m\omega^2 + ik\eta)} \tag{4.62}$$

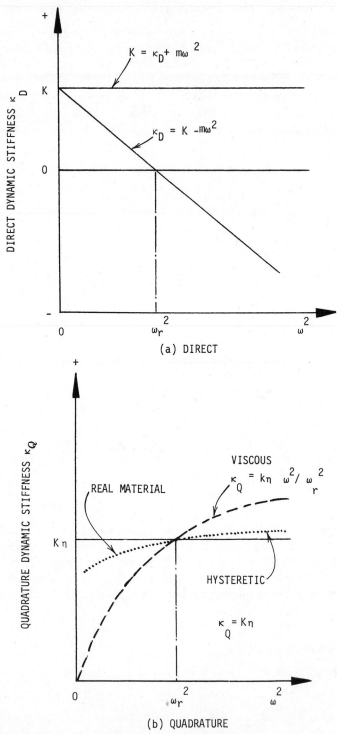

**FIGURE 4.20.** Direct and quadrature dynamic stiffness for single degree of freedom system.

This can also be written in the form

$$\alpha = |\alpha|e^{i\varepsilon} = \alpha_D + i\alpha_Q \qquad (4.63)$$

where $|\alpha|$ is the receptance amplitude, $\alpha_D$ is the direct receptance, $\alpha_Q$ is the quadrature receptance, and $\varepsilon$ is the phase angle, as given by equation (4.56). If $|\alpha|$ and $\varepsilon$ are measured in a test program, it is difficult, though perhaps not impossible, to determine $k$ and $\eta$ by solving equations (4.56) for given $m$. On the other hand, if we let $\kappa = 1/\alpha$ be the dynamic stiffness $(F/W_p^*)$, then

$$\kappa = k - m\omega^2 + ik\eta = \kappa_D + i\kappa_Q \qquad (4.64)$$

where $\kappa_D$ is the direct dynamic stiffness and $\kappa_Q$ is the quadrature dynamic stiffness, and

$$\kappa_D = k - m\omega^2 = \frac{\cos \varepsilon}{|\alpha|} \qquad (4.65)$$

$$\kappa_Q = k\eta = \frac{\sin \varepsilon}{|\alpha|} \qquad (4.66)$$

Hence a plot of $\kappa_D + m\omega^2$ versus frequency will give $k$ directly, and a plot of $\kappa_Q/(\kappa_D + m\omega^2)$ will give a direct measure of $\eta$. This direct approach is very helpful when $\eta$ is high. In summary,

$$k = \kappa_D + m\omega^2 \qquad (4.67)$$

$$\eta = \frac{\kappa_Q}{\kappa_D + m\omega^2} \qquad (4.68)$$

Figure 4.20 illustrates the variation of $\kappa_D$ with $\omega^2$ (from equation 4.65), of $k$ with $\omega^2$ (from equation 4.67), and of $\kappa_Q$ with $\omega^2$ for hysteretic and viscous damping mechanisms. Most real viscoelastic materials would be somewhere between these two extremes, as indicated.

## 4.4.  TRANSIENT RESPONSE

### 4.4.1.  Viscous Damped System

The viscously damped system has long been recognized as the only type of damping mechanism for which analytical solutions of the equations of motion can be obtained by one or more simple analytical methods, including the direct, Fourier transform and Laplace transform approaches.

**Direct method.**   The equation of motion of the single degree of freedom viscously damped system, subject to an impulsive force $F\delta(t)$ at $t = 0$, is written in the homogeneous form

$$m\ddot{w} + kw + C\dot{w} = 0 \qquad (4.69)$$

with the initial conditions $w(0) = 0$ and $\dot{w}(0) = F/m$. We assume a solution of the form $w = Ae^{\lambda t}$ and substitute in (4.69) to obtain the equation for $\lambda$

$$m\lambda^2 + C\lambda + k = 0$$

$$\therefore \quad \lambda = \frac{-C}{2m} \pm \sqrt{\left(\frac{C}{2m}\right)^2 - \frac{k}{m}} \qquad (4.70)$$

If $C/2\sqrt{km} < 1$, $\lambda$ will be complex, and an oscillatory decaying motion will occur. The complete solution, incorporating the initial conditions, is then

$$w(t) = \frac{F/m}{\sqrt{\dfrac{k}{m}(1 - \zeta^2)}} \sin\left(t\sqrt{\frac{k}{m}(1 - \zeta^2)}\right) e^{-Ct/2m} \qquad (4.71)$$

where $\zeta = C/2\sqrt{km}$. Now $\sin(\sqrt{(k/m)(1 - \zeta^2)}\, t)$ is equal to unity when

$$t\sqrt{\left(\frac{k}{m}\right)(1 - \zeta^2)} = \frac{(2n + 1)\pi}{2} \qquad n = 0, 1, 2, \ldots \qquad (4.72)$$

So the ratio of the amplitudes of the $n$th and $m$th peaks is

$$\frac{w_m}{w_n} = \frac{e^{-[C(2m+1)\pi/[4m\sqrt{(k/m)(1-\zeta^2)}]}}{e^{-[C(2n+1)\pi]/[4m\sqrt{(k/m)(1-\zeta^2)}]}} = \frac{e^{-[\zeta\pi(2m+1)]/[2\sqrt{(1-\zeta^2)}]}}{e^{-[\zeta\pi(2n+1)]/[2\sqrt{1-\zeta^2}]}}$$

$$= e^{-\zeta\pi(m-n)/\sqrt{1-\zeta^2}}$$

$$\therefore \quad \ln\left(\frac{w_m}{w_n}\right) = -\frac{\zeta\pi(m - n)}{\sqrt{1 - \zeta^2}}$$

$$\frac{-\zeta}{\sqrt{1 - \zeta^2}} = \frac{1}{\pi(m - n)} \ln\left(\frac{w_m}{w_n}\right) \qquad (4.73)$$

In particular, for $m = n + 1$

$$\delta = \ln\frac{w_m}{w_n} = \frac{-\pi\zeta}{\sqrt{1 - \zeta^2}} \qquad (4.74)$$

where $\delta$ is known as the logarithmic decrement and $\zeta$ is the damping ratio.

**Fourier transform method.**   If the equation of motion is written in the formal way

$$m\ddot{w} + kw + C\dot{w} = F\delta(t)$$

then one may perform a Fourier transform on each term, so as to obtain an algebraic equation instead of a differential equation. To do this, we factor each term by $e^{-i\omega t}$ and integrate over the limits $-\infty < t < \infty$;

$$\therefore \quad m\int_{-\infty}^{\infty} e^{-i\omega t}\left(\frac{d^2 w}{dt^2}\right) dt + K\int_{-\infty}^{\infty} e^{-i\omega t} w \, dt$$

$$+ C\int_{-\infty}^{\infty} e^{-i\omega t}\frac{dw}{dt} \, dt = \int_{-\infty}^{\infty} Fe^{-i\omega t}\delta(t) \, dt$$

$$\therefore \quad (-m\omega^2 + iC\omega + k)\bar{W} = F$$

$$\therefore \quad \bar{W} = \frac{F}{k - m\omega^2 + iC\omega} \tag{4.75}$$

where $\bar{W} = \int_{-\infty}^{\infty} w e^{-i\omega t} \, dt$. This equation can be inverted to recover $w$ by the method of residues, by the method of partial fractions, or simply from a recognition of the fact that the Fourier transform of $e^{-at}\sin(bt)H(t)$ is $b/[b^2 + (a + i\omega)^2]$. Since equation (4.75) can be rewritten in the form

$$\bar{W} = \frac{F}{m[(i\omega + C/2m)^2 + (k/m)(1 - C^2/4km)]} \tag{4.76}$$

one sees at once the recovery of equation (4.71). Obviously the Fourier transform method is less simple than the direct method, but for viscous damping no difficulties arise for any of the methods of solution. The Laplace transform may also be used without difficulty for this problem.

### 4.4.2.  Viscoelastically Damped System

When the damping element in the system is nonviscous, as for the complex stiffness $k(1 + i\eta)$ of a viscoelastic system, direct solutions of the equations of motion are not obtainable, and transform methods must be used. For stiffness representations using the standard viscoelastic model or the fractional derivative model, Laplace transforms may be used. For a complex modulus representation of real material behavior, for which $k$ and $\eta$ are functions of frequency, only a Fourier transform approach appears to be feasible, since the equations of motion can be written only in the frequency domain:

$$\left[-m\omega^2 + k\left(1 + \frac{i\eta\omega}{|\omega|}\right)\right]\bar{W} = \int_{-\infty}^{\infty} F\delta(t)e^{-i\omega t} \, dt = F \tag{4.77}$$

where $k(\omega) = k(-\omega)$ and $\eta(\omega) = \eta\omega/|\omega| = -\eta(-\omega)$. It is necessary to define $k(\omega)$ as a symmetric function of $\omega$ and $\eta(\omega)$ as a nonsymmetric function of $\omega$ because the integrals in the inverse Fourier transform encompass the frequency range $-\infty < \omega < \infty$ and fallacious results are obtained if these restrictions are not imposed. For presentation in the real-world frequency domain, $0 < \omega < \infty$, of course the question does not arise. The inversion of equation (4.77) becomes

$$w(t) = \frac{1}{2\pi} \int_{-\infty}^{\infty} \frac{F e^{i\omega t}\, d\omega}{k - m\omega^2 + ik\eta\omega/|\omega|} \tag{4.78}$$

which, using the symmetric and antisymmetric properties discussed before, leads to the real integral:

$$\frac{w(t)}{F} = \frac{1}{\pi} \int_{0}^{\infty} \frac{[(k - m\omega^2)\cos \omega t + k\eta \sin \omega t]}{(k - m\omega^2)^2 + (k\eta)^2}\, d\omega \tag{4.79}$$

This integration must, in general, be performed numerically. If $k$ and $\eta$ are assumed to be constants, the integration leads to very perplexing and confusing results, including noncausality of the form $w(t) \neq 0$ for $t < 0$. This happens because for all physically real materials or systems, $k$ and $\eta$ must vary with frequency, over a sufficiently wide range, and can never be constants. Hence the dilemma is not a real-world problem at all but only a consequence of an unduly extreme approximation. If $k(\omega)$ and $\eta(\omega)$ are described by functions representing actual measured data, no questions of noncausality can or do arise, except to the extent that the selected functions or the experimental data are themselves in error. To illustrate this matter, we have integrated equation (4.79) numerically for several cases, as summarized in Table 4.2.

These cases illustrate some of the matters discussed in this section. Cases A and B, for example, are for $k$ and $\eta$ constant; Cases C and D represent typical viscoelastic material behavior, C being low damped and D high damped; and Cases E and F represent viscous damping. Figure 4.21 illustrates the variation of $k$ and $\eta$ with frequency for these several cases. Figures 4.22 to 4.24 show

TABLE 4.2.   Transient Response Cases ($m = 0.007382$ kg, $f$ in Hz)

| Case | Type | $k$ N/m | $\eta$ |
|------|------|---------|--------|
| A | Hysteretic | 2910 | 0.15 |
| B | Hysteretic | 2910 | 1.40 |
| C | Viscoelastic | $1158(1 + 100f)^{0.1}$ | 0.15 |
| D | Viscoelastic | $113(4 + f)^{0.7}$ | $1.40 e^{-0.175|\log f/100|^{1.5}}$ |
| E | Viscous | 2910 | $2.28 \times 10^{-4} \times 2\pi f$ |
| F | Viscous | 2910 | $2.28 \times 10^{-3} \times 2\pi f$ |

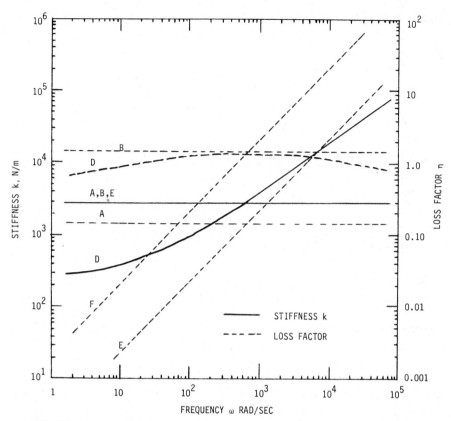

**FIGURE 4.21.** Variation of $k$ and $\eta$ with frequency for various cases (See Table 4.3).

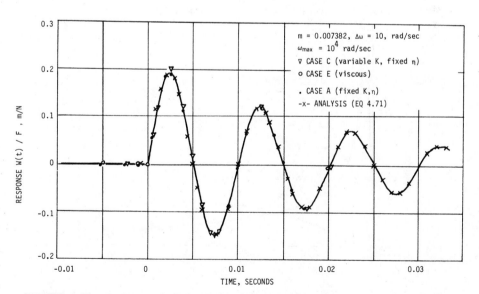

**FIGURE 4.22.** Typical transient response behavior for low damping elastomer (see Table 4.3).

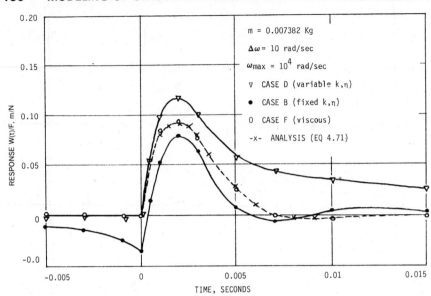

**FIGURE 4.23.** Typical transient response behavior for high damping polymer (see Table 4.3).

typical predicted responses. For $\eta = 0.15$ (Case A) the noncausality is quite small, but for $\eta = 1.4$ it is very significant. As $\eta$ approaches zero, the constant $k$, $\eta$ model does become a reasonably usable representation of real system behavior, but for values of $\eta$ greater than 0.1 or so, errors grow rapidly. Cases E and F show that the numerical integration gives the same results as the analytical model, equation (4.71), for viscous damping.

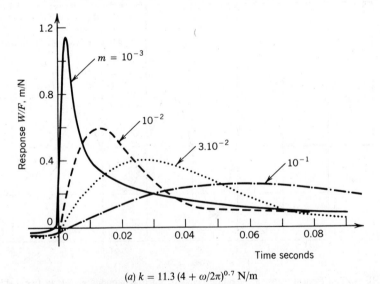

(a) $k = 11.3 \, (4 + \omega/2\pi)^{0.7}$ N/m

**FIGURE 4.24.** Typical transient response behavior (Case D, Table 4.3).

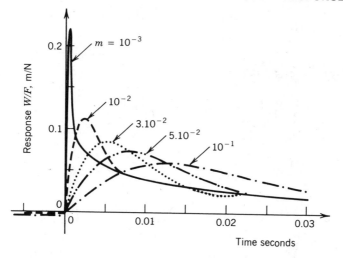

(b) $k = 113 (4 + \omega/2\pi)^{0.7}$ N/m

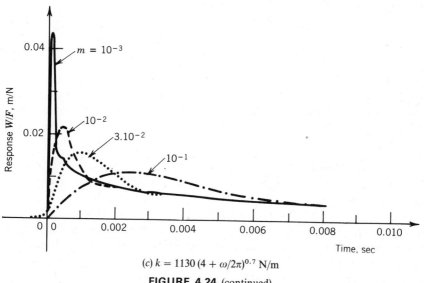

(c) $k = 1130 (4 + \omega/2\pi)^{0.7}$ N/m

**FIGURE 4.24** (continued)

## 4.5.  RANDOM RESPONSE

### 4.5.1.  Force Excitation

When a single degree of freedom system is excited by a random force, the driving force $F(t)$ varies continually with time in an unpredictable way, as illustrated in Figure 4.25a, and exact, deterministic, predictions of response are no longer possible or meaningful. If the time signal $F(t)$ is analyzed by performing a Fourier transform, the time interval $\Delta T$ over which the integration

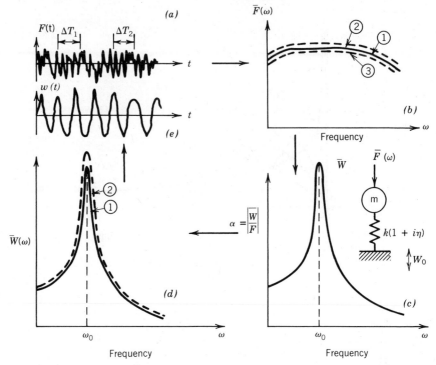

**FIGURE 4.25.** Random response. (*a*) Time domain. (*b*) Fourier transform of excitation. (*c*) Transfer function. (*d*) Fourier transform of response. (*e*) Time domain response (inverse Fourier transform).

is performed cannot be infinite in practice but must be limited. The selection of the maximum required interval $\Delta T$ depends on the minimum frequency at which significant excitation occurs, and $\Delta T$ is of the order of $1/\omega_{min}$—for example, if $\omega_{min} = 10$ Hz, $\Delta T$ is 1/10 sec; if $\omega_{min} = 1$ Hz, $\Delta T$ is 1 sec; and if $\omega_{min} = 100$ Hz, $\Delta T$ is 1/100 sec. If the signal is analyzed over an interval $\Delta T_1$ to obtain the Fourier transform, the spectrum level expected for broadband random excitation would look somewhat as in Figure 4.25*b*. If the analysis is repeated over another interval $\Delta T_2$, a similar spectrum is obtained if the excitation is stationary; that is, the statistical properties do not vary with time. In this case then each spectrum obtained will vary around an average spectrum, and one may think not in terms of deterministic predictions but rather in terms of averages, and expecially in terms of a root mean square average:

$$[\bar{F}(\omega)]^2 = \frac{1}{N} \sum_{n=1}^{N} [\bar{F}_n(\omega)]^2 \tag{4.80}$$

where $\bar{F}_n(\omega)$ is the spectrum for the nth sample and $\bar{F}(\omega)$ is the root mean square spectrum level. If the random excitation spectrum $\bar{F}_n(\omega)$ is applied to a single degree of freedom system, the spectrum level of the response is

$$\bar{W}_n(\omega) = \frac{\bar{F}_n(\omega)}{k(1 + i\eta) - m\omega^2} \tag{4.81}$$

for each sample $n$, as illustrated in Figure 4.25d. The response data therefore will also be random but will be dominated by the large resonance peak near $\omega_0 = \sqrt{k/m}$, so the response can be expected to be a harmonic signal of frequency $\omega_0$ having a randomly varying amplitude for each sample interval $\Delta T_n$. The root mean square amplitude is

$$
\begin{aligned}
(W_{rms})^2 &= \frac{1}{2N} \sum_{n=1}^{N} \int_{-\infty}^{\infty} \frac{[\bar{F}_n(\omega)]^2}{(k - m\omega^2)^2 + (k\eta)^2} \, d\omega \\
&= \frac{1}{2} \int_{-\infty}^{\infty} \frac{[\bar{F}(\omega)]^2}{(k - m\omega^2)^2 + (k\eta)^2} \, d\omega
\end{aligned} \tag{4.82}
$$

If $[\bar{F}(\omega)]^2$ is constant, as for a "white noise" spectrum, then

$$
\begin{aligned}
(W_{rms})^2 &= \frac{[\bar{F}(\omega)]^2 \sqrt{k}}{k^2 \sqrt{m}} \int_0^{\infty} \frac{d(\omega\sqrt{m/k})}{(1 - m\omega^2/k)^2 + \eta^2} \\
&= \frac{[\bar{F}(\omega)]^2}{k^{3/2}\sqrt{m}} \left[ \frac{\pi}{2\sqrt{2\eta}} \frac{1 + \sqrt{1 + \eta^2}}{\sqrt{1 + \eta^2}} \right]^2 \\
\therefore \quad W_{rms} &= \frac{\pi \bar{F}(\omega)}{2\sqrt{2} m^{1/4} k^{3/4} \eta^{1/2}} \left[ \frac{1 + (1 + \eta^2)^{1/2}}{\sqrt{1 + \eta^2}} \right]
\end{aligned} \tag{4.83}
$$

since

$$\int_0^{\infty} [(1 - \theta^2)^2 + \eta^2]^{-1} \, d\theta = [\pi(1 + \sqrt{1 + \eta^2})/2\sqrt{2\eta(1 + \eta^2)}]^2. \tag{4.84}$$

This very simple result shows that for random vibrations of a single degree of freedom system, the root mean square (rms) amplitude of the response depends on the rms excitation $\bar{F}(\omega)$, the mass in proportion to $m^{-1/4}$, the stiffness as $k^{-3/4}$, and the loss factor as $\eta^{-1/2}$. This means that increasing mass, stiffness, and damping will all reduce the rms level of response but at quite different rates, the stiffness effect being the strongest and the mass effect the weakest. Equation (4.83) can be rewritten in the form

$$\frac{W_{rms}}{\bar{F}(\omega)} \doteq \frac{\pi}{\sqrt{2}} \frac{\sqrt{\omega_0}}{k\sqrt{\eta}} = \frac{\pi}{\sqrt{2}} \frac{1}{m\omega_0^{3/2}\sqrt{\eta}} \tag{4.85}$$

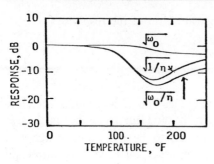

**FIGURE 4.26.** Variation of $\sqrt{1/\eta}$, $\sqrt{\omega_0}$, and $\sqrt{\omega_0/\eta}$ with temperature (force excitation).

for small values of $\eta$, showing how the resonant frequency $\omega_0$, the stiffness and the loss factor control the response. To illustrate the application of this equation, consider Figure 4.26 which shows the variation of $\sqrt{\omega_0}$, $\sqrt{\omega_0/\eta}$, and $1/\sqrt{\eta}$, with temperature from a specific set of tests on a typical stiffened panel used in an aircraft fuselage and damped by a constrained layer treatment. The plots are normalized at low temperatures and plotted on a decibel scale. Clearly $\sqrt{\omega_0}$ decreases as temperature increases, and $1/\sqrt{\eta}$ goes through a minimum.

### 4.5.2.  Base Excitation

The response of a one degree of freedom system to a random excitation $W_0(t)$ of the base to which it is attached can be found in the same way as before:

$$W_{rms}^2 = \frac{1}{2} \int_{-\infty}^{\infty} \frac{[\bar{W}_0(\omega)]^2 k^2 (1 + \eta^2)}{(k - m\omega^2)^2 + (k\eta)^2}\, d\omega \qquad (4.86)$$

If $[\bar{W}_0(\omega)]^2$ is constant, then

$$W_{rms}^2 = \frac{[\bar{W}_0(\omega)]^2 \sqrt{k}(1 + \eta^2)}{\sqrt{m}} \int_0^{\infty} \frac{d(\omega\sqrt{m/k})}{(1 - m\omega^2/k)^2 + \eta^2} \qquad (4.87)$$

$$W_{rms} = \frac{\pi\sqrt{\omega_0}}{2\sqrt{2}} \frac{\bar{W}_0(\omega)}{\eta^{1/2}} [1 + \sqrt{1 + \eta^2}] \qquad (4.88)$$

This means that $W_{rms}/\bar{W}_0(\omega)$ is proportional to $\sqrt{\omega_0/\eta}$ for a base-excited system, such as a damped valve cover used in a diesel engine.

## 4.6.  STEADY STATE RESPONSE OF MULTIPLE DEGREE OF FREEDOM SYSTEMS

### 4.6.1.  Discrete Element Modeling

Most machines and structures, when excited by an oscillating force, will exhibit many resonant response peaks in the frequency range of engineering interest. It is often difficult to predict the response of a complex structure directly, so one approach is to break it up conceptually into a number of discrete subelements and treat each element as a single mass joined to the others by various springlike links. This breakdown can be done in a number of ways; however, the following points must be considered:

1.  The number of masses must be equal to, or greater than, the number of important resonant peaks.
2.  The model may not be unique because several configurations of masses and stiffnesses might reproduce the observed behavior to an acceptable degree of accuracy.

A great deal of judgment must therefore be exercised in fitting such a physical model to observed response behavior. Note that this is not the same as a finite element representation in which the system is divided into a large number of masses at the centers or at the nodes of appropriate plate, beam, or shell elements selected according to the structure geometry. This usually results in the number of masses being far greater than the number of observed peaks in a given frequency range.

The advantage of the discrete mass representation is that for systems with only a few significant modes, a model can readily be generated to provide a simple analytical basis for predicting the effect of various structural modifications on the response.

### 4.6.2.  Free Vibrations of Beam by the Classical Approach

Consider the system shown in Figure 4.27 which consists of a beam supported by three torsional and three extensional springs. One of the "classical" approaches to analyzing this system is to write the differential equations, separate variables, determine a solution for each span of the beam, and solve for the arbitrary constants from equations of constraint for each point at which the springs are attached. For instance, to determine the free vibration natural frequencies and normal modes, the homogeneous Euler-Bernouli beam equation is written as

$$\frac{\partial^4 w(x, t)}{\partial x^4} + \frac{\rho H b}{EI} \frac{\partial^2 w}{\partial t^2} = 0 \qquad (4.89)$$

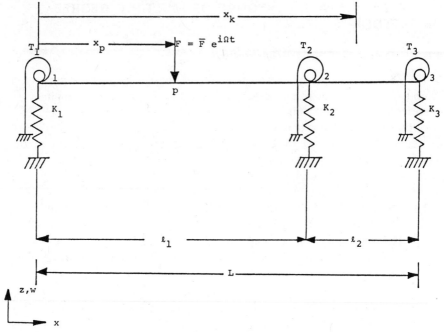

**FIGURE 4.27.**   Beam-spring system.

We seek a solution by separating variables, letting $w(x, t) = W(x)T(t)$. If we let

$$\beta^4 = \frac{\rho H b}{EI} \tag{4.90}$$

Then we can write

$$\frac{W^{iv}}{\beta^4 W} = \frac{-T''}{T} = \text{a constant } (\Omega^2)$$

which yields the solution

$$T = C_1 e^{i(\Omega t + \varepsilon)}$$

$$\therefore \quad w(x, t) = W(x) e^{i(\Omega t + \varepsilon)}$$

where

$$W(x) = W_1(x) = A_1 \sin \lambda x + A_2 \cos \lambda x + A_3 \sinh \lambda x + A_4 \cosh \lambda x \tag{4.91}$$

for $0 \le x \le l_1$, and

$$W(x) = W_2(x) = B_1 \sin \lambda x + B_2 \cos \lambda x + B_3 \sinh \lambda x + B_4 \cosh \lambda x \tag{4.92}$$

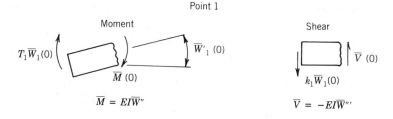

Point 1

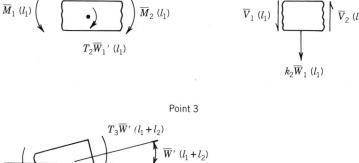

Point 2

Point 3

FIGURE 4.28.    Equilibrium diagrams for beam-spring system.

for $l_1 \leq x \leq l_1 + l_2$, with $\lambda = \sqrt{\Omega\beta}$ and the constant $C_1$ incorporated into the constants $A_n$ and $B_n$ of equations (4.91) and (4.92). Thus, to calculate natural frequencies and normal modes, it is necessary to write eight equations of constraint to determine the eigenvalues and constants in equations (4.91) and (4.92). The equilibrium conditions at points 1, 2, and 3 are illustrated in Figure 4.28 and can be expressed as follows:

At point 1, from moment equilibrium,

$$EIW_1'' = T_1 W_1'|_{x=0} \qquad (4.93)$$

and from equilibrium in the $z$ direction,

$$EIW_1''' = -k_1 W_1|_{x=0} \qquad (4.94)$$

Likewise at point 3

$$EIW_2'' = -T_3 W_2'|_{x=l_1+l_2} \tag{4.95}$$

and

$$EIW_2''' = k_3 W_2|_{x=l_1+l_2} \tag{4.96}$$

Finally, at point 2

$$EIW_2'' = EIW_1'' + T_2 W_1'|_{x=l_1} \tag{4.97}$$

and

$$EIW_2''' = EIW_1''' - k_2 W_1|_{x=l_1} \tag{4.98}$$

Continuity of displacement and slope at point 2 yields

$$W_1(l_1) = W_2(l_1) \tag{4.99}$$

and

$$W_1'(l_1) = W_2'(l_1) \tag{4.100}$$

Substituting equations (4.91) and (4.92) into equations (4.93) to (4.100) yields a matrix equation representing the eight homogenous equations.

$$
\begin{bmatrix}
-T_1\lambda & -EI\lambda^2 & -T_1\lambda & -EI\lambda^2 & 0 & 0 & 0 & 0 \\
-EI\lambda^3 & +k_1 & +EI\lambda^3 & k_1 & 0 & 0 & 0 & 0 \\
0 & 0 & 0 & 0 & d_1 & d_2 & d_3 & d_4 \\
0 & 0 & 0 & 0 & d_5 & d_6 & d_7 & d_8 \\
d_9 & d_{10} & d_{11} & d_{12} & d_{13} & d_{14} & d_{15} & d_{16} \\
d_{17} & d_{18} & d_{19} & d_{20} & d_{21} & d_{22} & d_{23} & d_{24} \\
d_{25} & d_{26} & d_{27} & d_{28} & -d_{25} & -d_{26} & -d_{27} & -d_{28} \\
d_{29} & d_{30} & d_{31} & d_{32} & -d_{29} & -d_{30} & -d_{31} & -d_{32}
\end{bmatrix}
\begin{Bmatrix}
A_1 \\ A_2 \\ A_3 \\ A_4 \\ B_1 \\ B_2 \\ B_3 \\ B_4
\end{Bmatrix} = \{0\}
$$

$$\tag{4.101}$$

where the $d$ parameters are summarized in Table 4.3.

TABLE 4.3.    Coefficients of Equation (4.101)

$d_1 = T_3\lambda \cos \lambda(l_1 + l_2) - EI\lambda^2 \sin \lambda(l_1 + l_2)$

$d_2 = -T_3\lambda \sin \lambda(l_1 + l_2) - EI\lambda^2 \sin \lambda(l_1 + l_2)$

$d_3 = T_3\lambda \cosh \lambda(l_1 + l_2) + EI\lambda^2 \sinh \lambda(l_1 + l_2)$

$d_4 = T_3\lambda \sinh \lambda(l_1 + l_2) + EI\lambda^2 \cosh \lambda(l_1 + l_2)$

$d_5 = -k_3 \sin \lambda(l_1 + l_2) - EI\lambda^3 \cos \lambda(l_1 + l_2)$

$d_6 = -k_3 \cos \lambda(l_1 + l_2) + EI\lambda^3 \sin \lambda(l_1 + l_2)$

$d_7 = -k_3 \sinh \lambda(l_1 + l_2) + EI\lambda^3 \cosh \lambda(l_1 + l_2)$

$d_8 = -k_3 \cos \lambda(l_1 + l_2) + EI\lambda^3 \sinh \lambda(l_1 + l_2)$

$d_9 = -T_2\lambda \cos \lambda l_1 + EI\lambda^2 \sin \lambda l_1$

$d_{10} = T_2\lambda \sin \lambda l_1 + EI\lambda^2 \cos \lambda l_1$

$d_{11} = -T_2\lambda \cosh \lambda l_1 - EI\lambda^2 \sinh \lambda l_1$

$d_{12} = +T_2\lambda \sinh \lambda l_1 - EI\lambda^2 \cosh \lambda l_1$

$d_{13} = -EI\lambda^2 \sin \lambda l_1$

$d_{14} \doteq -EI\lambda^2 \cos \lambda l_1$

$d_{15} = EI\lambda^2 \sinh \lambda l_1$

$d_{16} = EI\lambda^2 \cosh \lambda l_1$

$d_{17} = k_2 \sin \lambda l_1 + EI\lambda^3 \cos \lambda l_1$

$d_{18} = k_2 \cos \lambda l_1 - EI\lambda^3 \sin \lambda l_1$

$d_{19} = k_2 \sinh \lambda l_1 - EI\lambda^3 \cosh \lambda l_1$

$d_{20} = k_2 \cosh \lambda l_1 - EI\lambda^3 \sinh \lambda l_1$

$d_{21} = -EI\lambda^3 \cos \lambda l_1$

$d_{22} = EI\lambda^3 \sin \lambda l_1$

$d_{23} = EI\lambda^3 \cosh \lambda l_1$

$d_{24} = EI\lambda^3 \sinh \lambda l_1$

$d_{25} = \sin \lambda l_1$

$d_{26} = \cos \lambda l_1$

$d_{27} = \sinh \lambda l_1$

$d_{28} = \cosh \lambda l_1$

$d_{29} = \lambda^2 \sin \lambda l_1$

$d_{30} = -\lambda^2 \cos \lambda l_1$

$d_{31} = \lambda^2 \sinh \lambda l$

$d_{32} = \lambda^2 \cosh \lambda l$

Natural frequencies of this system may now be found by determining values of $\lambda$ for which the determinant of the matrix in equation (4.101) is equal to zero. Once the natural frequencies of interest are found, then the normal mode associated with the $n$th natural frequency can be determined by substituting the appropriate value of $\lambda_n$ back into equation (4.101) and solving for any seven of the constants in terms of the eighth. Thus a mode shape for equations (4.91) and (4.92) is determined, apart from a constant factor which may be chosen to normalize in any desired fashion.

Although, in principle, this procedure is straightforward, it is not without numerical difficulties even for this simple system. For instance, it is necessary, to evaluate the determinant of an $8 \times 8$ matrix, whose elements are trigonometric and hyberbolic functions of various arguments. This is usually done on a computer by essentially a trial and error technique. See reference [4.5] and [4.6] for some interesting aspects of this problem.

### 4.6.3. Forced Vibration by the Normal Mode Method

Consider now the problem of calculating the response of the example beam when excited by a harmonic force, $F = Fe^{i\omega t}$, applied at point $p$. In order to solve this problem, we could proceed in the classical manner and treat the applied force as an additional boundary condition, thereby adding to the difficulties already discussed in the classical treatment of the vibration problem, or one could use another near classical technique for analyzing forced vibration problems, namely the normal mode method, in which the forcing function and the response are expanded in term of the undamped normal modes of the system, already presumed known. Thus

$$F(t) = Fe^{i\omega t} = \sum_{n=1}^{\infty} A_n \phi_n(x) e^{i\omega t} \tag{4.102}$$

where $\phi_n(x)$ is the $n$th normal mode. In the example problem one must solve the nonhomogeneous differential equation

$$\frac{\partial^4 w(x, t)}{\partial x^4} + \frac{\rho H b}{EI} \frac{\partial^2 w(x, t)}{\partial t^2} = \frac{F}{EI} \delta(x - x_p) e^{i\omega t} \tag{4.103}$$

where $\delta(x - x_p)$ is the Dirac delta functional, for which $\delta(x - x_p) = 0$, $x \neq x_p$ and $\int_0^L \delta(x - x_p) \, dx = 1$.

The complete solution to the differential equation (4.103) consists of both the complementary and a particular solution. Usually it is assumed that the particular solution is the only one of interest, since it represents the steady state response of the system, and that the complementary solution is either not excited because of an appropriate choice of initial conditions or would eventually decay in a real system due to damping, even though damping is not specifically accounted for in the differential equation (4.103). Thus the steady state response of the beam occurs at the forcing frequency $\omega$ and equation (4.103) can be written as

$$\sum_{n=1}^{\infty} B_n \left[ \frac{d^4 \phi_n(x)}{dx^4} - \frac{\rho b H \omega^2}{EI} \phi_n(x) \right] = \frac{1}{EI} \sum_{n=1}^{\infty} A_n \phi_n(x) \tag{4.104}$$

where

$$F\delta(x - x_p) = \sum_{n=1}^{\infty} A_n \phi_n(x)$$

$$w(x, t) = \sum_{n=1}^{\infty} B_n \phi_n(x) e^{i\omega t}$$

Recall that each normal mode satisfies the equation

$$\frac{\partial^4 \phi_n}{\partial x^4} = +\frac{\rho H b}{EI} \omega_n^2 \phi_n \tag{4.105}$$

Hence equation (4.104) can be written as

$$\sum_{n=1}^{\infty} B_n(\omega_n^2 - \omega^2) \frac{\rho H b}{EI} \phi_n(x) = \frac{1}{EI} \sum_{n=1}^{\infty} A_n \phi_n(x) \tag{4.106}$$

Utilizing the important property of orthogonality of the normal modes,

$$\int_0^L \rho H b \phi_n(x) \phi_m(x)\, dx = \begin{cases} 0 & \text{for } m \neq n \\ M_n & \text{for } n = m \end{cases} \tag{4.107}$$

It can be shown that

$$A_n = F\phi_n(x_p) \tag{4.108}$$

Equation (4.106) therefore represents a set of uncoupled equations, and each $B_n$ can be obtained by the relationship

$$B_n = F \frac{\phi_n(x_p)}{M_n(\omega_n^2 - \omega^2)} \tag{4.109}$$

Thus the forced response can be written in the form

$$w(x, t) = F \sum_{n=1}^{\infty} \frac{\phi_n(x)\phi_n(x_p)}{M_n(\omega_n^2 - \omega^2)} e^{i\omega t} \tag{4.110}$$

Of course the response determined in equation (4.110) becomes unbounded when $\omega = \omega_n$ because damping has not been accounted for in the equations of motion. Damping can be introduced into the differential equation in the form of viscous damping, or as hysteretic damping. In the case of viscous damping the differential equation may be written as

$$\frac{\partial^4 w(x, t)}{\partial x^4} + \frac{C(x)}{EI} \frac{\partial w(x, t)}{\partial t} + \frac{\rho H b}{EI} \frac{\partial^2 w(x, t)}{\partial t^2} = F(x, t) \tag{4.111}$$

Using hysteretic damping, the differential equation of the beam may be written as

$$E*I \frac{\partial^4 w(x, t)}{\partial x^4} + \rho H b \frac{\partial^2 w(x, t)}{\partial t^2} = F(x, t) \qquad (4.112)$$

where $E* = E'(1 + i\eta) = E' + iE''$. In either the viscous or hysteretic damping cases expansion in terms of the undamped normal modes will result in uncoupled equations only if the damping is distributed along the beam in certain ways, such as in proportion to the mass or stiffness of the beam. Lin [4.7] has pointed out that proportional viscous damping results in an orthogonal relationship for the damping terms, which in the case of the beam would be

$$\int_0^L C(x)\phi_n(x)\phi_m(x)\, dx = \begin{cases} C_n & \text{for } n = m \\ 0 & \text{for } n \neq m \end{cases} \qquad (4.113)$$

and also that in practical examples the normal mode approach is widely used for structures with light damping by disregarding the coupling due to damping and assuming that the integral in equation (4.113) is zero for $n \neq m$, even though it often is not exactly so. Equation (4.113) corresponds to a case where the normal mode method would result in uncoupled equations, since the entire beam is considered to be made from material with the same complex Young's modulus.

However, if the case considered is that where a beam having negligible damping is suspended on springs having appreciable damping, such as would occur if the springs were made of an elastomer having a complex modulus with a loss factor $\eta \geq 0.2$, then the normal mode method becomes more unwieldy. The damping forces at each spring must be introduced as externally applied forces proportional to the displacement of the spring and in phase with and opposing the velocity of the springs. These terms couple the equations and make solution in terms of the undamped normal modes more tedious.

Other methods are available for the solution of forced vibration problems having generally distributed viscous or hysteretic damping. It has been shown, for instance, that the equations of motion for a viscous or hysteretically damped linear system can be uncoupled through the use of complex-valued "damped normal modes" and complex eigenvalues. However, these damped normal modes are not the classical normal modes of the system discussed here and are not so readily determined [4.5, 4.6].

### 4.6.4. Transfer Matrix Method

The transfer matrix method, the third of the "exact" methods of solution, is not usually regarded as a "classical" technique, yet it is exact in precisely the same sense that solutions can be obtained to specific types of problems without any further approximations beyond the physical approximations involved

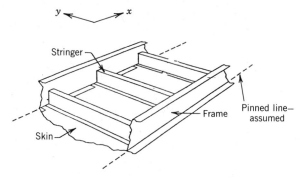

**FIGURE 4.29.**  Sketch of multispan skin-stringer structure with one set of parallel pinned boundaries.

in deriving the original equations of motion. One limitation is that the structure being analyzed must be essentially one dimensional. For example, the method is readily applicable to beams with many internal boundary conditions, such as beams on many supports, but is applicable to complex plate geometries only if one set of parallel boundaries is pinned, as shown in Figure 4.29. In this system, provided that the response in the $y$ direction can be represented adequately by a series of the form $\sum_{n=1}^{N} W_n(x) \sin(n\pi y/l)$, then the two-dimensional Euler–Bernoulli plate equation can be reduced to one involving the $x$ and $t$ variables only. The essence of the transfer matrix method is to solve the equations of motion step by step using a matrix notation, regarding the values of displacement, slope, bending moment and shear, as variables to be estimated simultaneously in the solution process, not separately as in the direct classical or normal mode methods. A transfer matrix analysis is written in terms of a state vector, which in this example problem (Figure 4.27) is a four-element vector of forces and displacements:

$$\{Z\}_k = \{W, W', M, V\}_k \tag{4.114}$$

and where it is understood that $\{\ \}$ signifies a column vector and

$$\text{Displacement} = w(x, t) = W(x)e^{i\omega t}$$
$$\text{Slope} = w'(x, t) = W'(x)e^{i\omega t}$$
$$\text{Moment} = M(x, t) = M(x)e^{i\omega t} \tag{4.115}$$
$$\text{Shear} = V(x, t) = V(x)e^{i\omega t}$$

and the $k$ subscript indicates that the state vector $\{Z\}_k$ is evaluated at $x = x_k$. If $W, W', M, V$ are assumed to be given at station $i$, then the state vector at any other point $j$ to the right of $i$, with no external loads or boundary supports between $i$ and $j$, can be obtained by solving the fourth-order Euler–Bernoulli equation and imposing $W$, $W'$, $M$, $V$ as boundary conditions at station $i$. This gives

$$\{Z\}_j = \begin{bmatrix} C_0 & C_1 & -\dfrac{C_2}{EI} & -\dfrac{C_3}{EI} \\[2ex] \lambda^4 C_3 & C_0 & -\dfrac{C_1}{EI} & -\dfrac{C_2}{EI} \\[2ex] -\lambda^4 EIC_2 & -\lambda^4 EIC_3 & C_0 & C_1 \\[2ex] -\lambda^4 EIC_1 & -\lambda^4 EIC_2 & \lambda^4 C_3 & C_0 \end{bmatrix} \{Z\}_i \qquad (4.116)$$

where

$$\left. \begin{aligned} C_0 &= \tfrac{1}{2}[\cosh(\lambda l_{ij}) + \cos(\lambda l_{ij})] \\[1.5ex] C_1 &= \frac{1}{2\lambda}[\sinh(\lambda l_{ij}) + \sin(\lambda l_{ij})] \\[1.5ex] C_2 &= \frac{1}{2\lambda^2}[\cosh(\lambda l_{ij}) - \cos(\lambda l_{ij})] \\[1.5ex] C_3 &= \frac{1}{2\lambda^3}[\sinh(\lambda l_{ij}) - \sin(\lambda l_{ij})] \end{aligned} \right\} \qquad (4.117)$$

and

$$\lambda = (\rho H b \omega^2 / EI)^{1/4} \quad \text{and} \quad l_{ij} = x_j - x_i \qquad (4.118)$$

Equation (4.116) can be rewritten as

$$\{Z\}_j = {}_j[U]_i \{Z\}_i \qquad (4.119)$$

where ${}_j[U]_i$ is the field transfer matrix between points $i$ and $j$ and $l_{ij}$ is the length between stations $i$ and $j$. If, on the other hand there is a spring attachment point at point $k$ the relationship between the state vector $\{Z\}_k^R$ on the right of $k$ to the state vector $\{Z\}_k^L$ on the left of $k$ is:

$$\{Z\}_k^R = \begin{bmatrix} 1 & 0 & 0 & 0 \\ 0 & 1 & 0 & 0 \\ 0 & -T & 1 & 0 \\ k & 0 & 0 & 1 \end{bmatrix} \{Z\}_k^L \qquad (4.120)$$

where $k$ and $T$ are the flexural and torsional spring constants, respectively. This equation can be written in the form

$$\{Z\}_k^R = {}_k^R[P]_k^L \{Z\}_k^L \qquad (4.121)$$

where ${}_k^R[P]_k^L$ is the point transfer matrix relating the right and left sides of point $k$. Finally, if there is an applied shear load $F$ at a point $p$, then

$$\{Z\}_p^R = \{Z\}_p^L + \begin{Bmatrix} 0 \\ 0 \\ 0 \\ F \end{Bmatrix} \tag{4.122}$$

where $F$ is the amplitude of the harmonic excitation. One can now see that the state vector at the right-hand end of a system can be calculated from field transfer matrices in terms of the state vector at the left-hand side:

$$\{Z\}_3^R = {}_3^R[T]_1^L \{Z\}_1^L + {}_3^R[T]_p^R \begin{Bmatrix} 0 \\ 0 \\ 0 \\ F \end{Bmatrix} \tag{4.123}$$

where the transfer matrices $[T]$ are products of point and field transfer matrices, that is,

$$
{}_3^R[T]_1^L = {}_3^R[P]_3^L \, {}_3[U]_2 \, {}_2^R[P]_2^L \, {}_2[U]_1 \, {}_1^R[P]_1^L \tag{4.124}
$$

and

$$
{}_3^R[T]_p^R = {}_3^R[P]_3^L \, {}_3[U]_2 \, {}_2^R[P]_2^L \, {}_2[U]_p \tag{4.125}
$$

Boundary conditions at the right- and left-hand ends of the beam can now be applied as

$$
\begin{aligned}
\{Z\}_1^L &= \{W(0), W'(0), 0, 0\} \\
\{Z\}_3^R &= \{W(l_1 + l_2), W'(l_1 + l_2), 0, 0\}
\end{aligned} \tag{4.126}
$$

This results in the following second-order homogeneous matrix equation formed by considering only the rows in which zero elements are located in $\{Z\}_3^R$ and eliminating the columns of ${}_3^R[T]_1^L$ multiplied by zeros in $\{Z\}_1^L$:

$$
\begin{Bmatrix} 0 \\ 0 \end{Bmatrix} = {}^R\begin{bmatrix} t_{31} & t_{32} \\ t_{41} & t_{42} \end{bmatrix}_1^L \begin{Bmatrix} W \\ W' \end{Bmatrix}_1^L + {}^R_3\begin{Bmatrix} t_{34} \\ t_{44} \end{Bmatrix}_p^R F \tag{4.127}
$$

where $t_{ij}$ are elements of $[T]$. Solving for the nonzero elements in $\{Z\}_1^L$,

$$
\begin{Bmatrix} W \\ W' \end{Bmatrix}_1^L = - {}^R\begin{bmatrix} t_{31} & t_{32} \\ t_{41} & t_{42} \end{bmatrix}_1^{L^{-1}} {}^R_3\begin{Bmatrix} t_{34} \\ t_{44} \end{Bmatrix}_p^R F \tag{4.128}
$$

As in the free vibration case, once a state vector is determined at any one location in the structure for a given frequency $\omega$, then the state vector, and hence the displacement, slope, moment, and shear response, can be calculated for any other location in the structure. The state vector at $k$ would be determined from

$$\{Z\}_k = {_k}[T]_1^L \{Z\}_1^L + {_k}[T]_p^R \begin{Bmatrix} 0 \\ 0 \\ 0 \\ F \end{Bmatrix} \qquad \text{if } x_k > x_p \qquad (4.129)$$

Using equation (4.128) and the fact that two of the elements in $\{Z\}_1^L$ are zero, equation (4.129) can be written, for $x_k > x_p$:

$$\{Z\}_k = - {_k}\begin{bmatrix} t_{11} & t_{12} \\ t_{21} & t_{22} \\ t_{31} & t_{32} \\ t_{41} & t_{42} \end{bmatrix}_1^L \left( {^R}\begin{bmatrix} t_{31} & t_{32} \\ t_{41} & t_{42} \end{bmatrix}_1^L \right)^{-1} \begin{Bmatrix} t_{34} \\ t_{44} \end{Bmatrix}_p^R F + {_k}[T]_p^R \begin{Bmatrix} 0 \\ 0 \\ 0 \\ F \end{Bmatrix} \qquad (4.130)$$

If $k$ is to the left of $p$, then the last term in equation (4.130) vanishes; that is,

$$ {_k}[T]_p^R = [0] \qquad \text{for } x_k < x_p \qquad (4.131)$$

and likewise

$$ {_k}[T]_p^R = [I] \qquad \text{for } x_k = x_p \qquad (4.132)$$

where $[I]$ is the identity matrix.

Some of the advantages of the transfer matrix method for analyzing the forced vibration of damped structures have been illustrated in this example. It is seen that transfer matrices provide a concise notation that allows for the convenient application of boundary conditions and readily accounts for the effects of applied harmonic forces.

Although in this example hysteretic damping is demonstrated by considering the stiffness of some elements to be complex, viscous damping could have been handled quite easily by expressing the damping in terms of equivalent complex stiffness. For instance, consider a spring with complex stiffness $k^*$, where

$$k^* = k'(1 + i\eta) = k' + ik'' \qquad (4.133)$$

The force across the spring is

$$F_d = w(t)k^*$$

where $w(t)$ is the relative displacement of the ends of the spring as a function of time. Since steady state harmonic motion is assumed

$$w(t) = We^{i\omega t}$$

Likewise, if the complex spring were replaced by a real spring and parallel viscous dashpot, the force across the spring dashpot combination would be

$$
\begin{aligned}
F_v &= w(t)k + \dot{w}C \\
&= W(k + i\omega C)e^{i\omega t}
\end{aligned}
\tag{4.134}
$$

Therefore $\omega C$ is the imaginary part of the complex stiffness, equivalent to viscous damping. Since transfer matrices across structures having distributed mass are functions of frequency, no special additional problems are introduced by incorporating frequency dependent complex stiffness terms.

**Numerical difficulties and limitations of transfer matrices:** Certainly transfer matrix analysis is not a universally applicable technique for all vibration problems. The most obvious limitation is that it is only applicable to one-dimensional analysis; that is, the transfer matrix must be written in terms of one space variable only. This method handles forces at a point quite well, but distributed forces must be simulated by summing a set of equivalent forces at discrete points, and although the matrix operations required might appear readily suitable for calculation on a digital computer, numerical difficulties can arise.

Numerical difficulties encountered in the analysis of a beam suspended on stiff springs have been discussed by Pestel and Leckie [4.8]. These problems become even more acute in the analysis of aircraft panels. One of the basic problems that occurs is that, in matrix chain multiplications of the type shown in equation (4.125), if the chain becomes long and the spring stiffnesses in $[P]$ are much larger than the beam bending stiffnesses in $[U]$, then numerical instabilities occur, essentially as a result of computing small differences between large numbers in a machine with finite precison. In the free vibration case this means that sometimes, particularly when dealing with higher-order problems involving shell equations, it is difficult to determine the frequencies at which the frequency determinant is sufficiently close to zero to allow accurate calculation of mode shapes. In the forced vibration problem the numerical evaluation of the inverse of the reduced matrix shown in equation (4.128) can be difficult. This has been shown by Lin and McDaniel [4.7] to be due to the fact that

$$
\left( {}^R_3\begin{bmatrix} t_{31} & t_{32} \\ t_{41} & t_{42} \end{bmatrix}^L_1 \right)^{-1} = \frac{1}{\Delta}\begin{bmatrix} t_{42} & -t_{32} \\ -t_{41} & t_{31} \end{bmatrix}
\tag{4.135}
$$

where the determinant $\Delta = t_{42}t_{31} - t_{32}t_{41}$. If no damping is present, $\Delta$ is the frequency determinant, which is equal to zero at each natural frequency. With damping in the system, $\Delta$ is a small difference between two large numbers near each resonance. Various methods have been developed for circumventing these numerical problems. One way is to convert the transfer matrices to "delta matrices." This technique is adequately discussed in the literature [4.8–4.10] and will not be covered here. The resulting delta matrix, for an $n \times n$ transfer matrix, is of order $(n!/\frac{1}{2}n!)^2$, by $(n!/\frac{1}{2}n!)^2$. Another technique, referred to by some authors as the "super matrix" approach, has been adapted for use on curved panels [4.11].

### 4.6.5. Finite Element Analysis

Modern finite element analysis offers an attractive approach to predicting the vibratory characteristics of complex structures exhibiting multimodal response. General purpose codes such as NASTRAN, ANSYS, and MARC [4.12] have been used by many investigators in the analysis of structural vibration problems. Usually finite element analysis is used to predict resonant frequencies and normal mode shapes. Many of these codes have the capability of accounting for damping in some form. However, when a finite element analysis is used for quantitative analysis of the effects of viscoelastic materials incorporated into a structure, great care has to be taken to avoid common pitfalls. These difficulties include unrealistic expansion of computer time and storage requirements to handle complex-valued stiffnesses, or oversimplification of the problem to a set of solvable equations that do not adequately model physical reality.

It has been demonstrated [4.13, 4.14] that a modal strain energy method can be used to estimate accurately the effects of viscoelastic layers on the damping of "composite" structures, that is, structures containing both elastic and viscoelastic components. Simply stated, the principle of the modal strain energy method is that the ratio of the composite structural loss factor to the loss factor of the viscoelastic material, for a given mode of vibration, can be estimated from the ratio of elastic strain energy in the viscoelastic material to the total strain energy in the structure when it deforms in a particular un-damped mode shape [4.13]. Expressed mathematically this is

$$\frac{\eta_S^{(r)}}{\eta_d} = \frac{U_d^{(r)}}{U_S^{(r)}} \tag{4.136}$$

where

$\eta_S^{(r)}$ = the loss factor of the $r$th mode of the composite structure

$\eta_d$ = the material loss factor for the viscoelastic material

$U_d^{(r)}$ = elastic strain energy stored in the viscoelastic material when the structure deforms in its $r$th undamped mode shape

$U_S^{(r)}$ = elastic strain energy of the entire composite structure in the $r$th mode shape.

This relationship yields

$$\frac{\eta_S^{(r)}}{\eta_d} = \sum_{e=1}^{n} \frac{\phi_e^{(r)} k_e \phi_e^{(r)}}{\phi^{(r)T} k \phi^{(r)}} \tag{4.137}$$

where

$\phi^{(r)} = r$th mode shape vector

$\phi_e^{(r)} =$ subvector found by deleting from $\phi^{(r)}$ all entries not corresponding to motion of nodes of the $e$th viscoelastic element

$k_e =$ element stiffness matrix of the $e$th viscoelastic element

$k =$ stiffness matrix of the entire composite structure

$n =$ number of viscoelastic elements in the model.

Although equation (4.137) is not precisely true, since it assumes that energy dissipated depends only on strain energies associated with the undamped mode shapes, it has been shown [4.13–4.16] to give accurate engineering data even for cases where $\eta_d$ is in excess of unity. This work utilized the MSC/NASTRAN code and was found to result in quite tractable computational efforts.

Another approach to finite element analysis of viscoelastically damped structures has been the development of special elements for direct solution of the complex structural equations of motion. Programs have been specially developed for steady state dynamic analysis of large three-dimensional structures vibrating about an initially stressed condition and are applicable to many different types of structures including viscoelastically damped turbine blades and damped skin-stringer panels [4.15–4.17].

### 4.6.6. Experimental Modal Analysis

Although analytical techniques for predicting structural vibration in complex structures have become increasingly sophisticated, the majority of practical problems involving vibrations of real structures are solved by experimentally based methods [4.18]. The advent of the minicomputer-based digital Fast Fourier analyzer has made it possible to estimate directly from measured data the mass, stiffness, and damping properties of a vibrating structure [4.19]. Furthermore three-dimensional animated mode shapes, generated from digitized frequency response functions measured at many points on a structure, are an invaluable aid in visualizing complicated vibration phenomena in structures with complex geometries.

The Fourier analyzer approximates the Fourier transform of the form

$$F(f) = \int_{-\infty}^{\infty} f(t)e^{-i2\pi ft} \, dt \tag{4.138}$$

by the truncated Fourier series

$$F(m\Delta f) = \Delta t \sum_{n=1}^{N-1} f(n\Delta t)e^{-i2\pi(m\Delta f)} \qquad (4.139)$$

where $N$ is the number of samples taken in the time window, and $m, n = 0, 1, 2, \ldots, N - 1$.

Unlike the Fourier transform integral, this discrete Fourier transform supplies amplitude and phase information only at evenly spaced intervals in the frequency domain. It is also important to realize that the discrete Fourier transform represented by equation (4.139) exists for periodic functions only. Therefore it is an inherent assumption that a time history observed from time $t = 0$ to $t = T$, and sampled at intervals $\Delta t$, repeats itself with a period $T$ for all time. With proper understanding of the limitations arising from this assumption, errors such as aliasing, leakage, and the "picket-fence" effect can be held within bounds, and data adequate for most engineering purposes can be obtained [4.20–4.22].

Primary data provided by a Fourier analyzer are frequency response functions obtained by simultaneously measuring and digitally storing the time histories of the structure excitation force and the acceleration response. These time-domain signals are then used by the Fourier analyzer to calculate in the frequency domain the acceleration, velocity, or displacement auto power spectrum $Gyy$, the force auto power spectrum $Gxx$, the cross power spectrum $Gyx$, and the frequency response function $H(f)$ where

$$H(f) = \frac{\overline{Gyx}}{\overline{Gxx}} \qquad (4.140)$$

where the $\overline{Gyx}$ and $\overline{Gxx}$ are averaged over several samples of $Gyx$ and $Gxx$. The analyzer can also be used to calculate the coherence function, $\gamma^2(f)$, where

$$\gamma^2(f) = \frac{|\overline{Gyx}|^2}{|\overline{Gxx}||\overline{Gyy}|} \qquad 0 < \gamma^2 < 1 \qquad (4.141)$$

The coherence function is a measure of the "quality" of the data [4.22], since a coherence value of unity indicates that the output (acceleration, velocity, or displacement) is completely causally related to the input (force). A coherence value other than unity indicates that something could be wrong in the test setup. Problems such as bad cables, ground loops, spurious excitation by ambient acoustic noise, nonlinearities in the test structure, or problems in the signal conditioning system all tend to reduce coherence values.

The usual method for measuring a frequency response function is to excite the test specimen with a measured input force, at a point, and measure the structure response function at another point. Typically the structural response is measured with an accelerometer, which will result in an accelerance form of frequency response function. However, strain gages, velocity transducers, eddy current probes, and so on, can also be used. Force input is usually provided by one of the following: (1) impact, (2) electromagnetic shaker excitation, or (3) noncontacting magnetic transducer excitation. This force is measured either directly, through a piezoelectric force gage, or by monitoring electrical current to the magnetic transducers [4.23].

Several types of excitation time histories can be utilized with shakers, or noncontacting transducer exciters, including periodic chirps, pseudorandom, periodic random, and pure random signals. Each of these techniques has its advantages and disadvantages, as discussed in references [4.24, 4.25].

Modal damping can be determined from digital Fourier analyzer data in a variety of ways, including measurement of the bandwidth of the frequency response function, observation of the Nyquist diagram, or utilization of curve fit algorithms to estimate modal mass, stiffness, and damping from the response curves [4.19, 4.27–4.32].

## 4.7  EXAMPLES AND ILLUSTRATIONS

### 4.7.1.  One Degree of Freedom High Damping System (Analysis of Test Data)

If the one degree of freedom system is lightly damped, one can determine the values of $k$, $m$, and $\eta$ (or $C$) at the resonant frequency by means of all of the methods outlined in Section 4.3. For example, one can determine the loss factor $\eta$ from the half-power bandwidth (equations 4.37 or 4.39), the resonant amplification (equation 4.42 or 4.44), the Nyquist diagram, the hysteresis loop, or the quadrature bandwidth (equation 4.61). Since $\kappa_Q$ is small, one may have difficulty using equation (4.68), based on dynamic stiffness, for very lightly damped structures. Each measure of damping so used will give essentially the same numerical quantities, so that

$$\eta \simeq 2\zeta \simeq \frac{\Delta\omega}{\omega_r} \simeq 1.554 \frac{\Delta\omega_Q}{\omega_r} \simeq \frac{1}{A} \tag{4.142}$$

However, nothing can be determined concerning the values of the parameters away from the resonance frequency. For example, if the damping is hysteretic, the loss factor $\eta$ at other frequencies above and below $\omega_r$ is about the same as at $\omega = \omega_r$; but if, on the other hand, the damping mechanism is viscous, then $\eta$ is equal to twice the viscous damping ratio $\zeta$ only at $\omega = \omega_r$. This can be very important in many situations. For example, if one has a structure isolated by a set of damped springs, then the value of $\eta$ at resonance can be determined from the relationships in equations (4.142), but at a frequency of, say, 10 times the resonant frequency, $\eta$ would be $10 \times 2\zeta$, that is, 10 times the value at the resonance frequency, if the damping mechanism is viscous.

Only the dynamic stiffness approach can, when errors are not excessive, give values of $\eta$ away from resonance. These problems can be resolved in various ways when conducting tests. For example, one could vary the mass $m$ in a series of tests, so as to vary the resonant frequency $\omega_r$ over a wide range, thereby obtaining a number of discrete estimates of $\eta$ at several frequencies. This should, in principle, correlate with the results obtained by the dynamic stiffness approach. In much the same way $k$ can be estimated. If $m$ is known, for example, $k$ can be estimated from $k = m\omega_r^2$, and if $m$ is varied, $k$ can be estimated at a number of frequencies. For a lightly damped system therefore the values of $k$ and $\eta$ can be estimated from a sufficient number of tests, even when they are strongly frequency dependent as for viscoelastic materials. When the material is highly damped, this is no longer possible.

Consider the test system and measured receptance amplitude and phase angle measurements illustrated in Figure 4.30. The specimen dimensions are indicated in the figure. The material was 3M-ISD-110, having complex modulus properties described in Data Sheet 017. The response of the mass $m$ (5.355 kg) was measured by an accelerometer, and excitation was by impact through a force gage, as described later in this chapter. A fast Fourier trans-

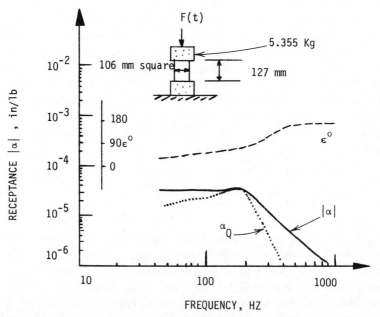

**FIGURE 4.30** Measured receptance amplitude and phase for high damping single degree of freedom system (106 mm × 106 mm square cross section).

**TABLE 4.4.  Measured Data for Single Degree of Freedom System**

| Freq, Hz | Measured Receptance | | Calculated Dynamic Stiffness | |
|---|---|---|---|---|
| | $\lvert\alpha\rvert$ in./lbf | Phase $\varepsilon°$ | $\kappa_D = \cos\varepsilon/\lvert\alpha\rvert$ N/m | $\kappa_Q = \sin\varepsilon/\lvert\alpha\rvert$ N/m |
| 56  | 3.5 E −5 | 33  | +4.21 E 6 | +2.74 E 6 |
| 75  | 3.5 E −5 | 45  | +3.55 E 6 | +2.74 E 6 |
| 100 | 3.3 E −5 | 52  | +3.28 E 6 | +4.20 E 6 |
| 125 | 3.4 E −5 | 61  | +2.51 E 6 | +4.51 E 6 |
| 150 | 3.6 E −5 | 73  | +1.43 E 6 | +4.67 E 6 |
| 175 | 3.6 E −5 | 91  | −8.52 E 4 | +4.88 E 6 |
| 200 | 3.3 E −5 | 112 | −2.00 E 6 | +4.93 E 6 |
| 225 | 2.7 E −5 | 128 | −4.00 E 6 | +5.13 E 6 |
| 250 | 2.1 E −5 | 143 | −6.67 E 7 | +5.04 E 6 |
| 275 | 1.6 E −5 | 151 | −9.61 E 6 | +5.32 E 6 |
| 300 | 1.3 E −5 | 157 | −1.24 E 7 | +5.29 E 6 |
| 350 | 8.5 E −6 | 164 | −1.98 E 7 | +5.69 E 6 |
| 400 | 6.1 E −6 | 171 | −2.84 E 7 | +4.50 E 6 |
| 500 | 3.2 E −6 | 177 | −5.48 E 7 | — |
| 600 | 2.1 E −6 | 169 | −8.20 E 7 | +1.60 E 7 |
| 700 | 1.7 E −6 | 169 | −1.01 E 8 | +1.97 E 7 |
| 800 | 1.3 E −6 | 174 | −1.34 E 8 | +1.41 E 7 |

form gave the receptances or compliances (displacement/force) shown. It is evident that neither $k$ nor $\eta$ can be measured by amplitude or bandwidth methods at any frequency, including the resonant frequency. The quadrature bandwidth might give a gross estimate of the value of $\eta$ at one frequency, but the plot of $\alpha_Q = \lvert\alpha\rvert\sin\varepsilon$ versus frequency in Figure 4.30 does not indicate that this would be a good estimate. Table 4.4 summarizes some of the measured receptance data and the corresponding values of the direct and quadrature dynamic stiffnesses.

Figure 4.31 shows the corresponding Nyquist diagram, namely a graph of $\alpha_Q = \lvert\alpha\rvert\sin\varepsilon$ versus $\alpha_D = \lvert\alpha\rvert\cos\varepsilon$. It is seen that no usable data can be obtained, except perhaps a gross estimate of the resonant frequency and of the loss factor, without any proper accounting for the variability of $k$ and $\eta$ with frequency.

On the other hand, with $\kappa_D$ and $\kappa_Q$ known for each frequency, $k$ and $\eta$ can be calculated directly from equations (4.67) and (4.68). The results are plotted in Figure 4.32. It is seen that the data are usable at all frequencies except for errors at low frequency resulting from instrumentation limitations and at high frequency resulting from the inertia term $m\omega^2$ becoming too great, a commonly encountered problem. Even with the success of the dynamic stiffness approach, a clear need is seen for varying $m$ so that bands of data centered on other frequency ranges can be obtained.

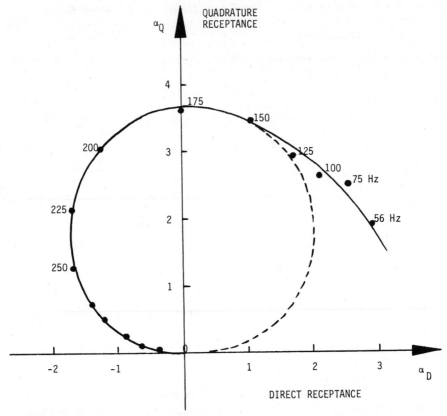

**FIGURE 4.31.** Measured hysteresis loop for highly damped single degree of freedom system.

### 4.7.2. Harmonic Response of Single Degree of Freedom System with Viscous or Hysteretic Damping and Fixed Mass and Stiffness (Force Excitation)

Consider a single degree of freedom system with $m = 0.0155$ kg and $k = 9.8 \times 10^4$ N/m. For the four cases (A) $\eta = 0.169$, (B) $\eta = 0.169$ ($f/400$), (C) $\eta = 1.00$, and (D) $\eta = 1.00$ ($f/400$), plot the receptance amplitude $|\alpha| = |W/F|$ and phase $\varepsilon$ versus frequency, and examine the possibility of recovering the system parameters from the variation of $|\alpha|$ and $\varepsilon$ or the dynamic stiffness $\kappa = 1/\alpha$. As usual

$$|\alpha| = \frac{1}{\sqrt{(k - m\omega^2)^2 + (k\eta)^2}}$$

$$\tan \varepsilon = \frac{-k\eta}{k - m\omega^2}$$

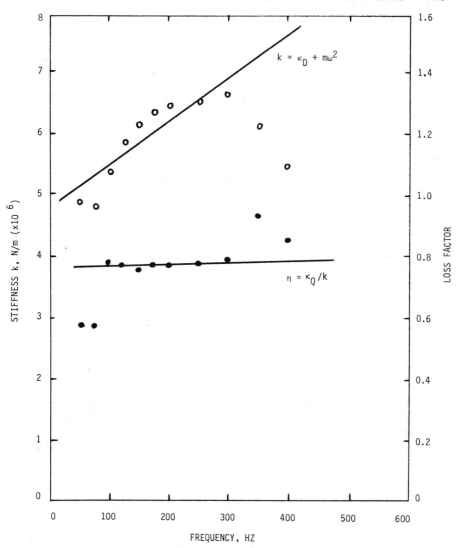

**FIGURE 4.32.** Stiffness and loss factor from dynamic stiffness data.

These expressions are plotted versus frequency for the four cases in Figure 4.33. Calculated values of $|\alpha|$ and $\varepsilon$ are summarized in Table 4.5. Also tabulated are the values of the direct and quadrature dynamic stiffnesses

$$\kappa_D = \frac{\cos \varepsilon}{|\alpha|}$$

$$\kappa_Q = -\frac{\sin \varepsilon}{|\alpha|}$$

calculated from the rounded-off values of $|\alpha|$ to two decimal places and $\varepsilon$ to the nearest degree. This would simulate to some extent the limitations of a

**TABLE 4.5. Receptance and Dynamic Stiffness Values for Single Degree of Freedom System**

| Case | $f$ Hz | $\lvert\alpha\rvert \times 10^{-5}$ N/m | $\varepsilon$ degree | $\kappa_D = \cos\varepsilon/\lvert\alpha\rvert$ $(\times 10^4)^a$ | $\kappa_Q = \sin\varepsilon/\lvert\alpha\rvert$ $(\times 10^4)^a$ | $\eta = -\kappa_Q/9.8 \times 10^4$ |
|---|---|---|---|---|---|---|
| A | 100 | 1.071 | -10.2 | 9.20 | -1.62 | 0.165 |
|   | 200 | 1.327 | -12.7 | 7.33 | -1.69 | 0.172 |
|   | 300 | 2.173 | -21.1 | 4.30 | -1.65 | 0.168 |
|   | 350 | 3.524 | -35.7 | 2.30 | -1.67 | 0.170 |
|   | 400 | 6.038 | -89.7 | 0 | -1.66 | 0.169 |
|   | 450 | 3.252 | -147.4 | -2.58 | -1.68 | 0.171 |
|   | 500 | 1.742 | -163.2 | -5.50 | -1.68 | 0.171 |
|   | 600 | 0.810 | -172.3 | -122 | -1.72 | 0.176 |
|   | 800 | 0.340 | -176.8 | -294 | -1.54 | 0.157 |
|   | 1000 | 0.194 | -178.2 | -515 | -1.80 | — |
| B | 100 | 1.087 | -2.5 | 9.16 | -0.48 | 0.049 |
|   | 200 | 1.352 | -6.4 | 7.37 | -0.77 | 0.079 |
|   | 300 | 2.238 | -16.1 | 4.29 | -1.23 | 0.126 |
|   | 350 | 3.674 | -32.2 | 2.31 | -1.44 | 0.147 |
|   | 400 | 6.038 | -89.7 | 0 | -1.66 | 0.169 |
|   | 450 | 3.133 | -144.3 | -2.58 | -1.88 | 0.192 |
|   | 500 | 1.702 | -159.4 | -5.49 | -2.11 | 0.215 |
|   | 600 | 0.801 | -168.5 | -12.3 | -2.39 | 0.244 |
|   | 800 | 0.338 | -173.6 | -19.3 | -3.07 | 0.313 |
|   | 1000 | 0.194 | -175.4 | -52.4 | -4.59 | 0.468 |

| | | | | | |
|---|---|---|---|---|---|
| C | 100 | 0.744 | −46.9 | 9.22 | −9.88 | 1.01 |
| | 200 | 0.816 | −53.1 | 7.34 | −9.74 | 0.99 |
| | 300 | 0.935 | −66.3 | 4.33 | −9.72 | 0.99 |
| | 350 | 0.993 | −76.8 | 2.27 | −9.86 | 1.01 |
| | 400 | 1.020 | −90.0 | 0 | −9.80 | 1.00 |
| | 450 | 0.987 | −104.8 | −2.61 | −9.76 | 1.00 |
| | 500 | 0.890 | −119.3 | −5.45 | −9.83 | 1.00 |
| | 600 | 0.638 | −141.3 | −12.1 | −9.83 | 1.00 |
| | 800 | 0.323 | −161.5 | −29.7 | −9.66 | 0.99 |
| | 1000 | 0.191 | −169.2 | −51.7 | −10.0 | 1.02 |
| D | 100 | 1.052 | −14.9 | 9.20 | −2.46 | 0.251 |
| | 200 | 1.132 | −33.7 | 7.30 | −4.95 | 0.505 |
| | 300 | 1.175 | −59.8 | 4.24 | −7.34 | 0.749 |
| | 350 | 1.126 | −75.0 | 2.29 | −8.55 | 0.872 |
| | 400 | 1.020 | −90.0 | 0 | −9.80 | 1.000 |
| | 450 | 0.883 | −103.2 | −2.56 | −11.1 | 1.133 |
| | 500 | 0.745 | −114.2 | −5.42 | −12.2 | 1.245 |
| | 600 | 0.523 | −129.8 | −12.4 | −14.7 | 1.500 |
| | 800 | 0.283 | −146.3 | −29.6 | −20.0 | 2.041 |
| | 1000 | 0.176 | −154.5 | −50.4 | −23.5 | 2.398 |

[a] Calculated using $|\alpha|$ to 2 decimal places, $\varepsilon$ to nearest degree.

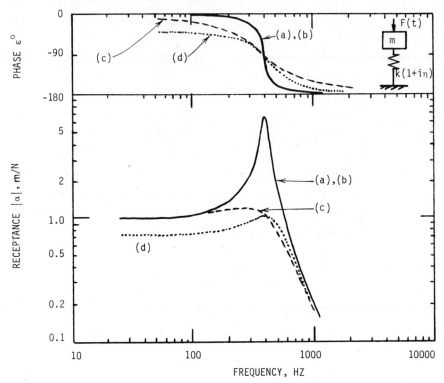

**FIGURE 4.33.** Receptance versus frequency.

test. From the receptance data in Figure 4.33, one can determine $\eta$ at 400 Hz, for the low damping cases, by the half-power bandwidth method:

$$\eta = \frac{435 - 365}{400} = 0.18$$

At resonance, $|\alpha| = 1/k\eta$, so that

$$6.2 \times 10^{-5} = \frac{1}{0.18k}$$

$$\therefore \quad k = 8.96 \times 10^4 \text{ N/m}$$

And

$$m = \frac{k}{\omega_r^2}$$

$$= \frac{8.96 \times 10^4}{(2\pi \times 400)^2}$$

$$= 0.0142 \text{ kg}$$

The errors in these values are 6.5% in $\eta$, 9.1% in $k$, and 9.2% in $m$. With care one could reduce these errors, but in a real experiment one usually has

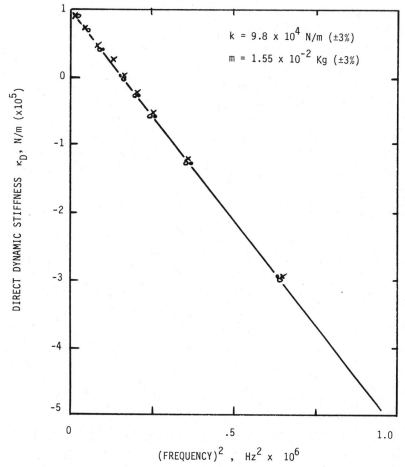

**FIGURE 4.34.**   Direct dynamic stiffness versus $f^2$ ($\bullet A$; $\times B$; $\bigcirc C$)

nothing to compare with to help reduce errors. Furthermore it is difficult to examine the variation of $k$ and $\eta$ with frequency, and for the higher values of damping, not even this degree of accuracy can be achieved.

Now look at Figures 4.34 and 4.35 where the calculated values of $\kappa_D = k - m\omega^2$ and $\kappa_Q = k\eta$ are plotted against $f^2$ and $f$, respectively. The graph of $\kappa_D$ versus $f^2$ is a straight line, from which $k$ is the intercept on the $f = 0$ axis and $m$ is the slope. It is seen from Figure 4.34 that

$$k = 9.8 \times 10^4 \text{ N/m}$$

$$m = 0.0155 \text{ kg}$$

From Figure 4.35 the graphs of $\kappa_Q$ versus $f$ are straight lines, whose equations, given the value of $k$, define the variation of $\eta$ with $f$. Errors are seen to

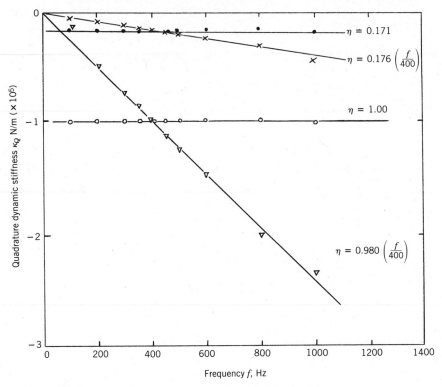

**FIGURE 4.35.**   Quadrature dynamic stiffness versus frequency ($\bullet$ A; $\times$ B; $\bigcirc$ C).

be quite small. This example shows that for constant $m$ and $k$, the dynamic stiffness approach allows one to determine $\eta$ as a function of frequency.

### 4.7.3. Harmonic Response of Single Degree of Freedom System with Variable Stiffness and Damping (Force Excitation)

To illustrate the problems associated with interpreting tests on systems with variable stiffness and damping, consider a system with several values of $m$, and stiffness and damping given by

$$k = 10^5 + 50f \text{ N/m}$$

$$\eta = 0.2 + 2 \times 10^{-5} f$$

With $f$ in Hertz, the values selected for $m$ are (A) 0.5 kg, (B) 0.1 kg, (C) 0.02 kg, and (D) 0.005 kg, which change the resonant frequency from about 70 Hz to 800 Hz as would occur in a series of resonance tests to measure material damping properties. In this example, $k$ changes from $10^5$ N/m at 0 Hz to $1.5 \times 10^5$ at 1000 Hz, and $\eta$ changes from 0.2 at 0 Hz to 0.22 at 1000 Hz.

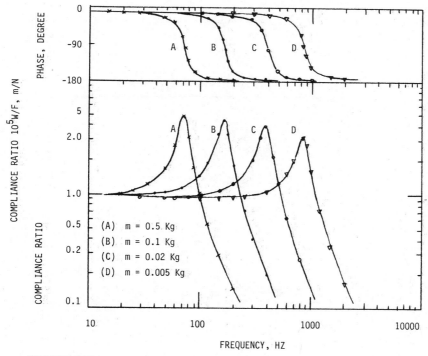

**FIGURE 4.36.**  Variation of compliance with frequency for various values of $m$.

typical for elastomeric materials in their rubbery region of temperature and frequency. Figure 4.36 shows the calculated variation of $|\alpha|$ and $\varepsilon$ for these cases as a function of frequency. Table 4.6 gives some tabulated values of $|\alpha|$ and $\varepsilon$ for $m = 0.5$ kg, and Table 4.7 for $m = 0.1$ kg, 0.02 kg, and 0.005 kg. From the tabulated and plotted values we can determine $k$ and $\eta$ at each resonance frequency as follows:

| Frequency Hz | $k = m\omega_r^2$ N/m | Percent error | $\eta = \Delta f/f$ | Percent error | $\eta = 1/\sqrt{A^2 - 1}$ | Percent error |
|---|---|---|---|---|---|---|
| 72.4 | 1.034 E 5 | −0.2 | 0.207 | +3.0 | 0.213 | +6.1 |
| 165.5 | 1.081 E 5 | −0.2 | 0.213 | +4.9 | 0.226 | +11.2 |
| 388 | 1.189 E 5 | −0.4 | 0.228 | +9.6 | 0.256 | +23.2 |
| 844 | 1.406 E 5 | −1.1 | 0.262 | +20.7 | 0.325 | +49.9 |

It is seen that the error in determining $k$ from the observed resonant frequencies is very small, but the error in $\eta$ is quite considerable and is due to the effect of variable $k$ on the apparent bandwidth of the peak at resonance. The values of $\eta$ determined from the amplification at resonance ($\eta = 1/\sqrt{A^2 - 1}$) are even more in error.

**TABLE 4.6. Receptance and Dynamic Stiffness Values for SDOF System with Variable $k$ and $\eta$ ($m = 0.5$ kg)**

| $f$ Hz | $|\alpha| \times 10^{-5}$ N/m | $\varepsilon$ degrees | $\kappa_D = \cos\varepsilon/|\alpha|$ $(\times 10^4)^a$ | $\kappa_Q = \sin\varepsilon/|\alpha|$ $(\times 10^4)^a$ | $k = \kappa_D + m\omega^2$ N/m $(\times 10^5)$ | $\eta = \kappa_Q/k$ |
|---|---|---|---|---|---|---|
| 0 | 0.981 | −11.3 | 10.00 | −1.95 | 1.00 | 0.20 |
| 10 | 0.994 | −11.5 | 9.88 | −2.10 | 1.01 | 0.21 |
| 20 | 1.050 | −12.3 | 9.32 | −1.98 | 1.01 | 0.20 |
| 25 | 1.096 | −12.9 | 8.86 | −2.05 | 1.01 | 0.20 |
| 50 | 1.754 | −21.2 | 5.33 | −2.05 | 1.03 | 0.20 |
| 60 | 2.627 | −33.0 | 3.19 | −2.07 | 1.03 | 0.20 |
| 64.5$^c$ | 3.376 | −44.6 | 2.09 | −2.09 | 1.03 | 0.20 |
| 70 | 4.562 | −72.0 | 0.68 | −2.09 | 1.04 | 0.20 |
| 72.4$^b$ | 4.791 | −89.6 | 0 | −2.09 | 1.03 | 0.20 |
| 75 | 4.517 | −109.2 | −0.72 | −2.09 | 1.04 | 0.20 |
| 79.5$^c$ | 3.388 | −134.7 | −2.09 | −2.09 | 1.04 | 0.20 |
| 80 | 3.265 | −136.8 | −2.24 | −2.09 | 1.04 | 0.20 |
| 90 | 1.687 | −159.2 | −5.52 | −2.12 | 1.05 | 0.20 |
| 100 | 1.055 | −167.1 | −9.19 | −2.12 | 1.05 | 0.20 |
| 120 | 0.557 | −173.1 | −17.7 | −2.18 | 1.07 | 0.20 |
| 150 | 0.296 | −176.3 | −33.3 | −2.33 | 1.11 | 0.21 |
| 200 | 0.147 | −178.1 | −66.6 | −2.33 | 1.24 | 0.19 |

[a] Calculated using $|\alpha|$ to 2 decimal places and $\varepsilon$ to nearest degree.
[b] Resonance frequency.
[c] Half-power points.

**TABLE 4.7. Receptance Values for SDOF System with Variable k and η (Force Excitation)**

| | m = 0.1 kg | | | m = 0.02 kg | | | m = 0.005 kg | |
|---|---|---|---|---|---|---|---|---|
| f Hz | $\|\alpha\| \times 10^{-5}$ | ε degrees | f Hz | $\|\alpha\| \times 10^{-5}$ | ε degrees | f Hz | $\|\alpha\| \times 10^{-5}$ | ε degrees |
| 0 | 0.981 | −11.31 | 0 | 0.981 | −11.31 | 0 | 0.981 | −11.31 |
| 50 | 1.054 | −12.54 | 100 | 1.066 | −12.30 | 200 | 0.957 | −12.40 |
| 100 | 1.452 | −17.94 | 200 | 1.226 | −15.97 | 400 | 1.088 | −15.76 |
| 150 | 3.482 | −49.45 | 300 | 2.000 | −28.33 | 600 | 1.537 | −25.06 |
| 160 | 4.345 | −72.46 | 400 | 3.883 | −104.23 | 800 | 3.013 | −65.68 |
| 170 | 4.392 | −104.22 | 500 | 1.299 | −160.00 | 850 | 3.234 | −90.21 |
| 180 | 3.430 | −130.43 | 350 | 3.126 | −49.49 | 900 | 2.862 | −115.20 |
| 75 | 3.961 | −118.77 | 360 | 3.443 | −57.34 | 1000 | 1.732 | −145.1 |
| 165 | 4.541 | −87.99 | 370 | 3.747 | −67.05 | 100 | 0.951 | −11.63 |
| 166 | 4.540 | −91.27 | 380 | 3.968 | −78.59 | 300 | 0.999 | −13.69 |
| 165.2 | 4.542 | −88.65 | 390 | 4.026 | −91.37 | 840 | 3.236 | −85.0 |
| 165.3 | 4.542 | −88.98 | 388 | 4.030 | −88.76 | 830 | 3.211 | −79.8 |
| 165.4 | 4.543 | −89.30 | 389 | 4.029 | −90.06 | 844 | 3.238 | −87.0 |
| 165.5 | 4.543 | −89.63 | 387 | 4.029 | −87.46 | 946 | 2.293 | −132.3 |
| 165.6 | 4.542 | −89.96 | 341 | 2.859 | −43.80 | 725 | 2.288 | −41.97 |
| 182.1 | 3.212 | −134.47 | 429.5 | 2.855 | −133.7 | 50 | 0.961 | −11.42 |
| 146.8 | 3.211 | −44.37 | 20 | 0.974 | −11.37 | 150 | 0.949 | −11.96 |
| 20 | 0.986 | −11.51 | 30 | 0.973 | −11.42 | 250 | 0.973 | −12.97 |
| 30 | 0.999 | −11.74 | 50 | 0.975 | −11.58 | 500 | 1.249 | −19.14 |
| 40 | 1.022 | −12.08 | 60 | 0.978 | −11.69 | 700 | 2.085 | −37.04 |
| 70 | 1.153 | −13.91 | 80 | 0.989 | −11.96 | 1200 | 0.773 | −163.91 |
| 120 | 1.865 | −23.58 | 150 | 1.083 | −13.67 | 1500 | 0.367 | −171.49 |
| 200 | 1.890 | −154.90 | 450 | 2.207 | −145.60 | 2000 | 0.169 | −175.35 |
| 250 | 0.734 | −170.25 | 600 | 0.638 | −169.87 | | | |
| 300 | 0.414 | −174.37 | 800 | 0.273 | −175.27 | | | |
| 400 | 0.195 | −177.21 | 1000 | 0.156 | −177.05 | | | |

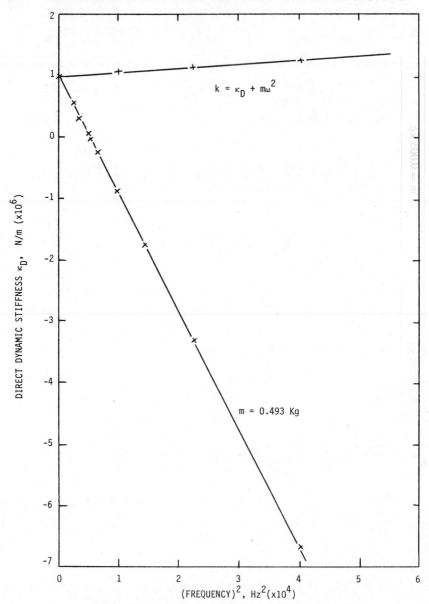

**FIGURE 4.37.**   Dynamic stiffness versus $f^2$ ($m = 0.5$ kg).

The dynamic stiffness approach seems to work better. Table 4.6 gives tabulated values of $\kappa_D = \cos \varepsilon/|\alpha|$, using values of $|\alpha|$ rounded off to two decimal places and values of $\varepsilon$ rounded off to the nearest degree. Figure 4.37 gives the corresponding plots of $\kappa_D$ and $k = \kappa_D + m\omega^2$, with $m = 0.5$ kg, versus $f^2$. From the slope of the $\kappa_D \to f^2$ line, $m$ was seen to be 0.493 kg, representing an error of about 1.4%. The error in $k$ is less than 1% except for $f = 200$ Hz. The error in $\eta$ is no greater, except for $f = 150$ Hz and 200 Hz.

### 4.7.4.  Harmonic Response of Single Degree of Freedom System with Variable Stiffness and Damping (Base Excitation)

If the single degree of freedom system is excited by a harmonic motion $W_0 e^{i\omega t}$ of its base, the transmissibility $T$ and phase angle between response $w(t) = We^{i(\omega t + \varepsilon)}$ and $W_0 e^{i\omega t}$ is given by

$$T = \left| \frac{W}{W_0} \right| = k \sqrt{\frac{1 + \eta^2}{(k - m\omega^2)^2 + (k\eta)^2}}$$

$$\varepsilon = \tan^{-1} \eta - \tan^{-1} \frac{k\eta}{k - m\omega^2}$$

If $k$ and $\eta$ vary in the same way with frequency as for example in Section 4.7.3, then $T$ and $\varepsilon$ may readily be calculated. Results are plotted for (A) $m = 0.5$ kg, (B) $m = 0.1$ kg, (C) $m = 0.02$ kg, and (D) $m = 0.005$ kg in Figure 4.38, and are summarized in Table 4.8. In this case there is no simple relationship between $\kappa_D$ and $\kappa_Q$, and the parameters $k$ and $\eta$, so the dynamic stiffness approach is not so useful. However, the plot of $T$ versus frequency gives

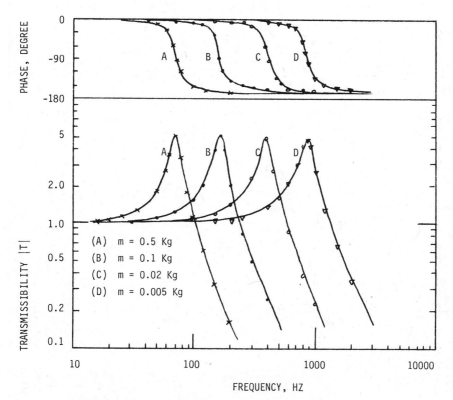

**FIGURE 4.38.**  Transmissibility versus frequency for various masses.

**TABLE 4.8. Receptance Values for SDOF System with Variable k and η (Base Excitation)**

| $m = 0.5$ kg | | | $m = 0.1$ kg (B) | | | $m = 0.02$ kg (C) | | | $m = 0.005$ kg | | |
|---|---|---|---|---|---|---|---|---|---|---|---|
| $f$ Hz | $\lvert\alpha\rvert \times 10^{-5}$ | $\varepsilon$ degrees | $f$ Hz | $\lvert\alpha\rvert \times 10^{-5}$ | $\varepsilon$ degrees | $f$ Hz | $\lvert\alpha\rvert \times 10^{-5}$ | $\varepsilon$ degree | $f$ Hz | $\lvert\alpha\rvert \times 10^{-5}$ | $\varepsilon$ degrees |
| 0 | 1.00 | 0 | 0 | 1.00 | 0 | 0 | 1.00 | 0 | 0 | 1.00 | 0 |
| 10 | 1.02 | −0.22 | 20 | 1.015 | −0.18 | 20 | 1.00 | −0.03 | 50 | 1.00 | −0.05 |
| 15 | 1.04 | −0.51 | 30 | 1.03 | −0.40 | 30 | 1.01 | −0.08 | 100 | 1.02 | −0.21 |
| 20 | 1.08 | −0.93 | 40 | 1.06 | −0.73 | 50 | 1.02 | −0.22 | 150 | 1.04 | −0.48 |
| 25 | 1.13 | −1.52 | 50 | 1.10 | −1.17 | 60 | 1.03 | −0.31 | 200 | 1.07 | −0.87 |
| 35 | 1.29 | −3.40 | 70 | 1.22 | −2.52 | 80 | 1.05 | −0.57 | 250 | 1.12 | −1.38 |
| 50 | 1.83 | −9.82 | 100 | 1.56 | −6.52 | 100 | 1.08 | −0.90 | 400 | 1.33 | −4.01 |
| 70 | 4.82 | −60.6 | 120 | 2.02 | −12.14 | 150 | 1.19 | −2.19 | 500 | 1.60 | −7.28 |
| 60 | 2.76 | −21.60 | 150 | 3.82 | −37.97 | 200 | 1.38 | −4.44 | 600 | 2.04 | −13.09 |
| 72 | 5.05 | −75.12 | 160 | 4.79 | −60.98 | 300 | 2.35 | −16.69 | 700 | 2.88 | −24.96 |
| 74 | 4.95 | −90.48 | 170 | 4.86 | −92.72 | 400 | 4.76 | −92.48 | 800 | 4.32 | −53.49 |
| 73 | 5.05 | −82.83 | 180 | 3.82 | −118.93 | 390 | 4.91 | −79.63 | 850 | 4.72 | −77.97 |
| 71 | 4.97 | −67.64 | 165 | 5.016 | −76.51 | 380 | 4.82 | −66.86 | 900 | 4.25 | −102.92 |
| 72.5 | 5.06 | −78.76 | 164 | 4.979 | −73.27 | 389 | 4.92 | −78.33 | 1000 | 2.66 | −132.74 |
| 64.5 | 3.56 | −37.17 | 166 | 5.018 | −79.77 | 343 | 3.49 | −33.28 | 1200 | 1.27 | −151.28 |
| 64.6 | 3.58 | −33.51 | 167 | 5.003 | −83.05 | 431 | 3.48 | −122.95 | 2000 | 0.35 | −161.85 |
| 75 | 4.78 | −97.81 | 165.7 | 5.019 | −78.79 | 450 | 2.76 | −133.79 | 1500 | 0.66 | −158.54 |
| 80 | 3.46 | −125.41 | 182.3 | 3.55 | −123.31 | 500 | 1.66 | −148.21 | 5000 | 0.08 | −161.99 |
| 79.5 | 3.59 | −123.36 | 147.2 | 3.55 | −33.49 | 600 | 0.85 | −157.90 | 840 | 4.70 | −72.72 |
| 40 | 1.80 | −147.75 | 200 | 2.12 | −143.37 | 800 | 0.39 | −163.08 | 860 | 4.69 | −83.25 |
| 100 | 1.13 | −155.65 | 250 | 0.84 | −158.67 | 1000 | 0.24 | −164.64 | 849 | 4.716 | −77.443 |
| 120 | 0.602 | −161.70 | 300 | 0.49 | −162.73 | 2000 | 0.07 | −165.57 | 736 | 3.34 | −32.40 |
| 150 | 0.325 | −164.82 | 400 | 0.24 | −165.46 | | | | 953 | 3.34 | −121.99 |
| 200 | 0.165 | −166.58 | | | | | | | | | |
| 1000 | 0.008 | −167.50 | | | | | | | | | |

estimates of $k$ and $\eta$ for four resonance peaks as for the preceding example. The summary of these results is as follows:

| Frequency Hz | $k = m\omega_r^2$ N/m | Percent error | $\eta = \Delta f / f$ | Percent error | $\eta = 1/\sqrt{A^2 - 1}$ | Percent error |
|---|---|---|---|---|---|---|
| 72.5 | 1.038 E 5 | −0.2 | 0.206 | +2.3 | 0.202 | −0.3 |
| 165.7 | 1.084 E 5 | −0.1 | 0.212 | +4.3 | 0.203 | −0.2 |
| 389 | 1.195 E 5 | 0 | 0.226 | +8.8 | 0.208 | −0.1 |
| 849 | 1.423 E 5 | +0.1 | 0.256 | +18.0 | 0.217 | 0 |

Note that the errors in predicting $k$ are very small, and $\eta$ calculated by the amplitude method ($\eta = 1/\sqrt{A^2 - 1}$) is also recovered with very little error. The bandwidth method involves more error. These results illustrate the effectiveness of the simple resonance technique, using base excitation, for determining damping material properties.

## REFERENCES

4.1. R. E. D. Bishop and D. C. Johnson, *The Mechanics of Vibration*, Cambridge Univ. Press, Cambridge, 1960.

4.2. B. J. Lazan, *Damping of Materials and Members in Structural Mechanics*, Pergamon, New York, 1968.

4.3. J. P. Henderson, "Energy dissipation in a vibration damper utilizing a viscoelastic suspension," Air Force Materials Lab. Report AFML-TR-67-403, 1965.

4.4. D. I. G. Jones and A. Muszynska, "On the modal identification of multiple degree of freedom systems from experimental data," *Shock Vib. Bull.*, **53**, 91–110 (1983).

4.5. R. L. Adkins, "Modal analysis of linear non-conservative system," Air Force Materials Lab. Report AFML-TR-75-2, 1975.

4.6. K. A. Foss, "Coordinates which uncouple the equation of motion of damped linear systems," *J. Appl. Mech.*, **25**, 361 (1958).

4.7. Y. K. Lin and T. J. McDaniel, "Dynamics of beam-type periodic structures," *J. Eng. Industry, ASME*, **B91**(4), 1133–1141, 1969.

4.8. E. Pestel and F. Leckie, *Matrix Methods in Elasto-Mechanics*, McGraw-Hill, New York, 1963.

4.9. Y. K. Lin and B. K. Donaldson, "A brief survey of transfer matrix techniques with special reference to the analysis of aircraft panels," *J. Sound Vib.*, **10**(1), 103–143 (1969).

4.10. Y. K. Lin, "Free vibration of continuous skin-stringer panels," *J. Appl. Mech.*, **27**(4), 669–676 (1960).

4.11. J. P. Henderson, "Vibration analysis of curved skin-stringer structures having tuned elastomeric dampers," Air Force Materials Lab. Report AFML-TR-72-240, WPAFB, October 1972.

4.12. K. Desai, "Summary of general purpose programs," *Shock and Vibration Computer Programs, Reviews and Summaries*, (ed. W. Pilkey and B. Pilkey), SVM-10, Shock and Vibration Information Center, Department of Defense, 1975, pp. 529–539.

4.13. C. D. Johnson, D. A. Kienholz, and L. C. Rogers, "Finite element prediction of damping in beams with constrained viscoelastic layers," *Shock Vib. Bull.* **51**, Pt. 1, 71–81 (May 1981).

4.14. C. D. Johnson and D. A. Kienholz, "Finite element prediction of damping in structures with constrained viscoelastic layers," *AIAA J.*, **20**(9), 1284–1290 (September 1982).

4.15. M. L. Soni, "Finite element analysis of viscoelastically damped sandwich structures," *Shock Vib. Bull.*, **51**, Pt. 1, 97–108 (May 1981).

4.16. M. F. Kluesener and M. L. Drake, "Damped structure design using finite element analysis," *Shock Vib. Bull.*, **52**, Pt. 5, 1–12 (May 1982).

4.17. R. A. Brockman, "MAGNA (Materially and Geometrically Nonlinear Analysis)," Part IV, Quick Reference Manual, Air Force Flight Dynamics Lab. Report AFWAL-TR-82-3098, WPAFB, December 1982.

4.18. D. J. Ewins, "Why's and wherefores of modal testing," *SEE J.*, 1–13 (September 1979).

4.19. J. M. Leuridan, D. L. Brown, and R. J. Allemang, "Direct system parameter identification of mechanical structures with application to modal analysis," AIAA paper 82-0767-CP, Bound Collection of Technical Papers, AIAA/ASME/ASCE/AHS 23rd Structures, Structural Dynamics and Materials Conf., New Orleans, La; pp. 548–556, May 10–12, 1982.

4.20. G. D. Bergland, "A guided tour of the fast Fourier transform," *IEEE Spectrum* **6**(7), pp. 41–52, July 1969.

4.21. R. B. Blachman and J. W. Tukey, *The Measurement of Power Spectra*, Dover, New York, 1958.

4.22. J. W. Cooley, P. A. Lewis, and P. D. Wleeh, "Application of the fast Fourier transform to computation of Fourier integrals, Fourier series, and convolution integrals," *IEEE Trans. Audio and Electrocoustics*, **AU-15**, 79–84 (June 1967).

4.23. J. P. Henderson and M. L. Drake, "Vibration control using additive damping and FFT analysis," SAE Paper 790220, March 1979.

4.24. K. A. Ramsey, "Effective measurements for structural dynamics testing, Part II," *Sound Vib.* 18–31, April 1976.

4.25. D. Brown, G. Carbon, and K. Ramsey, "Survey of excitation techniques applicable to the testing of automotive structures," Proc. International Automotive Engineering Congress and Exposition, Detroit, February 28–March 4, 1977.

4.26. J. A. Bendat and A. G. Piersol, *Measurement and Analysis of Random Data*, Wiley, New York, 1966.

4.27. A. L. Klosterman, "On the experimental determination and use of modal representations of dynamic characteristic," Ph.D. dissertation, Univ. of Cincinnati, 1971.

4.28. A. Berman and W. G. Flannelly, "Theory of incomplete models of dynamic structures," *AIAA J.*, **9**, 1481–1487 (1971).

4.29. H. G. D. Goyder, "Methods of application of structural modelling from measured structural frequency response data," *J. Sound Vib.*, **68**(2), 209–230 (1980).

4.30. I. R. Ibrahim and E. C. Mikulcik, "A Method for Direct Identification of Vibration Parameter from Free Response," *Shock Vib. Bull.*, **47**, Pt. 4, 183–198 (1977).

4.31. D. J. Ewins and P. R. Gleeson, "A Method for Modal Identification of Lightly Damped Structures," *J. Sound Vib.*, **84**(1), 57–79 (1982).

4.32. W. G. Halvorsen and D. L. Brown, "Impulse techniques for structural frequency response testing," *Sound Vib.*, 8–21 (November 1979).

# 5

# DISCRETE DAMPING DEVICES

## ADDITIONAL SYMBOLS

$D_s$        energy dissipated per cycle

$F_d$        force on structure due to damper

$\bar{F}_d$        amplitude of $F_d$

$H_1, H_2$   thickness of beams 1, 2, respectively

Im           imaginary part

$k, k'$      stiffness

$k''$        quadrature stiffness

$m$          mass

$\left.\begin{array}{l} M_{nn}^{(1)} \\ M_{nn}^{(2)} \end{array}\right\}$   modal masses of beams 1, 2, respectively

$n, m$       mode number $(n, m = 1, 2, \ldots)$

Re           real part

$W(x_j)$     transverse displacement amplitude at point $x_j$

$W_r$        amplitude of relative displacement between mass and base

$W_0$        amplitude of base motion

$x_j, y_j$   coordinates of point $j(j = 1, 2, \ldots)$

$\alpha_{ijx,y,z}$   transfer receptance in $x, y$, or $z$ directions

$\Delta$     determinant of matrix or $x/L$

$\Gamma$     nondimensional stiffness parameter $(kL^3/EI\xi_1^4)$

$\Gamma_e$   nondimensional stiffness parameter (eq. 5.19)

$\phi_{n1,2}$      $n$th modal function of beams 1, 2, respectively

$\kappa_d^*$      complex dynamic stiffness of damper as seen at point of attachment

$\psi$      nondimensional mass parameter ($m/\rho bHL$)

$\psi_e$      nondimensional mass parameter

$\rho_1, \rho_2$      density of beams 1, 2, respectively

$\omega_r$      resonant frequency of structure

$\omega_D$      resonant frequency of damper

$\omega_{n1}$      $n$th resonant frequency of beam 1

$\omega_{n2}$      $n$th resonant frequency of beam 2

## 5.1. INTRODUCTION

One method of increasing the damping of a structure is through the use of one or more tuned damping devices. Such a damping device could be in the form of a single degree of freedom system consisting of a mass on a linear spring with viscous damping [5.1] or a mass on a viscoelastic spring [5.2–5.5], a viscoelastically damped resonant beam [5.6–5.9], or a tuned viscoelastic link joining different elements of a complex structure [5.10]. It should be noted that all of these devices contribute to the damping of the structure through the dissipation of energy and differ from the undamped tuned resonator or "dynamic absorber," which functions as a discrete tuned resonant energy transfer device [5.11–5.12]. Since tuned damping devices dissipate energy, they are effective over a range of frequencies, and in some cases a single damper can even be effective in controlling the response of a structure in several modes of vibration [5.2, 5.13].

The devices considered in this chapter differ from layered damping treatments, discussed in the subsequent chapter, in that tuned damping devices dissipate energy depending on the local displacement in a structure, rather than as layered damping treatments that depend on surface strains. Thus for structures that have low vibratory surface strains, involving nonplatelike elements, as in the case of space-frame structures such as large antennas or highly curved elements, the tuned damper can offer advantages over other forms of damping treatments. Essential prerequisites for a single degree of freedom tuned damper to be of value are that the damper be located at a point of high displacement response, such as an antinode, and that the structure have a single resonance or a group of resonances with similar strain energies [5.2]. In addition tuned dampers can be designed with multiple resonances that generally occur at widely separated frequencies that can in turn be effective in damping structures over a broad range of frequencies.

The reader is encouraged, after reading the first sections of this chapter, to move occasionally to the section on applications and examples, in order to observe how the equations and general principles are applied in practice.

## 5.2.   TUNED DAMPER BEHAVIOR

### 5.2.1.   Energy Dissipated in a Single Degree of Freedom Tuned Damper

Consider the single degree of freedom tuned viscoelastic damper shown schematically in Figure 5.1. It has been shown in the previous chapter, equation (4.31), that the energy dissipated per cycle in this system is

$$D_S = \frac{\pi \eta k \xi^4 W_0^2}{(1 - \xi^2)^2 + \eta^2} \tag{5.1}$$

where $\xi^2 = (\omega/\omega_r)^2 = \omega^2 m/k$ and $W_0$ is the amplitude of harmonic displacement of the structure at the point of connection to the damper. If hysteretic damping is assumed, then Figure 5.2 illustrates how the nondimensionalized energy dissipated in a single degree of freedom damper relates to frequency of

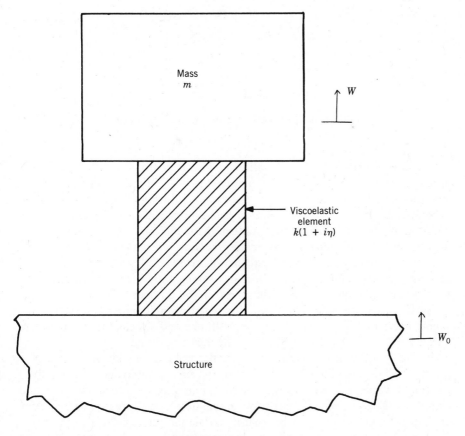

**FIGURE 5.1.**   Tuned damper.

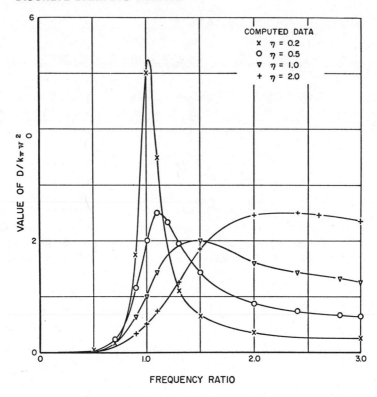

**FIGURE 5.2.** Energy dissipated in damper per cycle.

excitation for various assumed values of loss factor $\eta$. From this plot it is apparent that as the loss factor increases, the peak in energy dissipation decreases, but the frequency band associated with high energy dissipation becomes greater. The peak in energy dissipation occurs at a frequency slightly higher than the damper resonant frequency for materials with moderate loss factors, and the difference between the frequency of maximum energy dissipation and damper resonance frequency increases with an increase in $\eta$. In fact, as the excitation frequency becomes significantly greater than the resonant frequency of the damper, the mass of the damper tends to have very small displacements, and the energy dissipated in the damper approaches the energy dissipated in a viscoelastic link, with one end attached to the structure and the other end fixed in space (i.e., attached to ground).

Figures 4.6 and 4.7 illustrate how $R$, the ratio of energy dissipated in a tuned single degree of freedom damper to the energy dissipated in a link fixed to ground, varies with frequency for viscous damping, hysteretic damping, and complex stiffness values measured for real materials as a function of frequency at constant temperature. For the real materials the general shape of the curve for $R$ as a function of frequency is approximated by the hysteretic damping

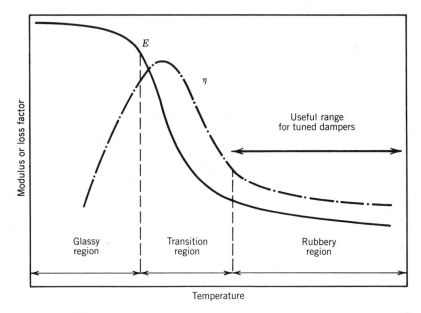

**FIGURE 5.3.**   Temperature zones of viscoelastic material behavior.

assumption, except that the effective frequency bandwidth is wider as a result of variation of the real part of the complex stiffness with frequency.

It has been shown in Chapter 3 that the properties of elastomeric materials vary with temperature in a manner idealized in Figure 5.3. For a tuned damper to be effective it must be operated near the frequencies of maximum energy dissipation illustrated in Figure 5.2. This tuning effect has a profound influence on the choice of elastomer used in the damper. If an elastomeric tuned damper were to be operated within the transition temperature range, where the loss factor is high and the modulus changes rapidly with temperature, then the internal heating of the damper due to energy dissipation would cause the resonant frequency to change and the damper to detune itself. Therefore elastomeric materials for tuned dampers should generally be used in the rubbery region, as shown in Figure 5.3, where small changes in temperature do not have a large effect on the stiffness. Therefore most practical elastomeric materials for tuned dampers have a maximum loss factor in the rubbery region, from 0.1 to 0.2, such as for some of the more highly damped silicone elastomers.

### 5.2.2.   Damping and Inertia Forces from a Single Degree of Freedom Tuned Damper

Assuming that the damper in Figure 5.1 acts as a single degree of freedom and that the frequency is such that the inertial mass of the spring can be neglected

[5.2], the force $F_d$ exerted on the structure during steady state vibration can be shown to be

$$F_d = k(1 + i\eta)W_r e^{i(\omega t - \varepsilon)} \tag{5.2}$$

$$= (k' + ik'')W_r e^{i(\omega t - \varepsilon)} \tag{5.3}$$

where $F_d = \bar{F}_d e^{i\omega t}$ and $W_r$ is the amplitude of relative displacement given in equation (4.30). This reduces to

$$\bar{F}_d(\omega) = \frac{m\omega^2 W_0}{1 - \omega^2 m/(k' + ik'')} \tag{5.4}$$

where $F_d$ and $\bar{F}_d$ are complex and $W_0$ is the magnitude of the harmonic displacement of the structure at the point of attachment of the damper. Equation (5.4) can be rewritten:

$$\bar{F}_d = \kappa_d^* W_0 \tag{5.5}$$

where

$$\kappa_d^* = \frac{\omega^2 mk'(k' + \eta^2 k' - \omega^2 m) + ik''m\omega^2(k' - m\omega^2)}{(k' - \omega^2 m)^2 + (k'')^2} \tag{5.6}$$

or $\kappa_d^* = \mathrm{Re}(\kappa_d^*) + i\,\mathrm{Im}(\kappa_d^*)$; $\mathrm{Re}(\kappa_d^*)W_0$ is the real part of (5.5) and represents the inertia force of the damper, and $\mathrm{Im}(\kappa_d^*)W_0$ is the imaginary part and represents the damping force, which is 90 degrees out of phase with the displacement.

### 5.2.3.  Damper Geometry and Design Considerations

Most tuned dampers are variations of a mass on a tension-compression spring or a mass on an elastomeric-shear spring, as shown in Figure 5.4. In either of these cases care must be taken when calculating the damper stiffness $k$ to correct for variation in the state of local stress caused by the end conditions. For a tension-compression elastomeric spring, as in Figure 5.4a, Snowdon [5.14] has shown that the stiffness of the spring can be calculated from

$$k = E_D \frac{S\kappa_T}{L} \tag{5.7}$$

where $E_D$ is the complex Young's modulus of the elastomer, $S$ is the cross-sectional area, $L$ is the length, and $\kappa_T$ is a shape factor given by

$$\kappa_T = 1 + \beta\left(\frac{S}{S'}\right)^2 \tag{5.8}$$

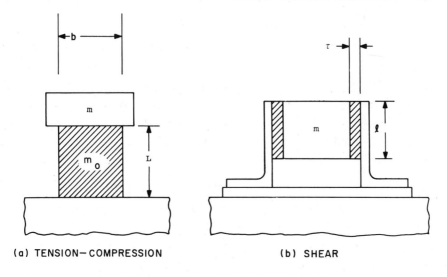

(a) TENSION–COMPRESSION                    (b) SHEAR

**FIGURE 5.4.**  Types of tuned dampers.

where $S'$ is the nonload-carrying area of the elastomeric spring and $\beta$ is a nondimensional constant equal to about 2.0 for an unfilled elastomer and 1.5 for a filled elastomer.

On the other hand, the stiffness of a shear-type damper, illustrated in Figure 5.4b can be calculated from the relation

$$k = \frac{G_D S \kappa_s}{\tau} \tag{5.9}$$

where $G_D$ is the shear modulus (approximately one-third of Young's modulus $E_D$), $S$ is the load-carrying area, and $\tau$ is the thickness of the shear layer. $\kappa_s$ is given [5.15] by

$$\kappa_s = \frac{1}{1 + L^2/36r^2} \tag{5.10}$$

and $L$ is the radius of gyration of the shear layer cross section around the neutral axis of bending. For a rectangular shear area $r = l/\sqrt{12}$.

Several other configurations of dampers have also been proposed, as illustrated in Figure 5.5 [5.8, 5.9], the crosshatched areas indicating viscoelastic materials. The "circular damper" was conceived as a way of fabricating a soft spring for low resonant frequency without unduly increasing the mass. The various beam dampers were conceived for use in rotating blades. Resonant beam damper 5 is designed to restrict creep deformation of the viscoelastic material when used on a rotating blade under high centrifugal loading [5.9].

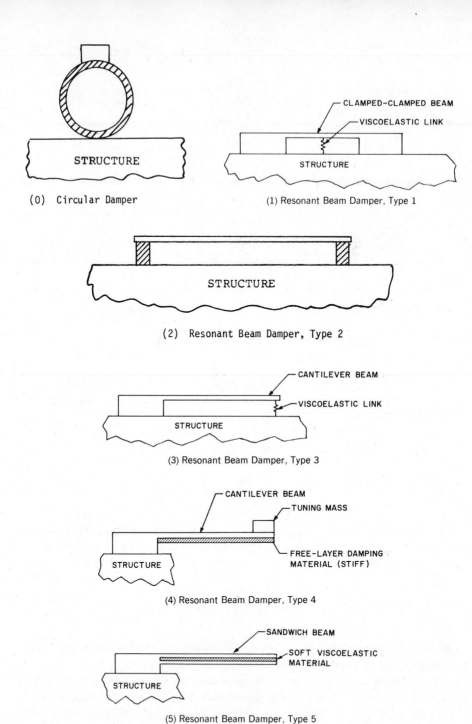

**FIGURE 5.5.** Damper configurations. (0) Circular damper. (1) Resonant beam damper, type 1. (2) Resonant beam damper, type 2. (3) Resonant beam damper, type 3. (4) Resonant beam damper, type 4. (5) Resonant beam damper, type 5.

**196**

Licari and Barhan [5.16] analyzed pendulum configurations of viscoelastic dampers, with the viscoelastic material in a cylindrical or spherical ball-joint suspension at the end of the pendulum. The pendulum configuration allowed for low resonant frequencies without excessive weight.

## 5.3. TUNED DAMPERS IN SIMPLE STRUCTURES

### 5.3.1. Introductory Remarks

If a structure has relatively widely separated resonances, a tuned damper can be designed simply by ensuring that the damper frequency $\omega_D = \sqrt{k/m}$ is close to the frequency of the mode to be damped.

The effect of the damper on the response of the structure illustrated in Figure 5.6 is to split the original single mode into two, as shown in Figure 5.7.

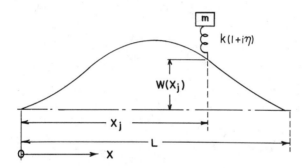

**FIGURE 5.6.** Tuned damper on simple structure.

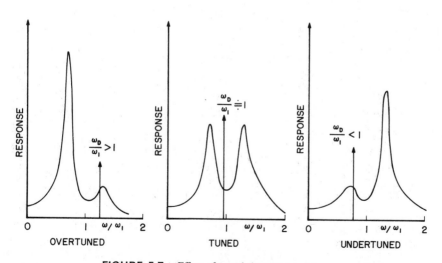

**FIGURE 5.7.** Effect of tuned damper on response.

The lower frequency branch corresponds to the damper mass and the structure surface moving essentially in phase with each other, whereas the higher frequency peak corresponds to the mass and surface moving out of phase. The effect of varying the damper frequency $\omega_D$ in relation to the frequency $\omega_n$ of the mode in question is to emphasize one of the peaks at the expense of the other, with an optimum damping case occurring when the two response peaks are of equal amplitude.

### 5.3.2. Modal Analysis of the Effect of Tuned Dampers on a Simple Beam with Force Excitation

Consider a beam of length $L$, with arbitrary boundary conditions and with tuned viscoelastic dampers of complex stiffness $k(1 + i\eta)$ and mass $m$ at a number of points $x = x_j (j = 1, \ldots, N)$, as shown in Figure 5.6. The amplitude of the harmonic force transmitted back to the structure $(F_j)$, by the damper at point $x_j$, can be obtained from equation (5.4) as

$$F_j = \frac{-m\omega^2 W(x_j)\delta(x - x_j)}{1 - m\omega^2/k(1 + i\eta)} \tag{5.11}$$

where $\delta$ is the Dirac delta function. The Euler–Bernoulli equation for a beam with a harmonic force excitation of amplitude $F(x)$ can therefore be written as

$$EI\left(\frac{d^4W}{dx^4}\right) - \rho bH\omega^2 W - \frac{m\omega^2}{1 - m\omega^2/k(1 + i\eta)} \sum_{j=1}^{N} W(x_j)\delta(x - x_j) = F(x) \tag{5.12}$$

If $W(x)$ and $F(x)$ are now expanded as series of normal modes of the undamped beam, then these modes must satisfy the homogeneous equation of motion:

$$\frac{d^4\phi_n(x)}{dx^4} - \left(\frac{\rho bH\omega_n^2}{EI}\right)\phi_n(x) = 0 \tag{5.13}$$

where $\phi_n$ is the $n$th normal mode of the undamped beam. Then

$$W(x) = \sum_{n=1}^{\infty} W_n \phi_n\left(\frac{x}{L}\right)$$

and

$$F(x) = \sum_{n=1}^{\infty} F_n \phi_n\left(\frac{x}{L}\right) \tag{5.14}$$

For vibrations in the vicinity of the fundamental mode, $\xi_n^4 \gg \xi_1^4$, allowing one to neglect all terms in the expansion apart from the first. It may be shown that the solution reduces to the form [5.17, 5.18]

$$\frac{EI\xi_1^4 W_1}{F_1 L^4} \simeq \left\{ 1 - (\xi^4/\xi_1^4) - \frac{\psi_e(\xi/\xi_1)^4}{1 - \psi_e(\xi/\xi_1)^4/\Gamma_e(1 + i\eta)} \right\}^{-1} \tag{5.15}$$

with

$$\xi_n = \left( \frac{\rho b H \omega_n^2 L^4}{EI} \right)^{1/4} \tag{5.16}$$

$$\xi = \left( \frac{\rho b H \omega^2 L^4}{EI} \right)^{1/4} \tag{5.17}$$

$\xi_n$ being the $n$th eigenvalue of the undamped beam and $\xi$ a nondimensional frequency parameter. In equation (5.15) $\psi_e$ is an effective mass parameter, and $\Gamma_e$ is an effective stiffness parameter, such that

$$\psi_e = \psi \frac{\sum\limits_{j=1}^{N} \phi_1^2(\Delta_j)}{\int_0^1 \phi_1^2(\Delta)\, d\Delta} \tag{5.18}$$

$$\Gamma_e = \Gamma \frac{\sum\limits_{j=1}^{N} \phi_1^2(\Delta_j)}{\int_0^1 \phi_1^2(\Delta)\, d\Delta} \tag{5.19}$$

where

$\Gamma = kL^3/EI\xi_1^4$, a nondimensional stiffness parameter

$\psi = m/\rho bHL$, a nondimensional mass parameter

$\Delta = x/L$ and $\Delta_i = x_i/L$

It is therefore apparent that the theory of the response of any simple beam, for which the resonant frequencies are sufficiently well separated for certain approximations to be made, can be reduced to a single expression if appropriate effective mass and stiffness parameters are defined for each particular set of boundary conditions. The integrals and summations in equation (5.18) are readily evaluated for most cases by using the tables of normal modes given by Bishop and Johnson [5.19]. Some of the integrals and summations are given in Table 5.1 for a number of boundary conditions.

**TABLE 5.1. Standard Properties for Various Beam Configurations in Fundamental Mode**

| Boundary Conditions | Clamped-free | Pinned-pinned | Clamped-pinned | Clamped-clamped | Free-free |
|---|---|---|---|---|---|
| $\xi_1$ | 1.875 | 3.142 | 3.927 | 4.730 | 4.730 |
| $\xi_1^4$ | 12.36 | 97.4 | 237.7 | 500.6 | 500.6 |
| $\Delta_j$ | 1.00 | 0.50 | 0.50 | 0.50 | 0.50 |
| $\phi_1(\Delta_j)$ | 1.000 | 1.000 | 0.957 | 1.000 | 1.000 |
| $\int_0^1 \phi_1(\Delta)\, d\Delta$ | 0.392 | 0.637 | 0.570 | 0.523 | 0.000 |
| $\int_0^1 \phi_1^2(\Delta)\, d\Delta$ | 0.250 | 0.500 | 0.439 | 0.397 | 0.250 |
| $\phi_1^2(\Delta_j)/\int_0^1 \phi_1^2(\Delta)\, d\Delta$ | 4.000 | 2.000 | 2.086 | 2.519 | 1.479 |
| $\int_0^1 \phi_1(\Delta)\, d\Delta/\int_0^1 \phi_1^2(\Delta)\, d\Delta$ | 1.568 | 1.274 | 1.298 | 1.317 | 0.000 |

If it is assumed that the beam is uniformly covered by a continuous distribution of tuned dampers, the effective mass parameter $\psi_e$ now becomes the
true mass ratio for the distributed dampers, that is, the ratio of the total mass
of the dampers to the total mass of the beam. This is apparent from equation
(5.18) when $N$ approaches infinity. Similarly $\Gamma_e$ is seen to be equal to $L^3/EI\xi_1^4$
times the total stiffness of all the damper springs in parallel. The theory of the
beam with distributed tuned dampers has been developed in reference [5.20].
The amplitude $|W_1|$ of the response can readily be determined from equation
(5.15) for various specific values of $\psi_e$, $\Gamma_e$, and $\eta$ as a function of $(\xi/\xi_1)^2$.
Typical graphs of $(EI\xi_1^4/F_1L^4)|W_1|$ are plotted against $(\xi/\xi_1)^2$, which is pro-
portional to the frequency $\omega$, in Figures 5.8 and 5.9.

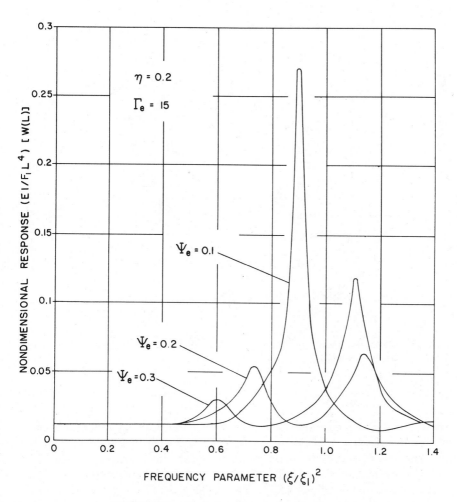

**FIGURE 5.8.**  Typical response spectra for $\eta = 0.2$.

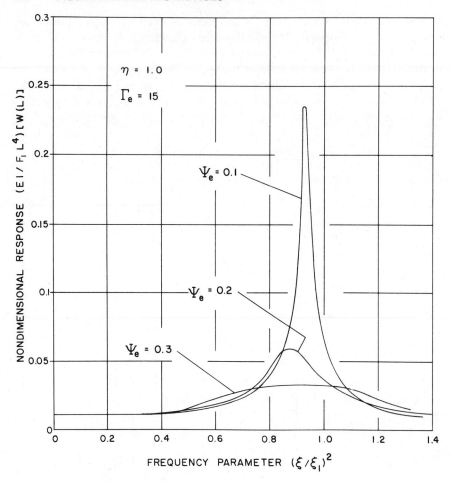

**FIGURE 5.9.** Typical response spectra for $\eta = 1.0$.

An effective loss factor $\eta_1$ can be defined for each of the two peaks in the response spectra for mode 1, namely

$$\eta_1 = (A^2 - 1)^{-1/2} \tag{5.20}$$

where $A$ is the amplification factor of each resonant peak. Typical graphs of $\eta_1$ against the effective stiffness parameter $\Gamma_e$ are plotted in Figure 5.10 for the high and low frequency resonant peaks at various mass ratios $\psi_e$ [5.20]. The dampers are said to be optimally tuned at the point where the two resonance peaks are of equal amplitude [5.21]. This is the point at which the curves of $\eta_1$ against $\Gamma_e$ intersect in Figure 5.10. At all other values of $\Gamma_e$, one or other of the two resonance peaks will have an amplification factor $A$ higher than at the

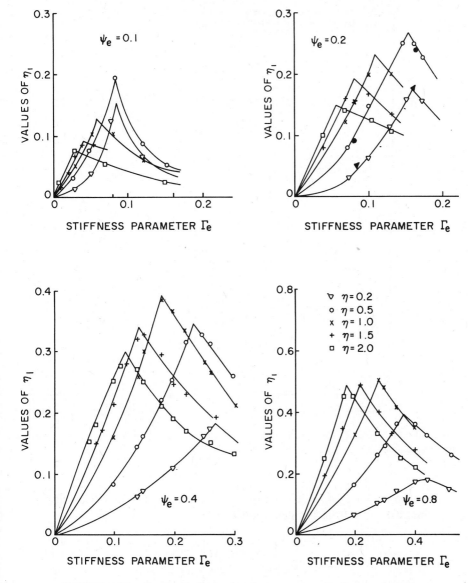

**FIGURE 5.10.**    Graphs of effective loss factor $\eta_1$ versus $\Gamma_e$ (force excitation).

point of optimal tuning. Typical graphs of the values of $\eta_1$ for optimally tuned dampers are plotted in Figure 5.11 as a function of the mass ratio $\psi_e$. The data are taken from reference [5.17], where values of $\eta_1$ and $\Gamma_e$ are summarized for various $\psi_e$ and $\eta$ for both the exact and approximate analyses. A cross-plot of the data in Figure 5.10 give $\eta_1$ as a function of $\eta$ for various $\psi_e$ and are plotted in Figure 5.12.

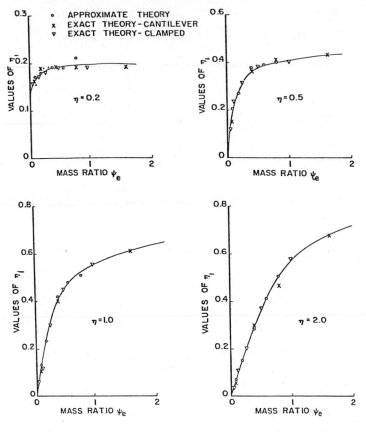

**FIGURE 5.11.** Graphs of optimum effective loss factor against effective mass ratio $\psi_e$ (base and force excitation).

The values of $\Gamma_e$ for which the dampers are optimally tuned are also of great interest, since from the definition of $\Gamma_e$

$$\Gamma_e = \frac{kL^3}{EI\zeta_1^4} \left\{ \frac{\sum_{j=1}^{N} \phi_1^2(\Delta_j)}{\int_0^1 \phi_1^2(\Delta) \, d\Delta} \right\} = \left(\frac{\omega_D}{\omega_1}\right)^2 \psi_e \tag{5.21}$$

$$\therefore \quad \frac{\omega_D}{\omega_1} = \left(\frac{\Gamma_e}{\psi_e}\right)^{1/2} \tag{5.22}$$

It is therefore a simple matter to determine the ratio of the natural frequency $\omega_D$ of the damper to the natural frequency $\omega_1$ of the undamped beam from the values of $\Gamma_e$ at the point of optimal tuning. Typical graphs of $\omega_D/\omega_1$

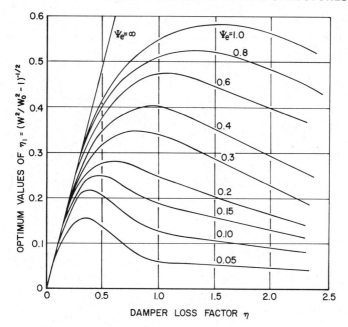

**FIGURE 5.12.**   Optimal damping versus damper loss factor.

against $\psi_e$ are shown in Figure 5.13. Of more interest, however, is the graph of $(\omega_D/\omega_1)(1 + \psi_e)^{1/2}(1 + \eta^2)^{1/4}$ against $\psi_e$ that is plotted for several values of the damper loss factor $\eta$ in Figure 5.14. An empirically derived representation collapses all the data onto a single line, so that the relationship between $\omega_D/\omega_1$ and $\psi_e$ and $\eta$ is

$$\frac{\omega_D}{\omega_1} = (1 + \psi_e)^{-1/2}(1 + \eta^2)^{-1/4} \tag{5.23}$$

Equation (5.23) implies that if $\psi_e$ and $\eta$ are known, it is possible to determine the natural frequency $\omega_D$ of the damper such that the beam-damper system is optimally damped. This simple relationship should therefore be of some value for simple structures exhibiting widely separated resonance frequencies.

### 5.3.3.  Modal Analysis of Tuned Dampers on a Beam with Base Excitation

Consider a beam excited at the base instead of by applied force $F(x)$. Such a case might occur if the beam were attached to the table of an electrodynamic shaker. With this type of excitation, one or more boundary points of the

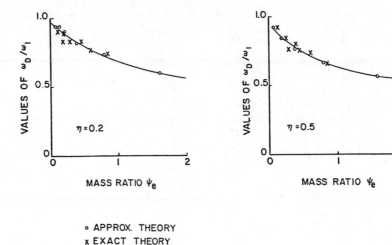

° APPROX. THEORY
x EXACT THEORY
(CANTILEVER)

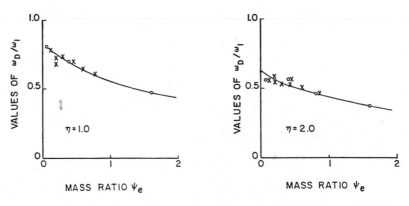

**FIGURE 5.13.**   Graphs of $\omega_D/\omega_1$ against $\psi_e$ for optimal tuning (force excitation).

beam-tuned damper combination, illustrated in Figure 5.6, oscillates with a vertical displacement, periodic in time, of amplitude $W_0$. If $W_r$ is the amplitude of transverse displacement of any point $x$ of the beam relative to the moving end or ends, the equation of motion may be written

$$EI\left(\frac{d^4 W_r}{dx^4}\right) - \rho b H \omega^2 W_r - \frac{m\omega^2}{1 - m\omega^2/k(1 + i\eta)} \sum_{j=1}^{N} W_r(x_j)\delta(x - x_j)$$

$$= \rho b H \omega^2 W_0 + \frac{m\omega^2 W_0}{1 - m\omega^2/k(1 + i\eta)} \sum_{j=1}^{N} \delta(x - x_j) \qquad (5.24)$$

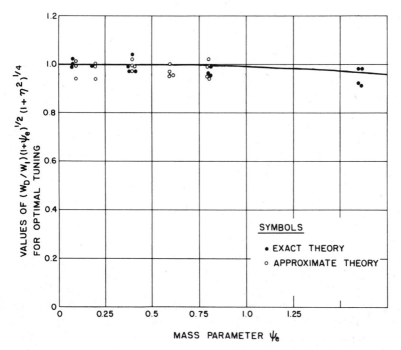

**FIGURE 5.14.** Graphs of $(\omega_D/\omega_1)(1 + \psi_e)^{1/2}(1 + \eta^2)^{1/4}$ against $\psi_e$ for optimal tuning (force excitation).

This equation is clearly different from equation (5.12) but may be solved in much the same way. Again the response $W_r(x)$ can be replaced by the appropriate expansion in normal modes:

$$W_r(x) = \sum_{n=1}^{\infty} W_{rn}\phi_n\left(\frac{x}{L}\right)$$

Hence, using equation (5.13) and the approximation neglecting all but the first term, equation (5.24) gives

$$\frac{W_{r1}}{W_0} = \frac{(\xi/\xi_1)^4\alpha_1 + \psi_e(\xi/\xi_1)^4\beta_1/[1 - \psi_e(\xi/\xi_1)^4/\Gamma_e(1 + i\eta)]}{1 - (\xi/\xi_1)^4 - \psi_e(\xi/\xi_1)^4\gamma_1/[1 - \psi_e(\xi/\xi_1)^4/\Gamma_e(1 + i\eta)]} \tag{5.25}$$

Therefore the absolute response $W$ is given by

$$\left.\begin{aligned}
\frac{W}{W_0} &= \frac{W_{r1} + W_0}{W_0} \\
&= 1 + \left(\frac{\xi}{\xi_1}\right)^4\left[\frac{\alpha_1 + \beta_1\psi_e(\xi/\xi_1)^4/\{1 - \psi_e(\xi/\xi_1)^4/\Gamma_e(1 + i\eta)\}}{1 - (\xi/\xi_1)^4 - \gamma_1\psi_e(\xi/\xi_1)^4/[1 - \psi_e(\xi/\xi_1)^4/\Gamma_e(1 + i\eta)]}\right]
\end{aligned}\right\} \tag{5.26}$$

where

$$\alpha_1 = \frac{\displaystyle\int_0^1 \phi_1(\Delta)\, d\Delta}{\displaystyle\int_0^1 \phi_1^2(\Delta)\, d\Delta}$$

$$\beta_1 = \frac{\displaystyle\sum_{j=1}^N \phi_1(\Delta_j)}{\displaystyle\int_0^1 \phi_1^2(\Delta)\, d\Delta}$$

$$\gamma_1 = \frac{\displaystyle\sum_{j=1}^N \phi_1^2(\Delta_j)}{\displaystyle\int_0^1 \phi_1^2(\Delta)\, d\Delta}$$

It is seen that the response is now governed by two additional parameters, namely $\alpha_1$ and $\beta_1$. In the special case where $N \to \infty$ (i.e., the dampers are uniformly distributed), $\alpha_1 \to \beta_1$. Also for the case where $N = 1$ and $\phi_1(\Delta_j) = 1$ (i.e., the single damper location and the point at which the mode shape is normalized are identical), then $\beta_1 = \gamma_1$. If $\beta_1 = \gamma_1$, then

$$\frac{W}{W_0} = \frac{1 + (\xi/\xi_1)^4(\alpha_1 - 1)}{1 - (\xi/\xi_1)^4 - [\psi_e(\xi/\xi_1)^4]/[1 - \psi_e(\xi/\xi_1)^4/\Gamma_e(1 + i\eta)]} \qquad (5.27)$$

In this particularly simple case, therefore, the problem of determining $|W/W_0|$ under base excitation reduces to that of multiplying the response under force excitation, given in equation (5.15), by $1 + (\xi/\xi_1)^4(\alpha_1 - 1)$. For the response determined in this way, two peaks are again observed. It has been shown in [5.17] that for shaker excitation the effective loss factor $\eta_1$ is defined by the relationship

$$\eta_1 = \frac{\alpha_1 \phi_1(\Delta)}{\sqrt{A^2 - 1}}$$

where $A$ is the amplification factor, that is, the value of $|W/W_0|$ at each resonance in the fundamental mode. Typical graphs of $\eta_1$ against $\Gamma_e$ for various $\psi_e$ are shown in Figure 5.15. From these graphs, the optimum loss factor corresponding to the point of crossover can be read off and plotted against $\psi_e$ for various values of $\eta$. The points, when plotted as in Figure 5.11, show that the variation of $\eta_1$ with $\psi_e$ is practically independent of whether the beam is force or shaker excited. On the other hand, the graphs of $\omega_D/\omega_1$ versus $\psi_e$ in

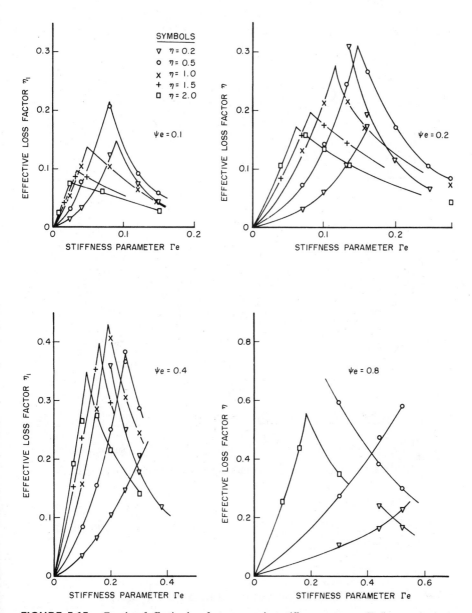

**FIGURE 5.15.** Graphs of effective loss factor $\eta_1$ against stiffness parameter $\Gamma_e$ (base excitation).

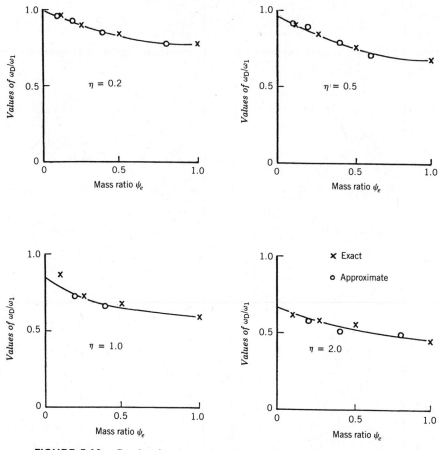

**FIGURE 5.16.**  Graphs of $\omega_D/\omega_1$ against $\psi_e$ for optimal tuning (base excitation).

Figure 5.16 and of $(\omega_D/\omega_1)(1 + \psi_e)^{+1/2}(1 + \eta^2)^{+1/4}$ against $\psi_e$ in Figure 5.17 do show some difference.

### 5.3.4.  Direct Solution for Tuned Dampers on Beams

Exact solutions of the Euler-Bernoulli equation for a cantilever beam with a tuned damper at the free end and forced by a harmonic excitation $F$ at the free end have been discussed by Young [5.22] and Nashif [5.23]. The Euler–Bernoulli equation for transverse vibration of a beam is

$$\frac{d^4W}{dx^4} - \frac{\rho b H \omega^2}{EI} W = 0 \tag{5.28}$$

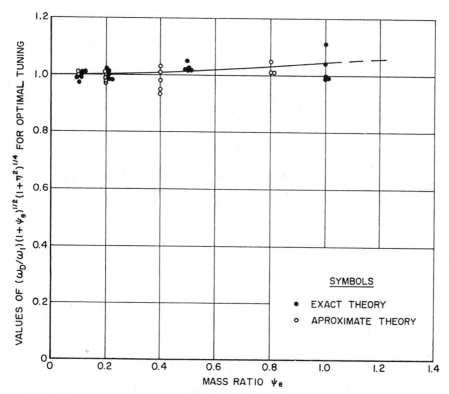

**FIGURE 5.17.**  Graph of $(\omega_D/\omega_1)(1 + \psi_e)^{1/2}(1 + \eta^2)^{1/4}$ against $\psi_e$ for optimal tuning base excitation).

If a cyclic force $Fe^{i\omega t}$ is applied at the free end of the beam, then the boundary conditions for the displacement $w = We^{i\omega t}$ become

at $x = 0$:

$$W = 0$$

$$\frac{dW}{dx} = 0$$

at $x = L$:

$$\frac{d^2 W}{dx^2} = 0$$

$$EI\left(\frac{d^3 W}{dx^3}\right) = F - F_D$$

$F$ is the amplitude of the applied cyclic load and $F_D$ is the amplitude of the force on the beam due to the damper which, from equation (5.4), is

$$F_D = \frac{-m\omega^2 W}{1 - m\omega^2/k(1 + i\eta)}$$

The general form of the response $W$ becomes

$$W = A \cosh \lambda x + B \sinh \lambda x + C \cos \lambda x + D \sin \lambda x \qquad (5.29)$$

where $\lambda^4 = \rho b H\omega^2/EI$ and the constants $A$, $B$, $C$, and $D$ in equation (5.29) are determined from the boundary conditions. After simplification the response can be written in the form

$$\frac{2EI}{FL^3} W = \frac{H}{Z} \qquad (5.30)$$

where

$$H = \left(\frac{1}{\xi^3}\right)[(\sinh \xi + \sin \xi)(\cosh \lambda x - \cos \lambda x)$$

$$- (\cosh \xi + \cos \xi)(\sinh \lambda x - \sin \lambda x)] \qquad (5.31)$$

$$Z = \frac{\psi_e}{1 - [\psi \xi^4/\Gamma(1 + i\eta)]} [\cos \xi \sinh \xi - \cosh \xi \sin \xi] + 1 + \cosh \xi \cos \xi$$

$$(5.32)$$

and $\xi = \lambda L$. Since $Z$ is a complex number, it can be expressed in the form

$$Z = Z_D + iZ_Q \qquad (5.33)$$

where

$$Z_D = \frac{\psi \xi \Gamma(\Gamma - \psi \xi^4 + \Gamma \eta^2)}{(\Gamma - \psi \xi^4)^2 + \Gamma^2 \eta^2} (\cos \xi \sinh \xi - \cosh \xi \sin \xi) + 1 + \cosh \xi \cos \xi$$

$$(5.34)$$

$$Z_Q = \frac{-\psi^2 \eta \Gamma \xi^4}{(\Gamma - \psi \xi^4)^2 + \Gamma^2 \eta^2} (\cos \xi \sinh \xi - \cosh \xi \sin \xi) \qquad (5.35)$$

The absolute value of $W$ can be written in the form

$$\frac{2EI}{PL^3} |W| = \frac{H}{\sqrt{Z_D^2 + Z_Q^2}} \qquad (5.36)$$

To compare this exact analysis with the approximate modal analysis, one notes that since there is only one damper in this case, $\sum_{j=1}^{N} \phi_1^2(\Delta_j) = 1$ and, as in Table 5.1, $\int_0^1 \phi_1^2(\Delta)\, d\Delta = 0.25$. Therefore for this case $\psi_e = 4\psi$ and $\Gamma_e = 4\Gamma$. The values of $\eta_1$ at optimum tuning as a function of stiffness ratio $\Gamma_e$, calculated from equation (5.36), are plotted in Figure 5.11. It is found that the computed points lie essentially along the same line given by the approximate theory. Furthermore the values of $(\omega_D/\omega_1)(1 + \psi_e)^{1/2}(1 + \eta^2)^{1/4}$, when plotted against $\psi_e$, lie on the same straight line as given by the approximate theory, as in Figure 5.14.

In another investigation Adkins [5.24] has given the direct analysis of a tuned damper, at the center of a clamped-clamped beam, on the response under base excitation. Again $\sum_{j=1}^{N} \Phi_1^2(\Delta_j) = 1$ and, from Table 5.1, $\int_0^1 \phi_1^2(\Delta)\, d\Delta = 0.439$. Thus for this case $\psi_e = 2.086\psi$ and $\Gamma_e = 2.086\Gamma$. From the values of $\Gamma$ and $\psi$ [5.24], $\Gamma_e$ and $\psi_e$ were deduced, and graphs of $\eta_1$ plotted against $\Gamma_e$, as in Figure 5.11. Again the computed points lie along the same curve as all the others.

## 5.4. TUNED DAMPERS IN COMPLEX STRUCTURES

### 5.4.1. Frequency Bandwidth of Effective Operation

Many practical structures exhibit modal response characteristics far more complex than the simple beams or single degree of freedom systems previously discussed. When tuned dampers are applied to complex structures possessing many closely spaced resonances, the simplicity of the foregoing is lost, and the effect of the dampers on the structural response is dependent on the exact nature of the structural geometry, so that no relatively general design concepts can now be formulated. However, as previously discussed and illustrated in Figure 5.2, a single degree of freedom tuned damper can dissipate energy over a significant band of frequencies. In fact it has been shown that on certain kinds of structures a single tuned damper can contribute significant damping to several different modes distributed over a significant frequency band.

To understand better why tuned dampers can be effective in controlling the multimodal response of some structures and not be effective in other structures, consider energy relationships in two different kinds of structures with attached tuned dampers. For the first type of structure, consider a clamped-clamped beam with a tuned damper attached at mid-span. If a damper is optimized for the fundamental mode, it has little effect on higher frequency modes of vibration. This would be true even if a damper could be designed that dissipated the same energy per cycle at the frequency of the third mode of vibration, which occurs at approximately 5.5 times the frequency of the first mode. A mid-span damper has no effect on the second mode of vibration since it is located on a node line. The loss factor $\eta_n$ for the beam with the damper attached is defined as $\eta_n = D_s/2\pi U_s$, where $U_s$ is the total strain energy in both the damper and the beam [5.25]. In the case of the clamped-clamped beam the strain energy in the third mode is much greater (about 30 times) than the strain energy in the fundamental mode for the same displacement at the center of the beam. Thus the loss factor $\eta_n$, which depends on the ratio of energy dissipated to the total strain energy in the system, is much lower in the third mode as compared to the fundamental mode.

For the second type of structure consider a continuous row of skin panels stiffened by stringers, typical of aircraft fuselage construction. It has been shown both experimentally [5.26] and analytically [5.27] that the most troublesome vibration modes occur in a frequency band bounded on the low side by the first "stringer twisting" mode and on the high side by the first "stringer bending" mode. These two limiting modes often occur at frequencies not more than an octave apart, and the strain energies associated with these modes are also quite close to each other. In fact the strain energy of the stringer bending mode is approximately four times that of the stringer twisting mode. Therefore it is possible for a single tuned damper to damp such modes effectively.

### 5.4.2.  Skin-Stringer Structures

Analyses of the effect of tuned dampers on the response of skin-stringer structures have illustrated the potential of using isolated elastomeric tuned dampers to control the response of several modes. These analytical studies include a normal mode approach to evaluate the effect of tuned dampers on a row of flat panels stiffened by stringers and frames, [5.28] and a transfer matrix approach to analyze the effect of tuned dampers on curved skin-stringer structure [5.13], as shown in Figure 5.18.

It has been shown that elastomeric tuned dampers with a loss factor of 0.2, which is typical of the more highly damped silicone materials in the rubbery temperature range, can be effective on skin-stringer structures. In one case a tuned damper was located in the center of each panel, and the mass of the damper was chosen to be 3% of the mass of the skin in the panel [5.13]. This amount of weight addition might be acceptable in an aeronautical system, particularly if the structural designer, faced with a resonant fatigue problem, could avoid increasing the skin thickness by one gage, with a resulting increase in panel weight of about 20%. As shown in Figure 5.19, these tuned dampers, having undamped resonant frequency slightly above 240 Hz, effectively controlled five modes of a cylindrically curved row of five stringer-stiffened panels. The resonant frequencies of the original modes ranged from approximately 210 to 445 Hz. Effects of damper location and total damper mass are illustrated in Figure 5.20, where the response of panel 1 of the flat structure with five tuned dampers, one in the center of each panel, is compared with the response of a structure with 15 tuned dampers located on the quarter-span points of each panel. The 15 dampers, each with mass equal to 1% of the mass of a panel, are less effective than the five dampers with the same total

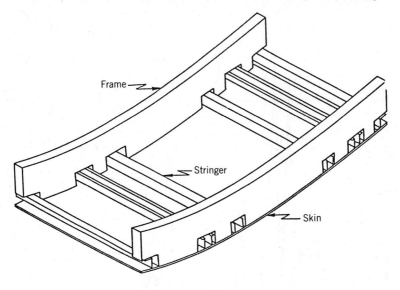

*(a)* Skin–stringer structural specimen

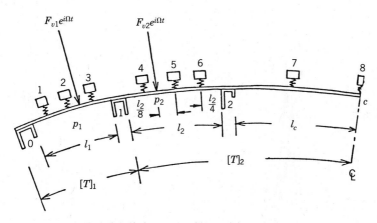

*(b)* Left half of structure with tuned dampers

**FIGURE 5.18.**   Multispan skin-stringer structure with tuned dampers.

mass located in the center of each panel where the displacements are greater. Doubling the mass of the 15 dampers results in slightly greater damping than for the 5 damper case. The effect of changing damper loss factor $\eta$ are shown in Figure 5.21, where the response of a flat structure, with dampers having loss factors of 0.1, 0.2, and 0.3, are compared. In this range of loss factors it appears that an increase in damper loss factor results in an increase in modal damping.

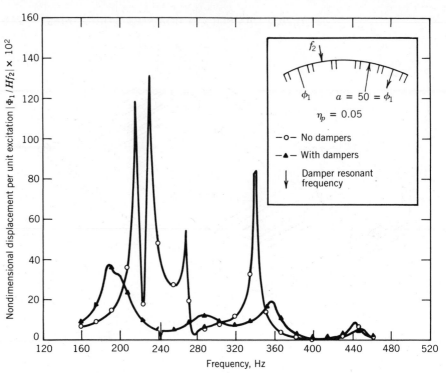

**FIGURE 5.19.** Dynamic response, panel 1, $a = 50$ in.

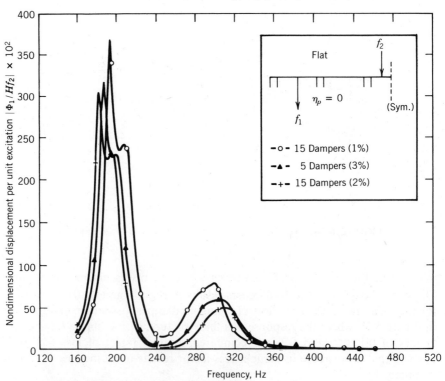

**FIGURE 5.20.** Effect of number of dampers and damper mass on response.

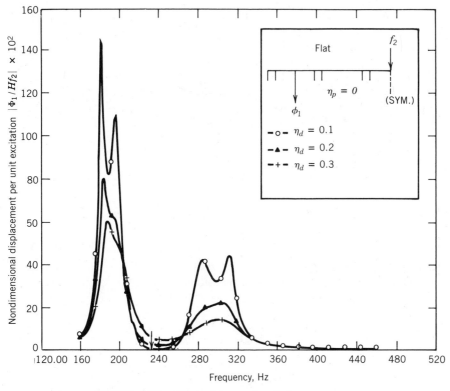

**FIGURE 5.21.**  Effect of damper loss factor on response.

## 5.5.  STRUCTURAL MODIFICATION BY LINKS JOINING TWO POINTS OF A STRUCTURE

### 5.5.1.  General Analysis

If any two points of a structural system are linked by a damped spring of complex stiffness $k(1 + i\eta)$, coupling will occur which may modify the behavior of the entire system. For example, for a two-dimensional system such as that illustrated in Figure 5.22, one may assume that the receptances of the system are known; that is,

$$
\left.
\begin{aligned}
\frac{\partial W_{x,i}}{\partial F_{y,j}} &= \alpha_{ij}^{xy} \\[2ex]
\frac{\partial W_{x,i}}{\partial F_{x,j}} &= \alpha_{ij}^{xx} \\[2ex]
\frac{\partial W_{y,i}}{\partial F_{y,j}} &= \alpha_{ij}^{yy} \\[2ex]
\frac{\partial W_{y,i}}{\partial F_{x,j}} &= \alpha_{ij}^{yx}
\end{aligned}
\right\}
\tag{5.37}
$$

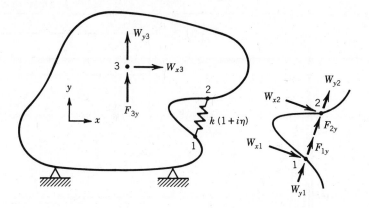

**FIGURE 5.22.** System with one-dimensional viscoelastic link.

where $x$, $y$, $z$ are local Cartesian coordinates and $i$, $j = 1, 2, 3, \ldots, n$ for $n$ particular points of interest. These receptances can be expressed in series form:

$$
\left.
\begin{aligned}
\alpha_{ij}^{xy} &= \sum_{n=1}^{N} \frac{\phi_{nix}\phi_{njy}}{M_{nn}[\omega_n^2(1 + i\eta_n) - \omega^2]} \\[6pt]
\alpha_{ij}^{xx} &= \sum_{n=1}^{N} \frac{\phi_{nix}\phi_{njx}}{M_{nn}[\omega_n^2(1 + i\eta_n) - \omega^2]} \\[6pt]
\alpha_{ij}^{yy} &= \sum_{n=1}^{N} \frac{\phi_{niy}\phi_{njy}}{M_{nn}[\omega_n^2(1 + i\eta_n) - \omega^2]} \\[6pt]
\alpha_{ij}^{yx} &= \sum_{n=1}^{N} \frac{\phi_{niy}\phi_{njx}}{M_{nn}[\omega_n^2(1 + i\eta_n) - \omega^2]}
\end{aligned}
\right\}
\tag{5.38}
$$

where $\phi_{niy}$ is the $n$th modal function at a particular point $i$, in the $y$ direction, for example. If in the particular system shown in Figure 5.22, the forces $F_{2y}$ and $F_{3y}$ result from the complex spring insertion, then

$$
-F_{2y} = +F_{1y} = k(1 + i\eta)(W_{y2} - W_{y1}) \tag{5.39}
$$

so the equation (5.37) may be used to write, by direct summation,

$$
\begin{aligned}
W_{y1} &= F_{1y}\alpha_{11}^{yy} + F_{2y}\alpha_{12}^{yy} + F_{3y}\alpha_{13}^{yy} \\
&= -k(1 + i\eta)(W_{y1} - W_{y2})\alpha_{11}^{yy} - k(1 + i\eta)(W_{y2} - W_{y1})\alpha_{12}^{yy} + F_{3y}\alpha_{13}^{yy} \quad (5.40)
\end{aligned}
$$

$$
\begin{aligned}
W_{y2} &= F_{1y}\alpha_{21}^{yy} + F_{2y}\alpha_{22}^{yy} + F_{3y}\alpha_{23}^{yy} \\
&= -k(1 + i\eta)(W_{y1} - W_{y2})\alpha_{21}^{yy} - k(1 + i\eta)(W_{y2} - W_{y1})\alpha_{22}^{yy} + F_{3y}\alpha_{23}^{yy} \quad (5.41)
\end{aligned}
$$

which may be simplified and written in matrix form:

$$\begin{bmatrix} 1 + k(1 + i\eta)(\alpha_{11}^{yy} - \alpha_{12}^{yy}) & -k(1 + i\eta)(\alpha_{11}^{yy} - \alpha_{12}^{yy}) \\ -k(1 + i\eta)(\alpha_{22}^{yy} - \alpha_{21}^{yy}) & 1 + k(1 + i\eta)(\alpha_{22}^{yy} - \alpha_{21}^{yy}) \end{bmatrix} \begin{Bmatrix} W_{y1} \\ W_{y2} \end{Bmatrix} = F_{3y} \begin{Bmatrix} \alpha_{13}^{yy} \\ \alpha_{23}^{yy} \end{Bmatrix}$$

(5.42)

from which $W_{y1}$ and $W_{y2}$ can be determined. With $W_{y1}$ and $W_{y2}$ known, $F_{2y}$ and $F_{1y}$ are known forces, and the response $W_{x3}$, $W_{y3}$ at point 3 is given by

$$W_{x3} = F_{1y}\alpha_{13}^{xy} + F_{2y}\alpha_{23}^{xy} + F_{3y}\alpha_{33}^{xy}$$

$$W_{y3} = F_{1y}\alpha_{13}^{yy} + F_{2y}\alpha_{23}^{yy} + F_{3y}\alpha_{33}^{yy}$$

(5.43)

and the responses at any other points can be obtained in the same manner. Therefore, given the initial receptance data for the unmodified structure, it is not difficult to predict the effect of a modification such as the complex link discussed here. The specific behavior of the modified structure is dependent on the characteristics of the receptance functions and is best illustrated through some simple examples rather than in abstract terms.

### 5.5.2. Two Parallel Beams Joined by a Link (Force Excitation/Receptance Analysis)

The particularly simple system shown in Figure 5.23, consisting of two parallel clamped beams joined at points 1 and 2 by a link $k(1 + i\eta)$, provides a useful illustration of the effect of such a link. The only relevant receptances are the point receptances $\alpha_{11}$ and $\alpha_{22}$, and all cross receptances between these two points are zero if the boundaries are completely rigid as assumed. If the beams are clamped,

$$\alpha_{11} = \sum_{n=1}^{\infty} \frac{\phi_{n1}^2(x_1/L)}{M_{nn}^{(1)}[\omega_{n1}^2 - \omega^2]}$$

(5.44)

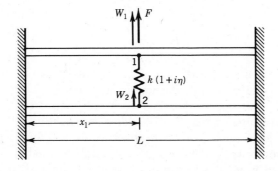

**FIGURE 5.23.** Two parallel clamped-clamped beams joined by viscoelastic link (force excitation).

and

$$\alpha_{22} = \sum_{n=1}^{\infty} \frac{\phi_{n2}^2(x_1/L)}{M_{nn}^{(2)}[\omega_{n2}^2 - \omega^2]} \tag{5.45}$$

if the initial damping is assumed to be zero. For the $n$th mode

$$\xi_n^2 = \left(\frac{\rho_1 b_1 H_1 \omega_{n1}^2 L^4}{E_1 I_1}\right)^{1/2} = \left(\frac{\rho_2 b_2 H_2 \omega_{n2}^2 L^4}{E_2 I_2}\right)^{1/2} \tag{5.46}$$

$$\left.\begin{aligned} M_{nn}^{(1)} &= 0.397\rho_1 b_1 H_1 \\ M_{nn}^{(2)} &= 0.397\rho_2 b_2 H_2 \end{aligned}\right\} \tag{5.47}$$

The equations obtained by summing receptances are

$$W_{y1} = F\alpha_{11} - k(1 + i\eta)(W_{y1} - W_{y2})\alpha_{11} \tag{5.48}$$

$$W_{y2} = k(1 + i\eta)(W_{y1} - W_{y2})\alpha_{22} \tag{5.49}$$

Therefore

$$\begin{bmatrix} 1 + k(1 + i\eta)\alpha_{11} & -k(1 + i\eta)\alpha_{11} \\ -k(1 + i\eta)\alpha_{22} & 1 + k(1 + i\eta)\alpha_{22} \end{bmatrix} \begin{bmatrix} W_{y1} \\ W_{y2} \end{bmatrix} = \begin{bmatrix} F\alpha_{11} \\ 0 \end{bmatrix} \tag{5.50}$$

and

$$\left.\begin{aligned} W_{y1} &= \frac{F\alpha_{11}[1 + k(1 + i\eta)\alpha_{22}]}{\Delta} \\ \\ W_{y2} &= \frac{F\alpha_{11}\alpha_{22}k(1 + i\eta)}{\Delta} \end{aligned}\right\} \tag{5.51}$$

where

$$\Delta = 1 + k(1 + i\eta)(\alpha_{11} + \alpha_{22}) \tag{5.52}$$

is the determinant of the matrix in Equation (5.50). One important feature that emerges from Equations (5.51) and (5.52) is that whenever $\alpha_{11}$ and $\alpha_{22}$ approach infinity at the same frequency for any mode, $W_{y1}$ and $W_{y2}$ also approach infinity, and the damped link does not control the resonant amplitudes. This can happen when the two beams are equal in all respects, so that $\omega_{n1} = \omega_{n2}$ for all $n$, or it can occur when $\omega_{m1} = \omega_{n2}$ for $n, m = 1, 2, \dots$. If $n = 1$, $m = 2$, for example, the first mode of beam 1 is equal in frequency to the second mode of beam 2. The entire set of conditions $\omega_{m1} = \omega_{n2}$ represents relationships for which the damped link is not effective in controlling vibration amplitudes. Conversely, the condition $\omega_{m1} \neq \omega_{n2}$ represents conditions for the link to be effective to some degree.


### 5.5.3.  Two Parallel Clamped-Clamped Beams Joined by a Link (Base Excitation/Direct Analysis)

To illustrate the direct analysis of two parallel beams joined by a viscoelastic link, consider the system illustrated in Figure 5.24 [5.10]. If the amplitude of the harmonic vibration of one beam relative to the clamped ends is $W_1(x)$ and that of the other beam is $W_2(x)$, and the supports are vibrating with harmonic displacement $W_0 e^{i\omega t}$, then the equations of motion of the two beams are

$$E_i I_i \left( \frac{d^4 W_i}{dx_i^4} \right) - \rho_i b_i H_i \omega^2 W_i = \rho_i b_i H_i \omega^2 W_0 \qquad i = 1, 2 \tag{5.53}$$

at all points apart from the points to which the link is attached. The general solution of Equation (5.53) is

$$W_i(x) = A_i \cosh (\lambda_i x) + B_i \sinh (\lambda_i x) + C_i \cos (\lambda_i x)$$
$$+ D_i \sin (\lambda_i x) - W_0 \qquad i = 1, 2 \tag{5.54}$$

where

$$\lambda_i^4 = \frac{\rho_i b_i H_i \omega^2}{E_i I_i} \qquad i = 1, 2 \tag{5.55}$$

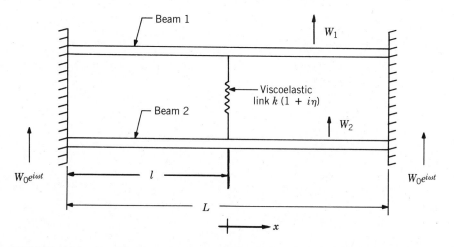

FIGURE 5.24.  Parallel clamped-clamped beams joined by viscoelastic link (base excitation).

The eight constants $A_i$, $B_i$, $C_i$, and $D_i$ are determined from the various boundary conditions. By symmetry, only the solution from $x = 0$ to $x = +L/2$ need be considered:

$$\left.\begin{array}{ll} W_i = \dfrac{dW_i}{dx} = 0 & \text{at } x = \dfrac{L}{2} \\[2ex] \dfrac{dW_i}{dx} = 0 & \text{at } x = 0 \\[2ex] 2E_1 I_1\left(\dfrac{d^3 W_1}{dx^3}\right) = k(1 + i\eta)(W_2 - W_1) & \text{at } x = 0 \\[2ex] 2E_2 I_2\left(\dfrac{d^3 W_2}{dx^3}\right) = k(1 + i\eta)(W_1 - W_2) & \text{at } x = 0 \end{array}\right\} \tag{5.56}$$

From these conditions the eight equations for the eight unknown constants are readily set up and solved. After some simplification the solution $W_1(x)$ is readily shown to be

$$\frac{W_1(x)}{W_0} = \frac{\Gamma(1 + i\eta)[b_1 c_2 + (b_2 c_1/\lambda)(\psi/\lambda)^{-3/4}] + 4c_1 a_2}{\Gamma(1 + i\eta)[a_2 b_1 + (a_1 b_2/\lambda)(\psi/\lambda)^{-3/4}] + 4a_1 a_2} \tag{5.57}$$

where the parameters are given by the following:

$$\Gamma = \frac{k}{E_1 I_1 \lambda_1^3} = \text{a nondimensional link stiffness parameter}$$

$$a_1 = \sinh \xi_1 \cos \xi_1 + \cosh \xi_1 \sin \xi_1$$

$$b_1 = a_1\left[\sinh\left(\frac{2\xi_1 x}{L}\right) - \sin\left(\frac{2\xi_1 x}{L}\right)\right]$$

$$- \cosh\left(\frac{2\xi_1 x}{L}\right)[\cosh \xi_1 \cos \xi_1 + \sinh \xi_1 \sin \xi_1 - 1]$$

$$- \cos\left(\frac{2\xi_1 x}{L}\right)[\cosh \xi_1 \cos \xi_1 - \sinh \xi_1 \sin \xi_1 - 1]$$

$$c_1 = \sin \xi_1 \cosh\left(\frac{2\xi_1 x}{L}\right) - \sinh \xi_1 \cos\left(\frac{2\xi_1 x}{L}\right) - a_1$$

$$a_2 = \sinh\left\{\left(\frac{\psi}{\lambda}\right)^{1/4}\xi_1\right\}\cos\left\{\left(\frac{\psi}{\lambda}\right)^{1/4}\xi_1\right\} + \cosh\left\{\left(\frac{\psi}{\lambda}\right)^{1/4}\xi_1\right\}\sin\left\{\left(\frac{\psi}{\lambda}\right)^{1/4}\xi_1\right\}$$

$$b_2 = a_2\left[\sinh\left\{\left(\frac{\psi}{\lambda}\right)^{1/4}\left(\frac{2\xi_1 x}{L}\right)\right\} - \sin\left\{\left(\frac{\psi}{\lambda}\right)^{1/4}\left(\frac{2\xi_1 x}{L}\right)\right\}\right]$$

$$- \cosh\left\{\left(\frac{\psi}{\lambda}\right)^{1/4}\left(\frac{2\xi_1 x}{L}\right)\right\}\left[\cosh\left\{\left(\frac{\psi}{\lambda}\right)^{1/4}\xi_1\right\}\cos\left\{\left(\frac{\psi}{\lambda}\right)^{1/4}\xi_1\right\}\right.$$

$$\left. + \sinh\left\{\left(\frac{\psi}{\lambda}\right)^{1/4}\xi_1\right\}\sin\left\{\left(\frac{\psi}{\lambda}\right)^{1/4}\xi_1\right\} - 1\right]$$

$$- \cos\left\{\left(\frac{\psi}{\lambda}\right)^{1/4}\left(\frac{2\xi_1 x}{L}\right)\right\}\left[\cosh\left\{\left(\frac{\psi}{\lambda}\right)^{1/4}\xi_1\right\}\cos\left\{\left(\frac{\psi}{\lambda}\right)^{1/4}\xi_1\right\}\right.$$

$$\left. - \sinh\left\{\left(\frac{\psi}{\lambda}\right)^{1/4}\xi_1\right\}\sin\left\{\left(\frac{\psi}{\lambda}\right)^{1/4}\xi_1\right\} - 1\right]$$

$$c_2 = \sin\left\{\left(\frac{\psi}{\lambda}\right)^{1/4}\xi_1\right\}\cosh\left\{\left(\frac{\psi}{\lambda}\right)^{1/4}\left(\frac{2\xi_1 x}{L}\right)\right\}$$

$$+ \sinh\left\{\left(\frac{\psi}{\lambda}\right)^{1/4}\xi_1\right\}\cos\left\{\left(\frac{\psi}{\lambda}\right)^{1/4}\left(\frac{2\xi_1 x}{L}\right)\right\} - a_2$$

where $\lambda = E_2 I_2/E_1 I_1$ is a nondimensional stiffness ratio and $\psi = \rho_2 b_2 H_2/\rho_1 b_1 H_1$ is a nondimensional mass ratio, relating beams 1 and 2. The transmissibility $T_1$ is defined as the ratio of the response at any point of beam 1, relative to a fixed point in space, to the input amplitude $W_0$, that is

$$T_1 = \frac{|W_1 + W_0|}{W_0} \tag{5.58}$$

For given values of the ratio $\psi$ of the beam masses per unit length, beam 1 being taken as reference and the ratio $\lambda$ of the flexural rigidities, the transmissibility $T_1$ can be expressed as a function of the frequency parameter $\xi_1$, the link loss factor $\eta$, and the link stiffness parameter $\Gamma$.

Calculations were performed for $\psi = 0.5$, $1.0$, and $2.0$, and a range of values of $\lambda$ between $0.01$ and $100$ at $x = 0$. Transmissibility spectra such as those illustrated in Figure 5.25 were computed. The characteristics of the response were found to depend on whether $\lambda/\psi = 1$, $<1$, or $>1$:

1.  If $\lambda/\psi = 1$, the first resonant frequency of beam 2 is equal to that of beam 1 and the two beams will always vibrate in phase with each other. In

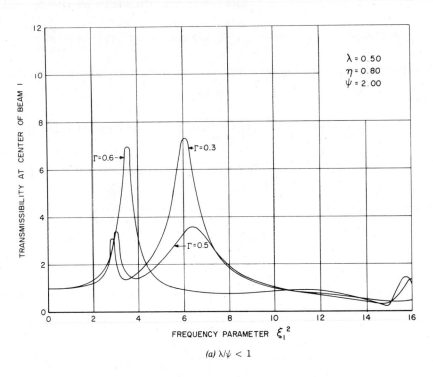

*(a)* $\lambda/\psi < 1$

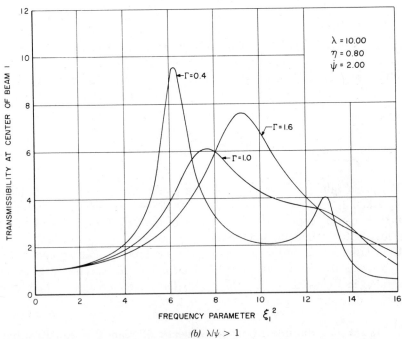

*(b)* $\lambda/\psi > 1$

**FIGURE 5.25.** Typical transmissibility spectra. *(a)* $\lambda/\psi < 1$. *(b)* $\lambda/\psi > 1$.

this case no deformation occurs in the viscoelastic link, and hence no damping can be introduced into the system by the viscoelastic link.

2.   If $\lambda/\psi < 1$, the first resonant frequency of beam 2 is always lower than that of beam 1 and the spectra shown in Figure 5.25a are typical. The figure shows that the amplitude of the low frequency resonance peak is smaller than that of the higher frequency peak for small values of $\Gamma$, and as $\Gamma$ increases, the amplitude of the second peak eventually becomes smaller than that of the first peak. The effective loss factor $\eta_1$ of a clamped-clamped beam under shaker excitation, for which the clamped ends are vibrated to give the excitation, has been shown earlier [5.17] to be

$$\eta_1 = 1.32(A^2 - 1)^{-1/2} \tag{5.59}$$

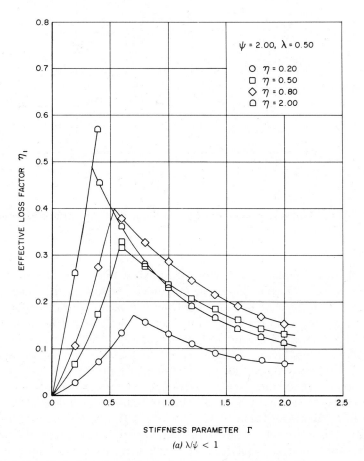

FIGURE 5.26.   Typical graphs of effective loss factor $\eta_1$ versus link stiffness parameter $\Gamma$. (a) $\lambda/\psi < 1$.

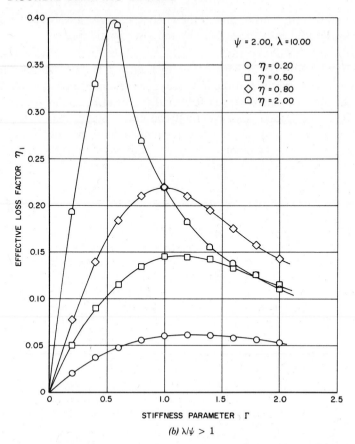

**FIGURE 5.26.**   Typical graphs of effective loss factor $\eta_1$ versus link stiffness parameter $\Gamma$. (b) $\lambda/\psi > 1$.

where $A$ is the amplification factor measured at the center of beam 1 at the resonant peaks corresponding to the first mode. Typical graphs illustrating the variation of the effective loss factors of the two peaks with the stiffness parameter $\Gamma$ are shown in Figure 5.26a for several values of the link loss factor $\eta$. It is seen that for each $\eta$ a value of $\Gamma$ exists for which both peaks will have the same effective loss factor. This loss factor corresponds to the case where the system is "properly tuned" for beam 1, since it represents the maximum loss factor obtainable for the given values of $\lambda$, $\psi$, and $\eta$ in the frequency range of the fundamental mode of beam 1. This procedure was followed for various values of $\lambda/\psi$ between 1 and 0.035. At $\lambda/\psi = 0.035$ the first natural frequency of beam 1 is identical to the third natural frequency of beam 2, and the effective loss factor is again zero. For values of $\lambda/\psi < 0.035$ analysis of the response spectra followed the procedure adopted for values of $\lambda/\psi$ between 1 and 0.035. However, in this case the natural frequency of the fundamental mode of beam 1 is higher than the natural frequencies of the first and third modes of beam 2, and hence the predominant peak due to beam 2 was compared instead with the first peak due to beam 1 to define the optimum effective loss factor.

3.   When $\lambda/\psi > 1$, the first natural frequency of beam 2 is greater than that of beam 1, and the spectra shown in Figure 5.25b are typical. It is seen that one peak now dominates the response for all values of the stiffness parameter $\Gamma$, even though two peaks still exist. Graphs of $\eta_1$ against $\Gamma$ for the predominant peak are illustrated in Figure 5.26b.

Depending on whether $\lambda/\psi < 1$ or $> 1$, therefore, one may define the point of optimum damping either as that at which the curves of $\eta_1$ against $\Gamma$ cross over or that at which the curve of $\eta_1$ against $\Gamma$ has a maximum, respectively. Graphs of the optimum damping so defined were determined in this manner for many values of $\lambda$ and $\psi$. Typical results are plotted in Figure 5.27 for $\psi = 0.5$ and various values of $\eta$. Figure 5.28 illustrates a comparison with experimental data [5.10] for this case. It is seen that the agreement is quite satisfactory. The test fixture used is also illustrated.

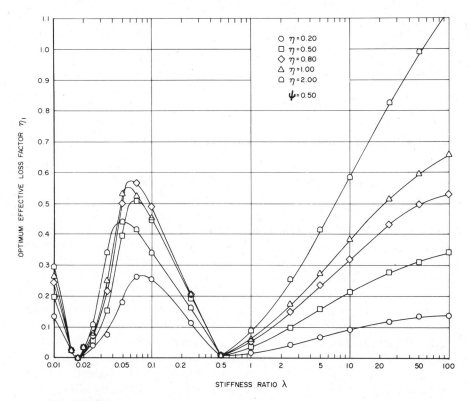

**FIGURE 5.27.**   Fundamental mode damping of beam 1 versus stiffness ratio $\lambda$.

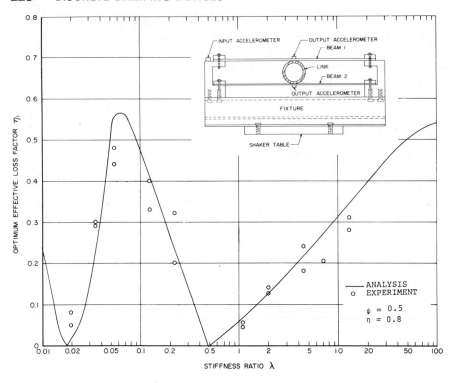

**FIGURE 5.28.** Measured and predicted fundamental mode damping for beam 1.

## 5.6. APPLICATIONS AND EXAMPLES

### 5.6.1. Development of a Tuned Damper for an Aircraft IFF Antenna

This example illustrates the process of designing a tuned damping device to control a resonant vibration problem for which a layered damping treatment was unsuitable. The problem involved high cycle resonant fatigue of an IFF antenna on the F100 aircraft and was first encountered in the mid-1960s when the antenna, placed between the four cannon on the aircraft, began to encounter failures when all four cannon were fired simultaneously, instead of two at a time as during training and practice usage. The power spectral density of the motion of the aircraft structure at the antenna location, as measured during the firing of the gun, was about $2\,g^2/Hz$ in the vicinity of the fundamental resonant frequency of the antenna, namely 485 Hz. This very severe environment, amplified by the resonance of the antenna, led to very high stresses and early failure within a few hours of operational use. Efforts by the manufacturer to stiffen the antenna did not lead to a sufficient decrease of dynamic stress to be useful, and the problem was eventually brought to the attention of the Materials Laboratory at Wright–Patterson Air Force Base, and the authors, in February 1966.

The antenna geometry is illustrated in Figures 5.29 to 5.31. The dish-shaped unit had a highly curved surface which did not experience high surface strains, even when the bulkhead area was vibrating at resonance with very high amplitudes, except around the rim where failure occurred. Since a tuned damper is one of the few devices in which damping depends on cyclic displacement, or more accurately acceleration, rather than surface strain, it was determined very early that such a device had to be designed to solve this particular problem.

In order to evaluate the character of the dynamic response behavior of the antenna, the antenna was supported by a circular fixture attached in turn to a shaker table, as illustrated in Figure 5.32. The response was monitored by an accelerometer placed in the dish, as in Figure 5.31, and the input acceleration by an accelerometer on the fixture at the edge of the antenna. The antenna was subjected to sine-sweep excitation at various input acceleration levels (1 g, 5 g, 10 g, 20 g) at the Materials Laboratory and in a parallel test series, simulating the operational vibration input spectrum on a shaker, at the Flight Dynamics Laboratory. Typical measured response spectra are shown in Figure 5.33. A single resonance peak is seen near 485 Hz, at which frequency

**FIGURE 5.29.**   IFF antenna.

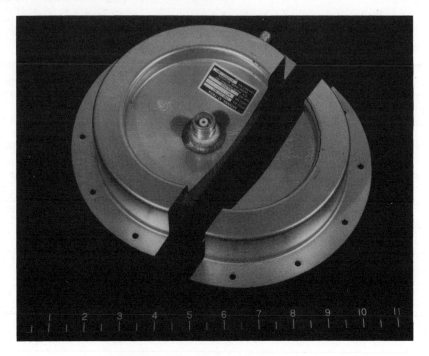

**FIGURE 5.30.** Section through antenna.

the antenna vibrated as a diaphragm. No other resonance peaks were observed up to 2000 Hz. The observed amplification factor at resonance was about 28 to 30 in the nominally undamped case.

**Preliminary damper design of prototype 1.** The initial efforts at damper design were based on the main principles discussed in this chapter. The active mass $m$ of the antenna (i.e., that involved directly in the single degree of freedom motion of the antenna) included the electrical connector, the bulkhead connector, and a fraction (about one-third) of the mass of the diaphragm and amounted to about 100 grams. Some uncertainty was involved (perhaps $\pm 20\%$) because a full modal survey of the antenna was not carried out and the accelerometer was not placed at the exact point of highest amplitude. The effective stiffness $k$ of the antenna, treated as a single degree of freedom system, was therefore given by

$$k = m\omega_r^2 = 0.100(2\pi \times 485)^2 = 9.29 \times 10^5 \text{ N/m}$$

Figure 5.11 shows the mass ratio needed to make an effective damper for various values of $\eta$. For $\eta < 0.2$, $\psi_e$ must be greater than 0.2 for over 50% of maximum possible amplitude reduction to be achieved. Therefore the early

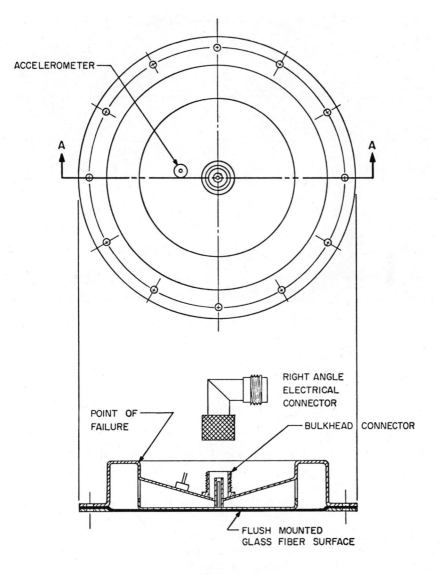

ACCELEROMETER

A ⟶    A

RIGHT ANGLE
ELECTRICAL
CONNECTOR

POINT OF
FAILURE

BULKHEAD CONNECTOR

FLUSH MOUNTED
GLASS FIBER SURFACE

SECTION A-A

**FIGURE 5.31.** Antenna geometry.

**FIGURE 5.32.** Base excitation of antenna on fixture.

tests were conducted with a damper mass $m_D$ of not less than 20 grams. The damper was turned in accordance with equation (5.32), so that

$$\omega_D = \sqrt{\frac{k_D}{m_D}} = 0.913 \times 485 = 443 \text{ Hz}$$

In view of the spatial restrictions in the vicinity of the antenna, and the fact that maximum acceleration occurred at the bulkhead connector location, the damper was designed so it fit snugly over the electrical connector, as illustrated in Figure 5.34. The detachable clamp was designed to hold the damper to the electrical connector, and the mass and stiffness elements sized to meet the mass and stiffness criteria discussed earlier. The design process was primarily experimental, using the general principles of tuned damper analysis as a background for the construction and shaker test evaluation of a series of progressively improving prototypes.

Prototype 1 was hand-built according to the geometry of Figure 5.35. The material selected was a polyvinyl chloride (PVC) copolymer made by Farbwercke Hoechst in Germany, labeled VPCO-15080 (Data Sheet 032). This material was readily available at the time of prototype damper design and was used to demonstrate initial feasibility. From the dimensions of the damper the stiffness $k_D$ was estimated to be

$$k_D = \frac{GS}{H} = 8.89G \frac{\text{N}}{\text{m}}$$

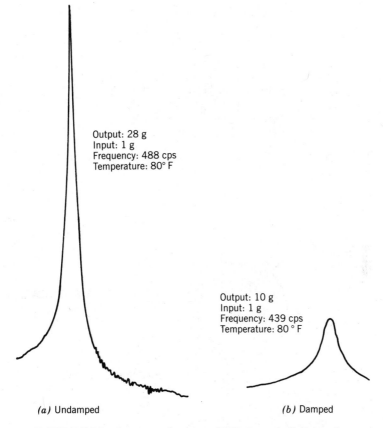

Output: 28 g
Input: 1 g
Frequency: 488 cps
Temperature: 80° F

Output: 10 g
Input: 1 g
Frequency: 439 cps
Temperature: 80 ° F

*(a)* Undamped                                      *(b)* Damped

**FIGURE 5.33.**   Response of antenna. (*a*) Undamped. (*b*) Damped.

where $G$ is the shear modulus in N/m$^2$. The tuning mass $m_D$ was selected to be 45 grams, which gave a value of $k_D$ for optimum tuning:

$$(k_D)_{\text{opt}} = m_D \omega_D^2$$

$$= 0.045(2\pi \times 443)^2$$

$$= 3.49 \times 10^5 \frac{\text{N}}{\text{m}}$$

Therefore $G = 3.92 \times 10^4$ N/m$^2$ or 223 lb/in$^2$. This corresponds to $E = 3 \times 223 = 669$ lb/in$^2$, which is seen to occur at about 130° F in Data Sheet 032. The damper was then attached to the connector of the antenna, and the system was subjected to base (shaker) excitation. A typical response spectrum is shown in Figure 5.33*b*. Tests were conducted at several temperatures from 75° F to 140° F and at several excitation input levels (1 g, 5 g, 10 g). Graphs of

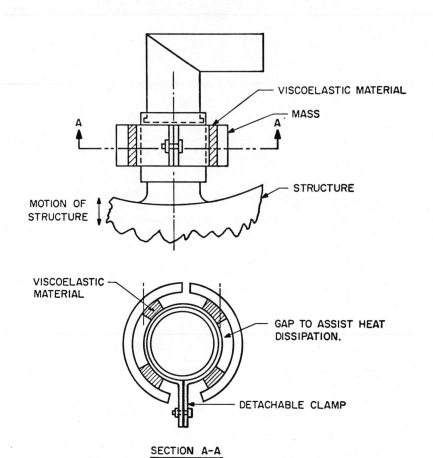

FIGURE 5.34. Prototype damper geometry.

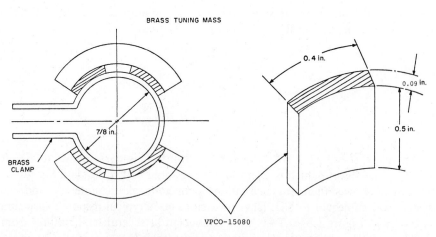

FIGURE 5.35. Prototype 1 damper geometry.

peak amplitude versus temperature are plotted in Figure 5.36. It is seen that significant reductions of peak amplitudes, and hence of peak stresses in the failure zone, were achieved between about 70° F and to over 150° F.

The damper was then tested under high level random excitation, simulating the operational environment, and performed well for about 15 minutes of continuous excitation without failure, although at that point the damper had heated up excessively, and the material had become too soft. It was later recognized that the test conditions were too severe, since the actual firing environment was 30 second bursts of random excitation interspersed with relatively long rest periods. Hence the damper worked effectively under very severe over-test conditions and was technically a complete success. However, potential supply problems in the material from Germany, lack of precise information on the best bonding techniques, and some concern for operations below 70° F and above 150° F, as well as an apparent lull in the urgency of the

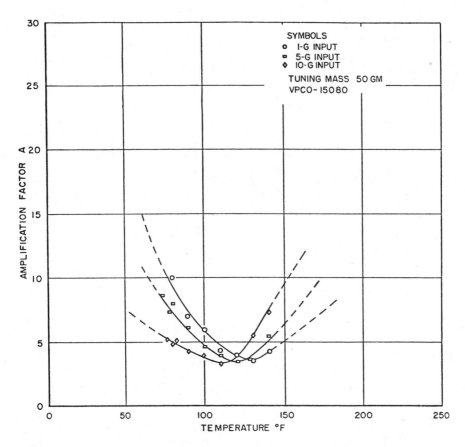

**FIGURE 5.36.**   Variation of amplification factor $A$ with temperature for antenna with prototype 1 attached.

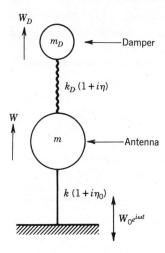

$W_D$

$m_D$ ←——Damper

$k_D(1+i\eta)$

$W$

$m$ ←——Antenna

$k(1+i\eta_0)$

$W_0e^{i\omega t}$

**FIGURE 5.37.** Antenna-damper model.

requirement, led to a search for other materials, which were used for later prototypes.

**Design analysis of prototype 1.** Although a design analysis was not carried out in great detail during the development of prototype 1, it is highly instructive to apply the analysis of this chapter to predict the response of the damped antenna and hence corroborate the experimental measurements.

Figure 5.37 shows the relevant two degree of freedom model of the antenna damper system with base excitation. For the antenna, $m = 0.1$ kg and $k = 9.29 \times 10^5$ N/m, with a probable error of $\pm 20\%$. For the damper, $m_D = 0.045$ kg and $k_D = 8.89G$. The values of $G$ and $\eta$ for the damper material are given in Data Sheet 032. On this basis the equations of motion of the system are

$$\left.\begin{array}{c} m\ddot{w} + k(1 + i\eta_0)(w - w_0) + k_D(1 + i\eta)(w - w_D) = 0 \\ m_D\ddot{w}_D + k_D(1 + i\eta)(w_D - w) = 0 \end{array}\right\} \tag{5.60}$$

These equations are solved in the usual way, for base excitation $w_0 = W_0e^{i\omega t}$, to give

$$\frac{\ddot{w}}{\ddot{w}_0} = \frac{Ak\sqrt{1 + \eta_0^2}}{\sqrt{B^2 + C^2}} \tag{5.61}$$

where

$$A = (k_D - m_D\omega^2)^2 + (k_D\eta)^2$$
$$B = A(k + k_D - m\omega^2) - k_D^2(k_D - m_D\omega^2)(1 - \eta^2) - 2k_D^3\eta^2 \tag{5.62}$$
$$C = A(k\eta_0 + k_D\eta) - 2\eta k_D^2(k_D - m_D\omega^2) + k_D^3\eta(1 - \eta^2)$$

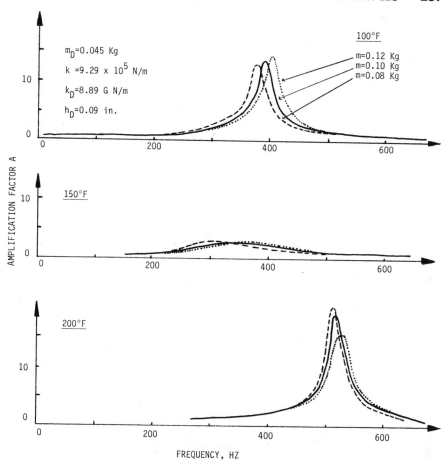

AMPLIFICATION FACTOR A

$m_D=0.045$ Kg

$k =9.29 \times 10^5$ N/m

$k_D=8.89$ G N/m

$h_D=0.09$ in.

100°F

m=0.12 Kg
m=0.10 Kg
m=0.08 Kg

150°F

200°F

FREQUENCY, HZ

**FIGURE 5.38.**   Predicted response behavior for antenna with prototype 1.

Figure 5.38 shows the predicted transmissibility $|\ddot{w}/\ddot{w}_0|$ versus frequency, for prototype 1 on the antenna, obtained at three temperatures using Equation (5.61). A comparison of analytical and experimental transmissibility at low excitation levels (1 g) is given in Figure 5.39. The agreement is sufficiently good to validate the analysis and also to demonstrate that the rapid rise of amplitudes below 100° F and above 200° F is due to the rapid variation of the properties, especially the shear modulus, with temperature. The same figure also shows the effect of changing the value of $m$ by $\pm 20\%$.

**Design of later prototypes.**   After the successful testing of prototype 1 some time passed during which uncertainties about the operational need for the damper were resolved, and more antennas were procured for testing purposes. The program resumed in August 1966, and the damper design evolved

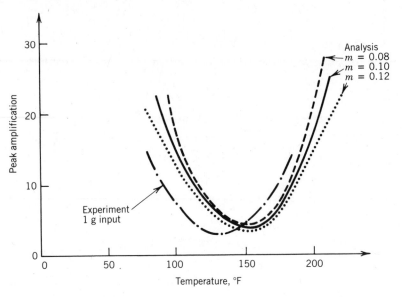

**FIGURE 5.39.**   Predicted and measured peak amplification factors for antenna with prototype 1.

through several prototypes, leading to a final configuration which was sent for field testing in October 1966. Prototype 2 was handmade, and prototype 3 was molded, using a nitrile rubber elastomeric element known as Paracril-D, made by the U.S. Rubber Company (Data Sheet 028). The dampers were successful in reducing the amplification factor over quite a wide temperature range. In order to extend the useful temperature range at the lower end, subsequent prototypes used a nitrile rubber having a lower transition temperature than Paracil-D, namely Paracil-BJ (Data Sheets 025 to 027). The formulations of the elastomeric elements in each of the prototypes, from 1 to 8, are summarized in Table 5.2. Prototype 4, using Paracil-BJ, was constructed in accordance with the design shown in Figures 5.40 and 5.41. The mold breakout is illustrated in Figure 5.42. Figure 5.43 shows the variation of the amplification factor with temperature for prototype 4. Tests under random excitation showed the damper to be too weak, so prototype 5 was made with 50 parts per hundred (PHR) of SAF carbon black added during processing of the elastomer for molding. Figure 5.44 shows the variation of amplification factor with temperature for this damper. Clearly the increased stiffness of the elastomeric material, resulting from the addition of the carbon black, caused excessive detuning of the damper and a compromise of 25 PHR of carbon black was used for prototype 7 (prototype 6 was not used). Figure 5.45 shows the measured variation of the amplification factor with temperature for prototype 7. The performance was improved, but further improvement was sought through slight modification of the damper geometry for prototype 8, as shown in

**TABLE 5.2. Nitrile Rubber Formulations**

| Material Composition | Prototype Number | | | | | | | |
|---|---|---|---|---|---|---|---|---|
| | 1 | 2 | 3 | 4 | 5 | 6 | 7 | 8 |
| VPCO-15080 | 100 | | | | | | | |
| Paracril-D | | 100 | 100 | | | | | |
| Paracril-BJ | | | | 100 | 100 | 100 | 100 | 100 |
| Polystyrene | | | | | | 10 | | |
| Zinc oxide | | 10 | 10 | 10 | 10 | 10 | 10 | 10 |
| Dicumyl peroxide | | 3 | | | | | | |
| Stearic acid | | | | 1.5 | | | | |
| Sulfur | | | 1.5 | 1.5 | 1.5 | 1.5 | 1.5 | 1.5 |
| Benzothiozyl disulphide | | | 1.5 | 1.5 | 1.5 | 1.5 | 1.5 | 1.5 |
| Antioxidant 2246[a] | | | | | | | | 1.0 |
| SAF black | | | | | 50 | | 25 | 25 |
| Cure temperature °F | | 280 | 310 | 310 | 310 | 310 | 310 | 310 |
| Cure time, minutes | | 60 | 30 | 30 | 30 | 30 | 30 | 30 |
| Mass grams | 50 | 45 | 35 | 35 | 35 | 35 | 25 | 25 |
| Mold | None | original | original | original | original | original | modified | modified |

[a] American Cyanamid Company trademark for 2,2′-methylenebis (4-methyl-6-tertiary-butyl phenol) [5.29].

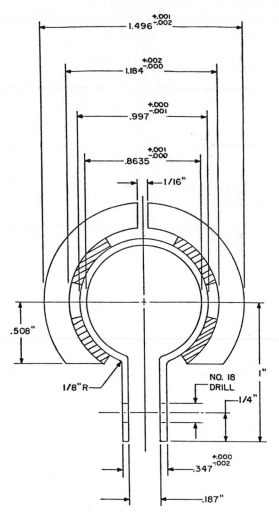

**FIGURE 5.40.** Geometry of prototypes 4–7.

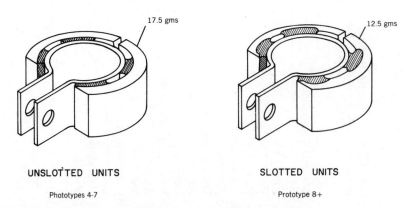

UNSLOTTED UNITS

Phototypes 4-7

SLOTTED UNITS

Prototype 8+

**FIGURE 5.41.** Damper geometries.

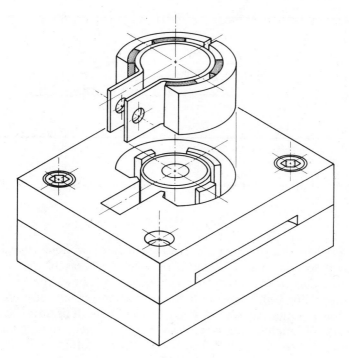

**FIGURE 5.42.** Mold for damper fabrication.

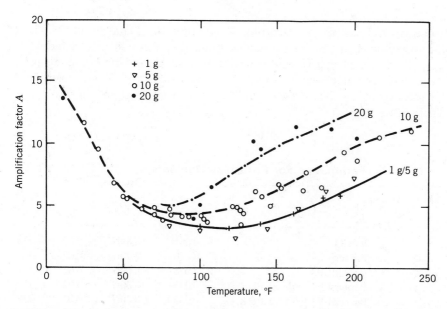

**FIGURE 5.43.** Variation of amplification factor $A$ with temperature for antenna with prototype 4 attached.

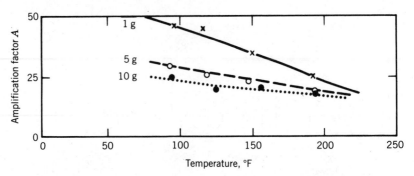

**FIGURE 5.44.**   Variation of amplification factor $A$ with temperature for antenna with prototype 5 attached.

Figure 5.46. This design change was accommodated without changing the mold, except for rounding of some corners at the edges of the elastomer elements to reduce stress concentrations. Figure 5.47 shows the variation of amplification factor with temperature for production versions of prototype 8, the final configuration, at several input accelerations. It is seen that satisfactory levels of amplification were achieved over a very wide temperature range, from 50° F to over 200° F. The damper also performed well in random vibration tests, and this design was used for the production dampers, several of which were sent for field testing at the end of 1966.

**Design analysis of prototypes 4 through 8.**   Data Sheet 026 shows the variation of the Young's modulus $E$ and loss factor $\eta$ with reduced frequency $f\alpha_T$, frequency $f$ Hz and temperature $T°$ F in the form of a nomogram. Using this data, and equation (5.61), the transmissibility $|\ddot{w}/\ddot{w}_0|$ was calculated for each geometry. The relevant input data for each case are summarized in Table 5.3.

**TABLE 5.3.   Parameters For Prototype Dampers**

| Prototype Number | Material | $m_D$ kg | $k_D$ N/m | $\eta_0$ | Data Sheet |
|---|---|---|---|---|---|
| 1 | VPCO-15080 | 0.045 | 8.89G | 0.02 | 032 |
| 2 | Paracril-D | 0.050 | 4.00G | 0.02 | 028 |
| 4 | Paracril-BJ | 0.035 | 8.00G | 0.02 | 025 |
| 5 | Paracril-BJ with 50 PHR C | 0.035 | 8.00G | 0.02 | 027 |
| 7 | Paracril-BJ with 25 PHR C | 0.035 | 8.00G | 0.02 | 026 |
| 8 | Paracril-BJ with 25 PHR C | 0.025 | 4.00G | 0.02 | 026 |

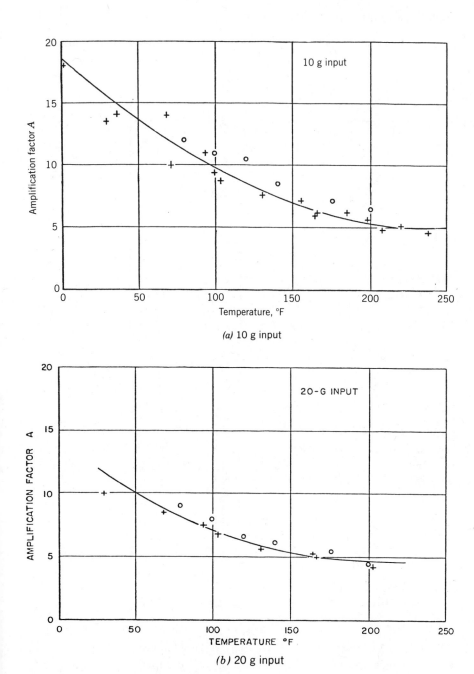

**FIGURE 5.45.** Variation of amplification factor $A$ with temperature for antenna with prototype 7 attached. (a) 10 g input. (b) 20 g input.

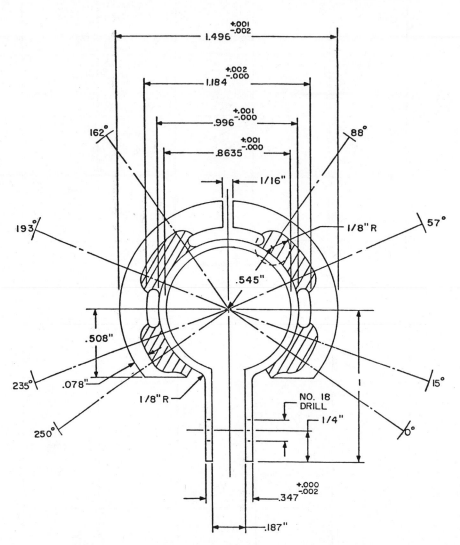

**FIGURE 5.46.** Geometry of prototype 8 and subsequent dampers.

Figure 5.48 shows the predicted variation of amplification factor $|\ddot{w}/\ddot{w}_0|$ with frequency and temperature for prototype 4 damper on the antenna. This damper, having Paracril-BJ with a O PHR C as the elastomeric element, is seen to exhibit high effectiveness over a wide temperature range. Other calculations show the effect of varying $k_D$ by a factor of $2(k_D = 4.00G)$, and of reducing $m$ by a factor of 2. These plots illustrate the effect of possible errors in $m$ (but not $k/m$), and $k_D$, on the damper effectiveness. The resulting peak

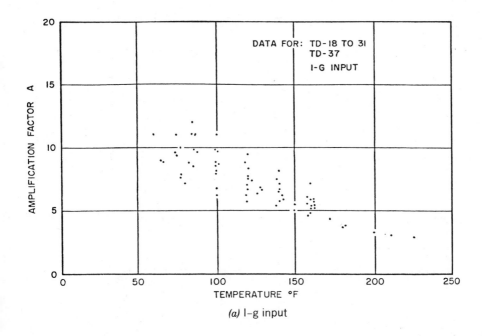

*(a)* 1–g input

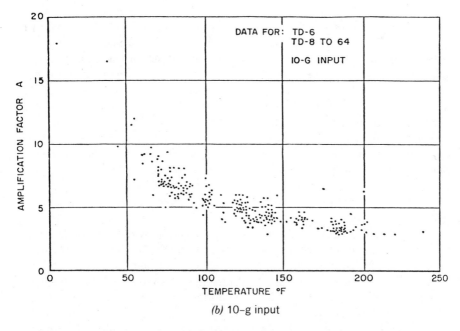

*(b)* 10–g input

**FIGURE 5.47.** Graphs of amplification factor $A$ against temperature for antenna with production dampers attached. (*a*) 1–g input. (*b*) 10–g input.

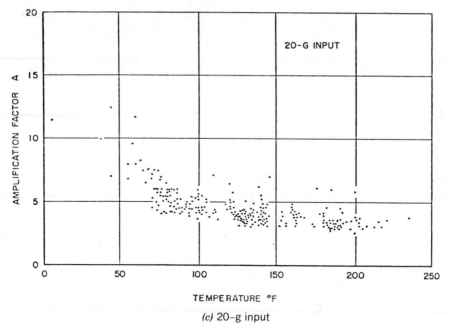

**FIGURE 5.47.**   Graphs of amplification factor $A$ against temperature for antenna with production dampers attached. (c) 20-g *input*.

amplification factors are plotted versus temperature in Figure 5.49 and compared with low excitation level experimental data. The nominal case ($m = 0.05Kg$, $k_D = 4G$) agrees quite well with measurements, indicating that the analysis worked reasonably well, that errors in geometry and material properties data are not too great, and that the damper design was satisfactory from the point of view of amplitude reduction. However, as described earlier, the design was inadequate from a strength point of view.

Figure 5.50 shows the predicted amplification $|\ddot{w}/\ddot{w}_0|$ versus frequency for prototype 5, having Paracril-BJ with 50 PHR carbon black as the elastomeric element. Amplifications were consistently too high. Figure 5.51 shows peak amplifications versus temperature, as compared with experiment. Agreement was again reasonably good. The damper was inadequate from an amplitude reduction point of view.

Figure 5.52 shows the predicted variation of amplification with frequency at various temperatures for prototype 7, having Paracril-BJ with 25 PHR carbon black as elastomeric element. Amplitudes were lower than for prototype 5 but were still too high. No experimental data at low excitation levels (1 g) were available for comparison with the calculations for prototype 7. However, the peak response is seen to be consistently high. Another calculation with $k_D$ reduced to 3/4 of its nominal value showed a great improvement. Hence prototype 8 was designed.

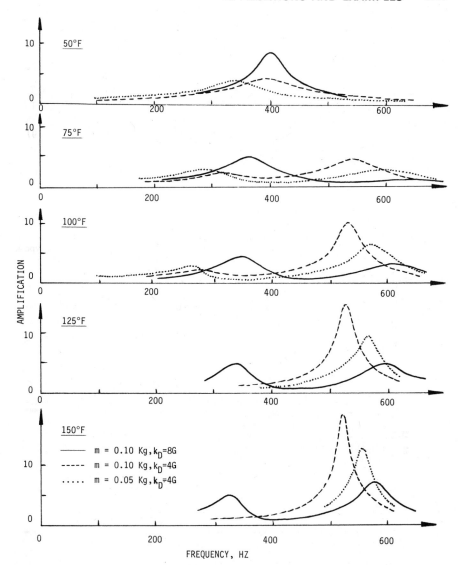

**FIGURE 5.48.** Predicted response for prototype 4 ($m_D = 0.035$ Kg, $k = 9.29 \times 10^5$ N/m, $\eta_0 = 0.02$).

Figure 5.53 shows the predicted variation of amplification factor with frequency at various temperatures for prototype 8, having Paracril-BJ with 25 PHR carbon black as the elastomeric element and a modified geometry ($k_D \simeq 3G$). The complex geometry of the damper made the estimate of the relationship between $k_D$ and $G$ somewhat imprecise, but the figure is probably not more than $\pm 25\%$ in error. Figure 5.54 shows a graph of peak amplification factor versus temperature, along with experimental data for 1 g input

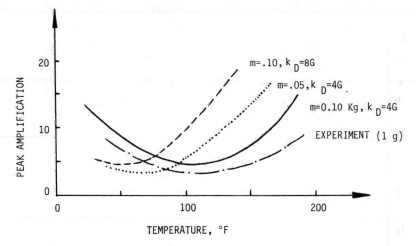

**FIGURE 5.49.** Predicted and measured peak amplification factors for antenna with prototype 4.

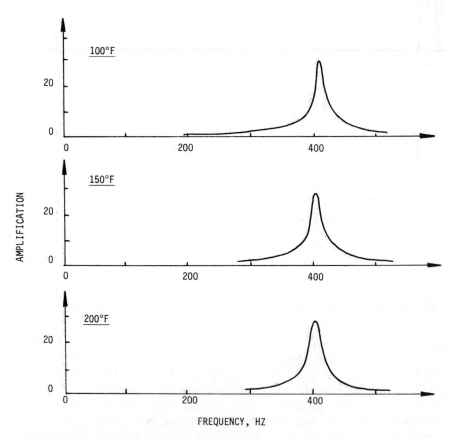

**FIGURE 5.50.** Predicted response for antenna with prototype 5 ($m_D = 0.035$ Kg, $m = 0.10$, $k_D = 8G$, Paracril-BJ with 50 PHR C).

**248**

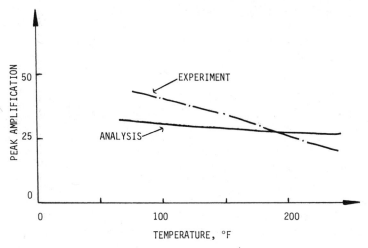

**FIGURE 5.51.** Predicted and measured peak amplification factors for antenna with prototype 5.

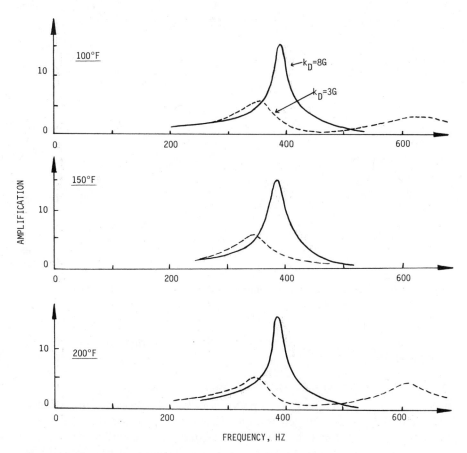

**FIGURE 5.52.** Predicted response for antenna with prototype 7 ($m_D = 0.035$ Kg, $m = 0.10$ g, Paracril-BJ with 25 PHR C).

**249**

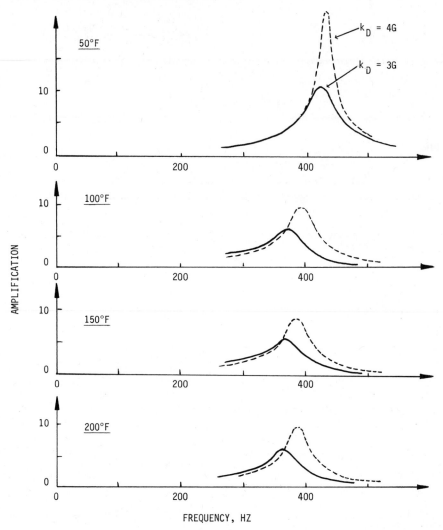

**FIGURE 5.53.**   Predicted response for antenna with prototype 8 ($m_D = 0.025$ Kg, $m = 0.1$ g, Paracril-BJ with 25 PHR C).

acceleration. The agreement is not too bad. This was the final damper configuration.

**Summary.**   The field tests were successfully completed by January 1967 and indicated that the dampers were performing well in service. This problem, though an old one, is an excellent example of the design procedure that might be adopted to develop a tuned damper to meet severe environmental and temperature requirements. Further details are available in references [5.29–31].

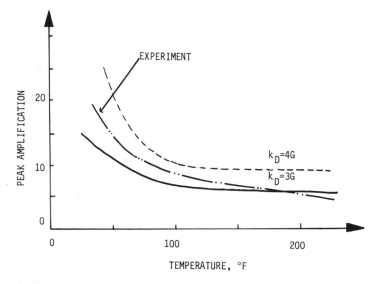

**FIGURE 5.54.**    Predicted and measured peak amplification factors for antenna with prototype 8.

## 5.6.2.  Tuned Damper for Vibration Control of Impeller Blades

A resonance corresponding to the first bending mode of a set of industrial impeller blades, at about 64 Hz, was believed to be excited under operating conditions at a speed of 640 RPM, thus leading to premature failure of the blades, illustrated in Figure 5.55. Although no strain measurements had been made on the blade when the problem was first investigated, it was believed that it would be beneficial to damp a set of existing blades so as to control the dynamic response around 64 Hz.

It was first thought that a constrained-layer damping treatment would be suitable for this application. However, after some initial analysis using the equations of Chapter 6 and some preliminary testing, it was concluded that an excessively thick and stiff constraining layer would be required to introduce sufficient damping into the system. It was not practical to apply this to the existing blade, and hence another approach was investigated. This was to use a tuned viscoelastic damper that could be placed inside the blade and tuned to the frequency of the first bending mode. When installed inside the blades, the dampers performed well; that is, they gave significant vibration reductions for the mode of vibration involved.

**Development of the dampers.**  The tuned damper was designed in the form of a cantilever beam with its first mode of vibration tuned to the frequency of the first bending mode of the blade. Sufficient damping was introduced into the cantilever beam by means of a sandwich configuration, so as to

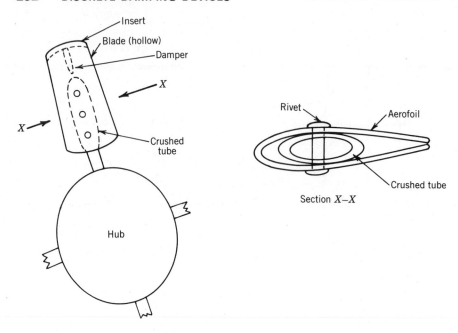

**FIGURE 5.55.** Blade geometry.

dissipate the vibrational energy at resonance. The beam was designed to fit inside the blade, ahead of the tubing, as shown in Figure 5.55.

The cantilever beam damper consisted of two sheets of steel 0.75 in. wide, 0.062 in. thick, 6.0 in. long with a 0.02 in. thick damper layer between them. The damping material was 3M-467 adhesive, manufactured by the 3M Company, St. Paul, Minnesota (Data Sheet 034). This material was selected because it has high damping capability near room temperature, which was the operating temperature of the blade. The beams were assembled with a thick root that could be bolted inside the blade. The root was epoxied to the insert. Because of the limited space inside the blade, ahead of the insert, the first resonant frequency of the beam was tuned by the addition of a small mass to its tip rather than by increasing its length. These masses were in the form of a small bolt, washer, and nut, which also served as a mechanical fastener to hold the tip of the beam and prevented it from separating in service. Figure 5.56 illustrates the variation of the damper stiffness and damping with temperature. These are analytical predictions of the resonant frequency of the first mode of the damper and of its modal damping in terms of temperature using the equations of Chapter 6. It is seen that the damper is tuned near to 64 Hz over a wide temperature range, covering the expected range seen by the blade. It can be seen that the modal damping is high; thus it should dissipate large amounts of vibrational energy when installed in the blade.

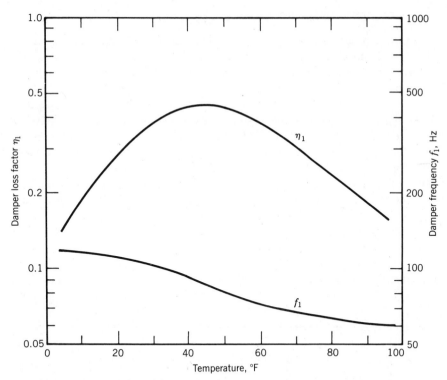

**FIGURE 5.56.** Variation of damper loss factor and resonant frequency with temperature.

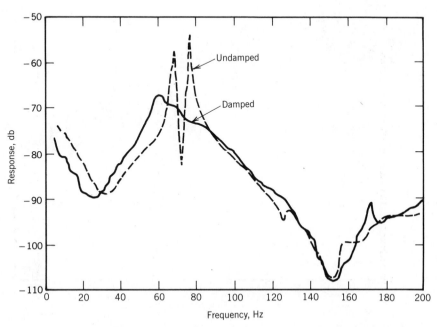

**FIGURE 5.57.** Damped and undamped vane response.

**253**

**Test evaluation of dampers.** Tuned dampers were made in the configuration just described and placed inside the blades of one completed wheel. The performance of the dampers was verified on a complete wheel by applying an oscillating force to one blade, close to the root, and measuring the response near to the same point. The wheel was not rotating for this test. This was done for both the original and the treated wheels. The results of these tests are illustrated in Figure 5.57. Figure 5.57 represents the compliance, the inverse of the stiffness, measured near the root, as a function of frequency at a temperature of approximately 75° F for the undamped and damped wheels. It can be seen that significant reductions, of the order of 14 dB, were achieved in the amplitude of vibration of the bending mode, at approximately 64 Hz. Because of the good results so achieved in reducing the amplitude of vibration of the first bending mode of the blades, the remaining sets of wheels were treated with this same type of damper.

The performance of the blades was verified under operating conditions by instrumenting the blades with strain gages. Results are given in Figures 5.58

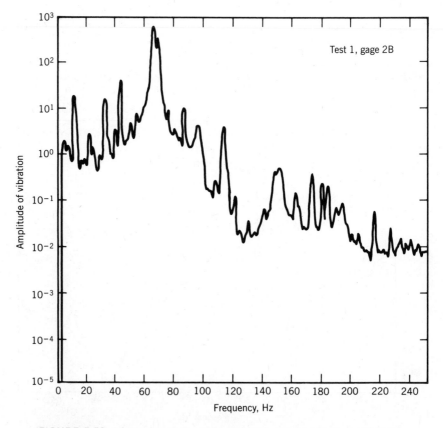

**FIGURE 5.58.** Operating strain spectrum of the untreated blade at 670 RPM.

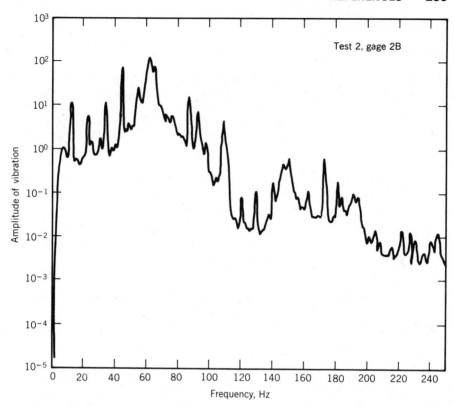

**FIGURE 5.59.** Operating strain spectrum of the treated blade at 670 RPM.

and 5.59 in terms of dynamic strain for the undamped and damped blade at a rotational speed of 670 RPM. The reductions of stress near 64 Hz were quite significant.

## REFERENCES

5.1.  F. E. Reed, "Dynamic vibration absorbers and auxiliary mass dampers," *Shock, and Vibration Handbook*, (eds. C. M. Harris and C. E. Crede), McGraw-Hill, New York, Chap. 6 1961.

5.2.  J. P. Henderson, "Energy dissipation in a vibration damper utilizing a viscoelastic suspension," Air Force Materials Lab., AFML-TR-65-403, WPAFB, November 1965.

5.3.  D. I. G. Jones, J. P. Henderson, and G. H. Bruns, "Use of tuned viscoelastic dampers in aerospace structures," Proc. 13th Annual Air Force Science and Engineering Symposium, Univ. of Tennessee, September 1966.

5.4.  D. I. G. Jones, A. D. Nashif, and R. L. Adkins, "Effect of tuned dampers on vibrations of simple structures," *AIAA J.*, **5**, (2) (1967).

5.5. D. I. G. Jones and J. P. Henderson, "A plural load viscoelastic damping device," U.S. Patent No. 3,419,111 (1968).

5.6. A. D. Nashif, and D. I. G. Jones, "A resonant beam tuned damping device," *Trans. ASME, J. Eng. Power*, 143–148 (July 1969).

5.7. A. D. Nashif, "Resonant beam/tuned damper," U.S. Patent No. 3,601, 228, 1971.

5.8. M. L. Parin and D. I. G. Jones, "Encapsulated tuned dampers for jet engine component vibration control," *J. Aircraft*, **12** (4), 293–295 (April 1975).

5.9. D. I. G. Jones, A. D. Nashif, and H. Stargardter, "Vibrating beam dampers for reducing vibrations in gas turbine blades," *Trans. ASME, J. Eng. Power*, 111–116 (January 1975).

5.10. D. I. G. Jones, "Damping of structures by viscoelastic links," *Shock Vib. Bull.* **36** (4), 9–24 (January 1967).

5.11. J. Ormondroyd and J. P. Den Hartog, "The theory of the dynamic vibration absorber," *Trans., ASME*, APM-50-7 (1928).

5.12. J. P. Den Hartog, *Mechanical Vibrations*, 4th ed., McGraw-Hill, New York, 1956.

5.13. J. P. Henderson, "Vibration analysis of curved skin-stringer structures having tuned elastomeric dampers," Air Force Materials Lab. AFML-TR-72-240, WPAFB, October 1972.

5.14. J. C. Snowdon, *Vibration and Shock in Damped Mechanical Systems*, Wiley, New York, 1968.

5.15. A. B. Davey and A. R. Payne, *Rubber in Engineering Practice*, Palmerton., New York, pp. 118–119, 1964.

5.16. J. P. Licari and E. Barhar, "A study of pendulum viscoelastic dampers," Air Force Materials Lab. Report AFML-TR-66-381, WPAFB, 1967.

5.17. D. I. G. Jones, "Analytical and experimental investigation of the effect of tuned viscoelastic dampers on the response of simple beams with various boundary conditions," Air Force Materials Lab., AFML-TR-67-214, WPAFB, June 1967.

5.18. D. I. G. Jones, "Response and damping of a simple beam with tuned damper," *J. Acoustical Soc. Am.*, **42**, (1), 50–53 (July 1967).

5.19. R. E. D. Bishop and D. C. Johnson, *The Mechanics of Vibration*, 1st ed., Cambridge Univ. Press, Cambridge, 375–387, 1960.

5.20. G. H. Bruns and A. D. Nashif, "Experimental verification of theory of damping of a simple structure by distributed tuned dampers," Air Force Materials Lab., AFML-TR-65-440, WPAFB, 1965.

5.21. J. C. Snowdon, "Vibration of Cantilever beams to which dynamic absorbers are attached," *J. Acoust. Soc. Am.*, **39** (5) 878–889 (1966).

5.22. D. Young, Proc. 1st U.S. National Congress on Applied Mechanics, pp. 91–96, 1952.

5.23. A. D. Nashif, "Effect of tuned dampers on vibrations of cantilever beams," Air Force Materials Lab., Report AFML-TR-66-83, WPAFB, 1966.

5.24. R. L. Adkins, "Effect of tuned viscoelastic dampers on response and damping of a clamped-clamped beam," Air Force Materials Lab., Report AFML-TR-66-100, WPAFB, 1966.

5.25. B. J. Lazan and L. E. Goodman, "Material and interface damping," *Shock and Vibration Handbook*, Vol. 2 (eds. C. M. Harris and C. E. Crede), McGraw-Hill, New York, Chap. 36, 1961.

5.26. B. L. Clarkson and R. D. Ford, "Experimental study of the random vibrations of an aircraft structure excited by jet noise," WADD Technical Report 61–70, March 1961.

5.27. Y. K. Lin, "Dynamic characteristics of continuous skin-stringer panels," *Acoustical Fatigue in Aerospace Structures* (eds. W. J. Trap and D. M. Forney, Jr.), Syracuse Univ. Press, Syracuse, N.Y., pp. 163–184, 1965.

5.28.   D. I. G. Jones, "Effect of isolated tuned dampers on response of multispan structures," *J. Aircraft*, **4** (4), 343–346 (July–August 1967)

5.29.   D. I. G. Jones, A. D. Nashif, G. H. Bruns, R. Sevy, F. S. Owens, J. P. Henderson, and R. L. Conner., "Development of a tuned damper to reduce vibration damage in an aircraft radar antenna," Air Force Materials Lab., Report AFML-TR-67-307, WPAFB, 1967.

5.30.   F. S. Owens, "Elastomers for damping over wide temperature ranges," *Shock Vib. Bull.*, **36**(4), 25–35 (1967).

5.31.   A. D. Nashif, "Development of practical tuned dampers to operate over a wide temperature range," *Shock Vib. Bull.*, **38**(3), 57–69 (November 1968).

# 6

## SURFACE DAMPING TREATMENTS

### ADDITIONAL SYMBOLS

| | |
|---|---|
| $a$ | length of plate |
| $b$ | width of plate |
| $D$ | measure of neutral axis position; also area of free layer treatment |
| $e$ | $E_D/E$ |
| $e_j$ | $E_j/E_1$ $(j = 1, 2, 3)$, modulus ratio |
| $E, E_3$ | Young's modulus |
| $E_c$ | modulus of constraining layers |
| $E_e$ | effective modulus of multiple layer treatment |
| $g, g_N$ | shear parameters |
| $G_2$ | shear modulus |
| $h$ | $H_D/H$ |
| $h_j$ | $H_j/H_1$ $(j = 1, 2, 3)$, thickness ratio |
| $H_D$ | net thickness of free layer treatment |
| $H_j$ | beam element thickness $(j = 1, 2, 3)$ |
| $H_{31}, H_{21}$ | thickness combinations |
| $I, I_1$ | second moment of area of beam cross section |
| $m$ | number of semiwavelengths in direction $b$ |
| $n$ | number of semiwavelengths in direction $a$, or mode number, according to context |
| $N$ | number of layer pairs |

**258**

| $p$ | wave number ($p = \xi_n/L$) |
|---|---|
| $\alpha_{nm}$ | nondimensional parameter |
| $\phi_{nm}$ | modal function |
| $\eta_1, \eta_1, \eta_2, \cdots$ | loss factors |
| $\eta_{nm}$ | modal loss factor in $n$, $m$ mode of plate |
| $v$ | Poisson's ratio |
| $\rho_D$ | density of free layer treatment |
| $\xi_n$ | $n$th eigenvalue |

## 6.1.  INTRODUCTORY REMARKS

Surface damping treatments are often used to solve a variety of resonant noise and vibration problems, especially those associated with sheet metal structure vibration. Such treatments can easily be applied to existing structures and provide high damping capability over wide temperature and frequency ranges. The surface damping treatments are usually classified in one of two categories, according to whether the damping material is subjected to extensional or shear deformation.

The analysis presented herein applies to both types of surface treatments and can be modified to take into account their various configurations and deformation patterns. Although the analysis is applicable to both beam and plate vibrations, most of the examples and discussions in this section will be restricted to beam applications.

## 6.2.  ANALYSIS FOR BEAMS AND PLATES

### 6.2.1.  Ross–Kerwin–Ungar Equations

A variety of approaches has been considered to describe the behavior of different types of surface damping treatments. Of these the analysis developed by Ross, Ungar, and Kerwin [6.1] has been the most widely used. This analysis (referred to hereafter as the RKU analysis) has been developed for a three-layer system and is usually used to handle both extensional and shear types of treatment. The analysis, within its limitations, can be extended to handle the response of damped plates as well as damped beams. Although this analysis has been developed to predict the response of damped three-layer systems, assuming that the properties of the damping material are known, it has on numerous occasions been used in reverse. In such cases the damping properties of the material are computed from the damped response of the system, most often a sandwich beam. A discussion of the basic analysis and how it can be extended for different types of treatments and objectives follows.

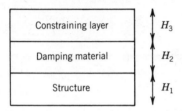

| Constraining layer | $H_3$ |
| Damping material | $H_2$ |
| Structure | $H_1$ |

**FIGURE 6.1.**   Constrained-layer damping treatment.

The flexural rigidity, $EI$, of the three-layer system of Figure 6.1 is [6.1]:

$$EI = E_1 \frac{H_1^3}{12} + E_2 \frac{H_2^3}{12} + E_3 \frac{H_3^3}{12} - E_2 \frac{H_2^2}{12}\left(\frac{H_{31} - D}{1 + g}\right)$$

$$+ E_1 H_1 D^2 + E_2 H_2 (H_{21} - D)^2 + E_3 H_3 (H_{31} - D)^2$$

$$- \left[\frac{E_2 H_2}{2}(H_{21} - D) + E_3 H_3 (H_{31} - D)\right]\left(\frac{H_{31} - D}{1 + g}\right) \qquad (6.1)$$

where

$$D = \frac{E_2 H_2 (H_{21} - H_{31}/2) + g(E_2 H_2 H_{21} + E_3 H_3 H_{31})}{E_1 H_1 + E_2 H_2/2 + g(E_1 H_1 + E_2 H_2 + E_3 H_3)} \qquad (6.2)$$

$$H_{31} = \frac{(H_1 + H_3)}{2} + H_2 \qquad (6.3)$$

$$H_{21} = \frac{(H_1 + H_2)}{2} \qquad (6.4)$$

$$g = \frac{G_2}{E_3 H_3 H_2 p^2} \qquad (6.5)$$

and $E$ is the Young's modulus of elasticity, $G$ is the shear modulus, $I$ is the second moment of area, $H$ is the thickness, $\xi_n$ is the $n$th eigenvalue. Subscript 1 refers to the base structure, subscript 2 to the damping layer, subscript 3 to the constraining layer, and no subscript refers to the composite system. $D$ is the distance from the neutral axis of the three layer system to that of the original beam, $H_1$.

### 6.2.2.   Assumptions and Precautions

The most important assumption to keep in mind while using the analysis, is that equations (6.1) through (6.5) were developed and solved using sinusoidal expansions for the modes of vibration, and the Ross-Kerwin-Ungar analysis therefore applies to simply supported beams or plates. For other types of boundary conditions, approximations must be used depending on the mode shape of the structure in question. The analysis also assumes rigid connections between the various layers of the system. However, in practice, because most damping materials are not self-adhesive, an additional adhesive layer is used for fastening purposes. In such cases the thickness of the adhesive layer must be kept to a minimum, and the modulus of elasticity of its material must be as high as possible. The loss factor of the adhesive should also be low.

**Beam equations.**   To use the RKU analysis to predict the damped response of simply supported beams, it is sufficient to keep in mind that the natural frequency, $\omega_n$, for such a beam is

$$\omega_n = \frac{\xi_n^2}{L^2} \sqrt{\frac{EI}{\rho Hb}} \tag{6.6}$$

where

$$\xi_n = n\pi \tag{6.7}$$

is the $n$th eigenvalue, $n$ is the mode number, $L$ is the length of the beam, and $\rho$ is the mass density.

**Plate equations.**   The RKU analysis can be extended to predict the damped response of a simply supported plate by recalling that its natural frequency can be written as

$$\omega_{nm} = \frac{\xi_{nm}^2}{a^2} \sqrt{\frac{EH^3}{12(1-v^2)\rho H}} \tag{6.8}$$

$$\frac{\xi_{nm}^2}{a^2} = \left(\frac{n\pi}{a}\right)^2 + \left(\frac{m\pi}{b}\right)^2 \tag{6.9}$$

$n$ is the number of semiwavelengths in direction $a$, $m$ is the number of semiwavelengths in direction $b$, $a$ is the length of the plate, $b$ is the width of the plate, $\rho$ is the density of the plate, and $v$ is Poisson's ratio for the plate material.

To incorporate the damping terms into equations (6.1), it is necessary to replace each modulus term by a complex modulus. For example, in the general case it is necessary to have

$$g \to g(1 + i\eta_2) \tag{6.10}$$

$$E_2 \to E_2(1 + i\eta_2) \tag{6.11}$$

$$E_3 \to E_3(1 + i\eta_3) \tag{6.12}$$

$$E_1 \to E_1(1 + i\eta_1) \tag{6.13}$$

$$E \to E(1 + i\eta) \tag{6.14}$$

where $\eta$ is the loss factor, assumed to be the same for the viscoelastic material when subjected to either extensional or shear deformation.

To solve equation (6.1), it is necessary to substitute the expressions given in equations (6.10) through (6.14) in equation (6.1) and then equate the real and imaginary parts of the equation to arrive at the expressions for $\eta$ and $EI$. Having obtained the stiffness for the damped system, then its natural frequency can be computed by using the appropriate beam or plate equations. A word of caution on using the plate equations (6.8) and (6.9) is in order, since very little information is available in the literature on Poisson's ratio of damping materials. For the proper analysis of damped plates Poisson's ratio must be known and available in the same way as the modulus and loss factor of the material are determined over the desired environment.

Since the material of the constrained layer is metallic, the loss factor $\eta_3$ can be assumed to be zero for most purposes. As far as the loss factor $\eta_1$ is concerned, it should be similar to the structural modal damping for the resonance of interest. For many cases, such as welded structures and integrally machined structures, the structural damping is almost the same as the material damping, and therefore $\eta_1$ can be assumed to be zero. However, for structures where the joint damping is high, such as riveted or bolted structures, the structural damping $\eta_1$ could be an important factor, and therefore it would have to be included in the analysis.

**Use of the RKU analysis for complex structures.** The foregoing analysis can be used for approximate prediction of the damping of complex structures. For this purpose it is necessary to know the frequency, damping, and mode shape of vibration for the resonance of interest. Such information could be obtained experimentally or analytically. The mode shape information is used to obtain the wavelength of vibration. This information is then used, along with either the beam or plate equations, to compute an equivalent thickness for the structure that will yield the same resonant frequency. The resultant equivalent thickness for the structure is then used to predict the effects of damping treatment that are applied to it.

The modal damping, if high, is used as an input to the equation for the loss factor term, $\eta_1$, and also to provide the basis for assessing how much improvement in vibration reduction can be achieved by the addition of the damping treatment.

More detailed information on this approach is given in the various case histories at the end of this chapter.

**Use of the RKU analysis for single constrained-layer treatment.** The RKU analysis can be used to predict the effect of different types of constrained-layer damping treatment on a given structure, assuming that the material properties and the structural modal properties are known.

For simply supported beams or plates it is necessary to know either the wavelength of vibration or the frequency, because either can be deduced from the other by using the appropriate equation. The wavelength is necessary as an input parameter to calculate the shear parameter $g$, so that the appropriate material properties can be selected, since they are frequency dependent.

For complex structures both the frequency and wavelength of vibration must be known, so that an equivalent thickness for the structure can be defined in addition to the information needed for the simply supported case.

## 6.3.  EXTENSIONAL DAMPING TREATMENT

### 6.3.1.  Equations

Extensional damping is one of the most commonly used treatments. Sometimes this treatment is referred to as the unconstrained- or free-layer damping treatment. The treatment is coated on one or both sides of a structure, so that whenever the structure is subjected to cyclic bending, the damping material will be subjected to tension-compression deformation. This type of damping treatment is illustrated in Figure 6.2.

The Ross–Kerwin–Ungar (RKU) equations can be applied to predict the performance of unconstrained-layer damping treatments. To use the analysis, it is necessary to consider the special case for which the constrained-layer thickness, $H_3$, is zero. This will reduce the earlier analysis of the three-layer system to a two-layer system. As $H_3$ approaches zero, the shear parameter $g$, approaches infinity, and eq. (6.1) can be simplified to

$$\frac{EI}{E_1 I_1} = 1 + e_2 h_2^3 + 3(1 + h_2)^2 \left( \frac{e_2 h_2}{1 + e_2 h_2} \right) \tag{6.15}$$

where $e_2 = E_2/E_1$ and $h_2 = H_2/H_1$.

If the loss factor of the damping material is $\eta_2$ and the system loss factor is $\eta$, equation (6.15) can be rewritten as

$$\frac{EI}{E_1 I_1}(1 + i\eta) = 1 + e_2 h_2^3(1 + i\eta_2) + 3(1 + h_2)^2 \left[ \frac{e_2 h_2(1 + i\eta_2)}{1 + e_2 h_2(1 + i\eta_2)} \right] \tag{6.16}$$

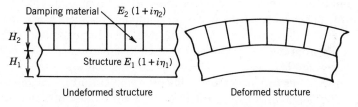

Damping material    $E_2 (1 + i\eta_2)$

$H_2$

$H_1$

Structure $E_1 (1 + i\eta_1)$

Undeformed structure          Deformed structure

**FIGURE 6.2.**   Unconstrained or free-layer damping treatment.

Equating the real and imaginary parts on both sides of equation (6.16), and noting that for all practical situations $(e_2 h_2)^2 \ll e_2 h_2$, the following two equations can be derived:

$$\frac{EI}{E_1 I_1} = \frac{1 + 4e_2 h_2 + 6e_2 h_2^2 + 4e_2 h_2^3 + e_2^2 h_2^4}{1 + e_2 h_2} \tag{6.17}$$

$$\frac{\eta}{\eta_2} = \frac{e_2 h_2 (3 + 6h_2 + 4h_2^2 + 2e_2 h_2^3 + e_2^2 h_2^4)}{(1 + e_2 h_2)(1 + 4e_2 h_2 + 6e_2 h_2^2 + 4e_2 h_2^3 + e_2^2 h_2^4)} \tag{6.18}$$

It is interesting to note that equations (6.17) and (6.18), for predicting the damping performance of free-layer treatments, are the same as those reported by Oberst [6.2]. Equation (6.17) is used to compute the frequency, whereas equation (6.18) is used to predict the damping performance.

In the following, most emphasis will be placed on the damping behavior of a beam coated on one side by damping material. In most of the discussions damping properties of a typical, commercially available, unconstrained-layer damping material will be used. The damping properties of the material are given in Data Sheet 010.

## 6.3.2.   Effects of Temperature

Since temperature is the most important single factor affecting the performance of any damping treatment utilizing viscoelastic materials, it is necessary to start discussion of the unconstrained-layer damping treatment by considering its dependence on temperature. By examining equation (6.18), it can be seen that the maximum loss factor is achieved by such a treatment only when the product of the real part of the modulus $(E_2)$ and the loss factor $(\eta_2)$ is a maximum. Since this product is a maximum in the transition region of the viscoelastic material, it is expected that the performance of the unconstrained-layer damping treatment, as a function of temperature, will be a maximum only in such a region.

Figure 6.3 illustrates the damping performance of a typical unconstrained-layer damping treatment, with temperature, for a fixed semiwavelength of vibration, for equal thickness of damping material and the aluminum structure. The particular material selected for this illustration has its optimum value around room temperature (Data Sheet 010). From Figure 6.3 it can be seen that the dependence of the free-layer damping treatment performance on temperature is very significant. Therefore it is very important to select and match the appropriate material for the desired environment over which it needs to operate. There are a number of commercially available materials whose transition regions occur over different temperatures, ranging from as low as $-200°$ F to as high as $2000°$ F. The properties of some of these materials are given in the next chapter.

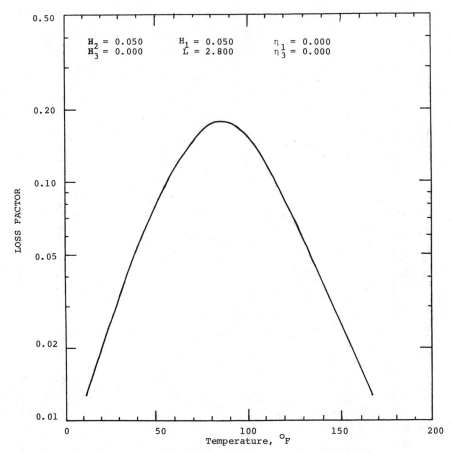

**FIGURE 6.3.** Variation of the damping performance of an unconstrained-layer treatment with temperature.

### 6.3.3. Effects of Thickness

Having selected a damping material with optimum damping capability over the desired temperature range of operation, the next variable to be investigated is the thickness of the damping material relative to that of the structure. Figure 6.4 illustrates the variation of the composite loss factor with temperature for different thicknesses of the damping material. It can be seen from this figure that by increasing the thickness of the viscoelastic material, higher damping can be achieved in the system. However, it is also evident that this increase in damping is not a linear function of thickness. This is because if the damping material thickness is much greater than that of the structure, the composite system damping approaches that of the viscoelastic material itself, and it is not possible to obtain higher composite damping than that of the viscoelastic material. On the other hand, for small thicknesss the increase in

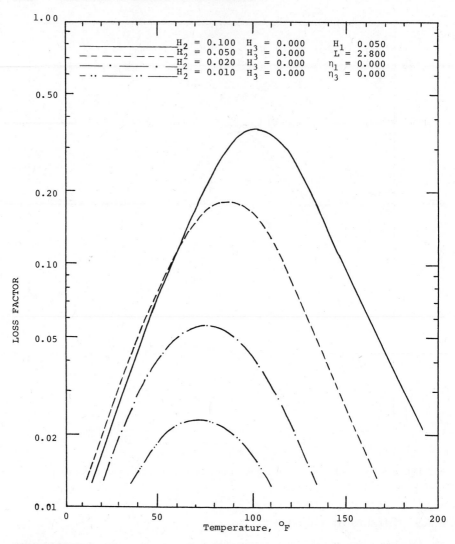

**FIGURE 6.4.** Variation of the damping performance of an unconstrained-layer treatment with temperature and thickness.

damping is almost linear with respect to the thickness; that is, if the thickness is doubled, the amount of damping in the system is also doubled. This rule is accurate only if the damping properties of the material do not change greatly with frequency, as will be discussed later.

Another interesting observation in Figure 6.4 concerns the temperature of peak damping for different thicknesses. As the thickness of the damping material increases, the temperature at which the damping is a maximum also increases. This can be explained by observing Figure 6.5, which illustrates the

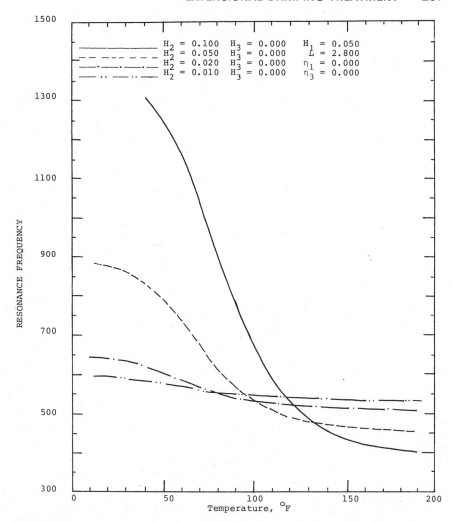

**FIGURE 6.5.** Variation of the resonant frequency of an unconstrained-layer treatment with temperature and thickness.

variation of the resonant frequency with temperature for the same thicknesses given in Figure 6.4. It can be seen that as the thickness increases, so does the resonant frequency, especially at low temperatures. Therefore, as the thickness of the unconstrained-layer treatment increases, the resonant frequency of the system increases, and the peak damping occurs at a higher temperature. This is because high frequency is equivalent to low temperature as far as the material properties are concerned. In other words, at higher frequencies the peak value of the product of the loss factor and the real part of the modulus of the material occurs at higher temperatures. Therefore the maximum damping will have to occur at those temperatures.

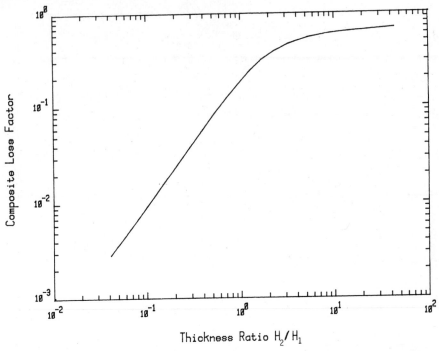

**FIGURE 6.6.** Variation of the maximum damping performance of an unconstrained-layer treatment with thickness.

Figure 6.6 illustrates the variation of the maximum loss factor achieved by an unconstrained-layer treatment with the ratio of damping material thickness to structure thickness. In this figure the maximum values of Figure 6.4 were used, as well as others. It should be noted that such maximum values occur at different temperatures and frequencies, as in Figures 6.4 and 6.5. Because of this some changes in the properties of the material occur between different configurations. For extremely low thickness ratios, the increase in damping is almost linear; however, at extremely high thickness ratios, the maximum damping tends to approach that of the viscoelastic material, which is approximately 0.8 for this example.

## 6.3.4. Effects of Initial Structural Damping

In the preceding discussion concerning the performance of an unconstrained-layer damping treatment with temperature, it was assumed that the initial structural damping was zero. However, in many cases the structural damping is significant and should be included in the analysis. This is particularly true for structures with many joints and/or subcomponents that are likely to contribute to the overall damping. For example, some aircraft structures of the riveted skin-stringer type, and some diesel engine components, could have modal loss factor values as high as 0.05. Hence, adding small amounts of an unconstrained-layer damping treatment to such a structure might not result in significant improvement beyond the 0.05 initial modal damping, and it would be necessary to utilize the material in much higher thicknesses for significant improvement to be achieved. This matter is illustrated in Figure 6.7 for $\eta_1 = 0.05$, as compared with Figure 6.4 for $\eta_1 = 0$.

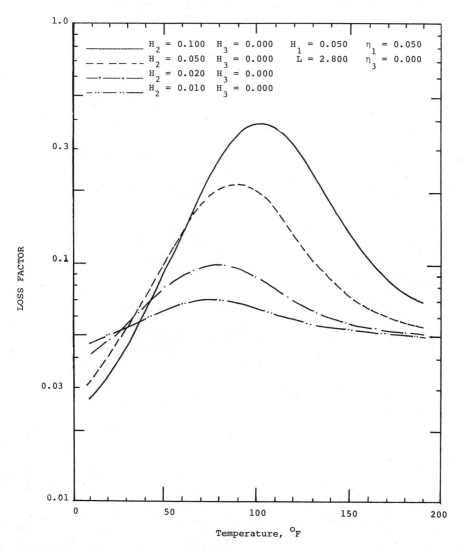

**FIGURE 6.7.** Effects of structural damping on the performance of an unconstrained-layer damping treatment.

### 6.3.5.   Effects of Frequency and Semiwavelength

The effect of frequency on the performance of an unconstrained-layer damping treatment is secondary to that of temperature. For many practical applications the effect is rather small and can be ignored unless there is a need to consider very wide frequency ranges.

Figure 6.8 illustrates the variation with temperature of the composite loss factor, for a given thickness beam, for two different semiwavelengths of vibration, 2.8 in. and 1 in. With a change of a semiwavelength of vibration from 2.8 in. to 1 in., the resonant frequency of the undamped system is increased

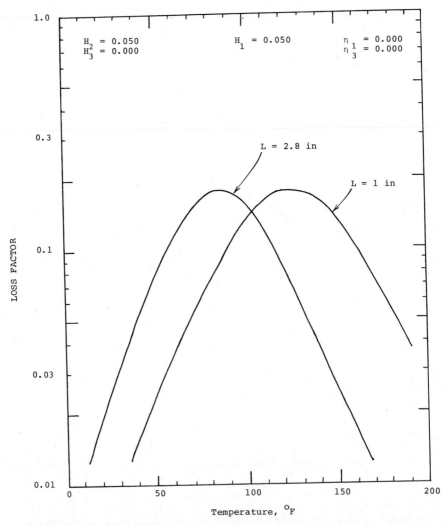

FIGURE 6.8.   Effect of frequency on the performance of an unconstrained-layer treatment.

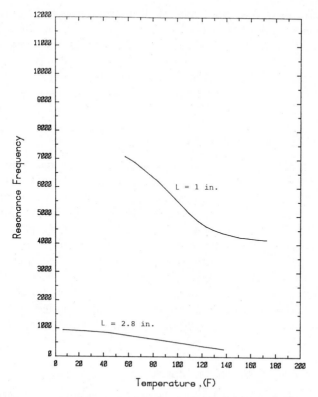

**FIGURE 6.9.** Variation of the resonant frequency with temperature for two different semi-wavelengths of vibration.

approximately eight times. However, it can be seen from Figure 6.8 that the damped performance for the 1 in. case is shifted to a higher temperature as compared to the 2.8 in. case. The reason for this shift, as before, is that higher frequencies are equivalent to lower temperatures as regards material properties. Figure 6.9 illustrates the true variation of the resonant frequency of the system with both temperature and wavelength of vibration. Note that for the extensional damping treatment, wavelength is important only to the extent that it affects modal frequencies.

## 6.3.6. Effects of Bonding Techniques

As might be expected, the performance of an unconstrained-layer damping treatment is only as good as the method of attachement to the structure. Some materials are self-adhesive and therefore are easy to attach to the structure. However, the majority of unconstrained-layer damping materials are not self-adhesive, and they require fastening to the structure. For most cases a structural epoxy layer is used.

A general rule of thumb for the adhesive layer between the unconstrained-layer damping material and the structure is that it be as thin and stiff as possible [6.3]. Otherwise, if a soft adhesive layer is used between the unconstrained layer and the structure, most of the shear deformation will occur in the adhesive layer, detracting from the performance of the damping material.

### 6.3.7.  Effects of Partial Coverage

Since high damping is usually best achieved when the damping material is subjected to large cyclic deformations, the location(s) of the damping material on the vibrating structure is important. Therefore a knowledge of the vibrational mode shapes of interest is necessary to determine the location where the maximum bending stresses occur. For the first mode of vibration of a simply supported beam, the maximum bending stress occurs around the center of the beam. Therefore, by concentrating the material around the center of the beam, better performance will be achieved than if the same amount of material is distributed uniformly over the beam [6.4]. Unfortunately for most cases it is not practical to do this because damping is usually required for many modes covering a wide frequency range, and conflicting optimum locations occur.

### 6.3.8.  Effects of Multiple Materials

Several attempts have been made to broaden the damping performance of the unconstrained-layer treatments with respect to temperature. Early work in this area has concentrated on slowing the rate of variation of the material properties in the transition region [6.5]. This is equivalent to widening the temperature range over which the material attains its maximum performance as an unconstrained-layer treatment, and this has been accomplished by varying the composition of the viscoelastic material, and/or the type of fillers used.

One of the easiest ways of broadening the temperature range over which high damping can be achieved, by an unconstrained-layer damping treatment, is by selecting several materials with different properties. These optimum properties must occur at two or more different temperatures. For instance, if the temperature range for which the treatment has to operate is from $-50°$ F to $150°$ F, it may be necessary to select one material that has its optimum value around $0°$ F and another material that has its optimum value at approximately $100°$ F. It will then be possible either to place one material layer on one side of the structure and one on the other or both on top of each other so that their performances are combined to give wide temperature coverage. To illustrate this point, the same unconstrained-layer damping material examined in previous sections will be considered and will be combined with another material having similar properties but shifted $75°$ F down in temperature. Figure 6.10 illustrates the combined performance of the two different materials.

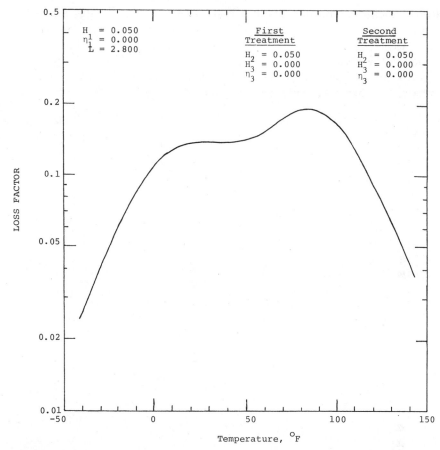

$H_1 = 0.050$
$\eta_1 = 0.000$
$L = 2.800$

First Treatment
$H_2 = 0.050$
$H_3 = 0.000$
$\eta_3 = 0.000$

Second Treatment
$H_2 = 0.050$
$H_3 = 0.000$
$\eta_3 = 0.000$

LOSS FACTOR

Temperature, $^\circ$F

**FIGURE 6.10.** Effect of multiple materials on the performance of an unconstrained-layer damping treatment.

As expected, good damping is achieved over a broader temperature range than for each material separately.

Other attempts have been made to combine the effects of multiple materials in one viscoelastic material. To do this, it is necessary to combine materials that are chemically incompatible. For example, nitrile-butadiene rubber, polyvinyl acetate, and polystyrene are three materials that are not chemically compatible. These materials, if mixed by a mechanical mill before curing, can produce a single macroscopically homogeneous material having three sets of properties merged in three adjoining temperature ranges. The reason for this is that the nitrile-butadiene rubber has its optimum damping below room temperature, the polyvinyl acetate has its optimum damping at approximately 150° F, and polystyrene has its optimum damping at approximately 300° F. Therefore, by combining these three materials, it is expected that three transition regions with temperature will occur, and high damping can be achieved around each of those temperatures [6.6]. Figure 6.11 illustrates the performance of such a blend with temperature. It can be seen that three distinct peaks exist where each of the materials goes through its transition region. Although the performance of such a treatment is very broad, the peak loss factor is not very high at any single temperature. This is the trade-off that has to be made when resorting to multiple materials.

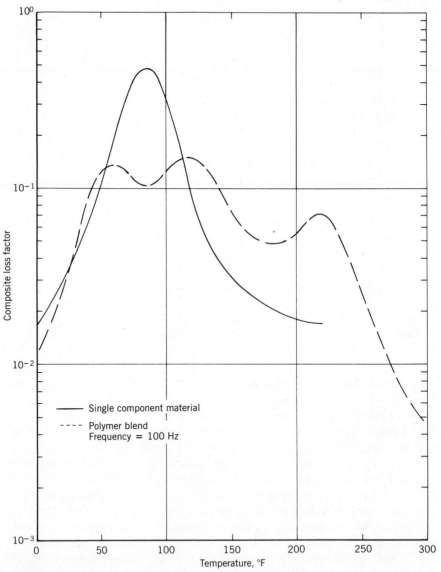

**FIGURE 6.11.** Effects of polymer blends on the performance of the unconstrained-layer damping treatment (refer to data sheets 010, 031).

## 6.4.  STIFFENED PLATE WITH EXTENSIONAL DAMPING TREATMENT

### 6.4.1.  Equations

If a beam or plate is stiffened by intermediate structural members, such as stringers, frames, or flexible supports, then the equations of Oberst do not adequately predict the modal damping and resonant frequencies. This is because of the significant amount of stored strain energy in each mode, contained by the additional structural members. This implies that the modal damping, defined as $\eta_{nm} = D_S/2\pi U_S$, must also be smaller, since the stored energy $U_S$ will be larger while the dissipated energy per cycle $D_s$ will not change greatly. Approximations for using an equivalent thickness of a simply supported plate that will yield the same frequency as that of a complex structure were discussed earlier. Another approach is to modify Oberst's equations to allow for the effect of stiffeneners, which can be done for some skin-stringer type structures [6.8–6.9]. The equations derived are

$$\eta_{nm} = \frac{\eta_2}{[1 + (A - 2 + \beta_{nm})/Be]} \tag{6.20}$$

$$\left(1 + \frac{\rho_2 H_2}{\rho H}\right)\left(\frac{f'_{nm}}{f_{nm}}\right)^2 = \frac{1 + (A - 2 + Be)}{\beta_{nm}} \tag{6.21}$$

where

$$A = \frac{(1 - h^2 e)^3 + (1 + [2h + h^2]e)^3}{(1 + he)^3} \tag{6.22}$$

$$B = \frac{(2h + 1 + h^2 e)^3 - (1 - h^2 e)^3}{(1 + he)^3} \tag{6.23}$$

$$\beta_{nm} = \frac{2\zeta_{nm}^4}{L^4 C} \iint_S \phi_{nm}^2 \, dx \, dy$$

$$C = \iint_D \left[\left(\frac{\partial^2 \phi_{nm}}{\partial x^2}\right)^2 + \left(\frac{\partial^2 \phi_{nm}}{\partial y^2}\right)^2 + 2v\left(\frac{\partial^2 \phi_{nm}}{\partial x^2}\right)\left(\frac{\partial^2 \phi_{nm}}{\partial y^2}\right)\right] dx \, dy \tag{6.24}$$

and $e = E_2/E$ and $h = H_D/H$. $\beta_{nm}$ is an extremely important parameter since it represents the effect of the structural stiffeners. Without going into details at this point, it is possible in principle to calculate $\beta_{nm}$ for a given mode of any given structure or to measure it by applying a known treatment and using the equations in reverse to calculate $\beta_{nm}$. Either way, the equations provide an

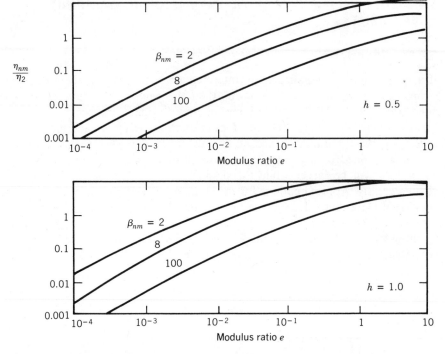

$\dfrac{\eta_{nm}}{\eta_2}$

$\beta_{nm} = 2$

8

100

$h = 0.5$

Modulus ratio $e$

$\beta_{nm} = 2$

8

100

$h = 1.0$

Modulus ratio $e$

**FIGURE 6.12.**   Variation of $\eta_{nm}/\eta_2$ with modulus ratio $e$ and $\beta_{nm}$.

extremely simple way of estimating the effects of uniform coverage of any treatment on any platelike structure.

When $\beta_{nm} = 2$, the equations (6.20) and (6.21) give essentially the same results as Oberst's equations, as would be expected. Figure 6.12 shows some calculated variation of $\eta_{nm}/\eta_2$ with $n$ and $e$ for various values of $\beta_{nm}$.

### 6.4.2.  Typical Test Results for a Multispan Stiffened Structure

In order to verify the foregoing analysis for multispan stiffened structures, a three-span skin-stringer structure was tested, as illustrated in Figure 6.13. The structure was first excited in the undamped state. The significant resonant

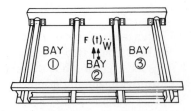

BAY ①   BAY ②   BAY ③   F (t) W

**FIGURE 6.13.**   Typical skin-stringer structure ($H_1 = 0.05$ in., 1.27 mm; $H_2 = 0.035$ in., 0.89 mm; $\beta_{nm} = 8$).

frequencies were observed to change only very slightly over the temperature range $-18°\text{C}$ $(0°\text{F})$ to $51°\text{C}$ $(125°\text{F})$.

The next tests were carried out with a single layer of damping material (LD-400) applied to the center bay only. Coupling between modes was quite strong, and hence the observed values of $\eta_{nm}$ for the first two modes behaved as shown in Figure 6.14, as a function of temperature. This variation bears little resemblance to that anticipated if no coupling were present and illustrates a problem involved in using partial treatment coverage in complicated structures.

The damping sheets were then attached to all three bays, and the tests were repeated. The variation of $\eta_{nm}$ with temperature is shown in Figure 6.15. The data are plotted for both modes and all three bays. If $\beta_{nm}$ is taken to be equal

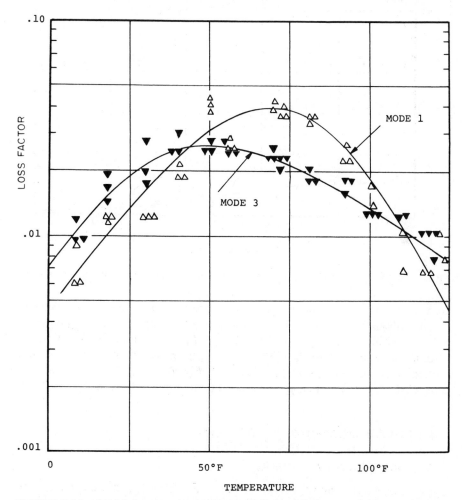

**FIGURE 6.14.** Modal damping versus temperature for partial coverage of three-panel system.

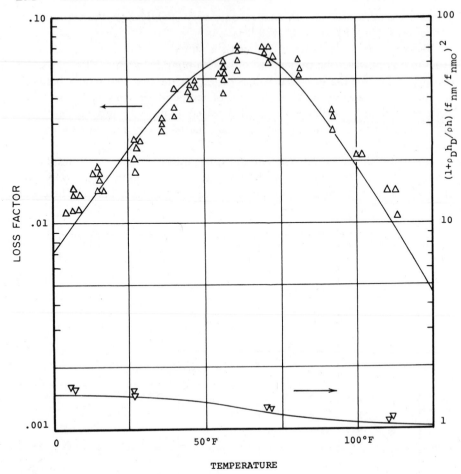

**FIGURE 6.15.** Modal damping versus temperature for full coverage of three-panel structure.

to 8, and if the same material properties are used, the analysis predicts the results shown in Fig. 6.15. The consistency of the agreement is quite satisfactory and indicates that a single parameter $\beta_{nm}$ for both modes, equal to 8.0, is sufficient to account for the observed damping and resonant frequencies. Also plotted is the nondimensional ratio $(1 + \rho_0 h_0/\rho h)(f_{nm}/f_{nm0})^2$ versus temperature, where $f_{nm0}$ in the undamped resonant frequency for the $n$, $m$ mode.

## 6.5. SHEAR DAMPING TREATMENT

### 6.5.1. Introductory Remarks

For a given weight, the shear type of damping treatment is more efficient than the unconstrained-layer damping treatment [6.4]. However, this efficiency is balanced by greater complication in analysis and application. The treatment is similar to the unconstrained-layer type, except the viscoelastic material is constrained by a metal layer. Therefore, whenever the structure is subjected to cyclic bending, the metal layer will constrain the viscoelastic material and force it to deform in shear. The shear deformation is the mechanism by which the energy is dissipated. This concept is illustrated in Figure 6.16.

To illustrate this case further, consider two extremes of the middle layer damping properties for Figure 6.16. At low temperatures, where the material is in its glassy region, both the structure and the constrained layer become rigidly coupled. In this case, whenever the system is subjected to cyclic bending, little shear deformation occurs in the middle layer, and hence the energy dissipation is also small. On the other hand, at high temperature, where the viscoelastic material is in its rubbery region and soft, both the structure and constrained layer become almost uncoupled. The energy dissipation in this case is also minimal, even though the shear deformation in the middle layer is high. This is because the shear modulus of the middle layer is low. Between

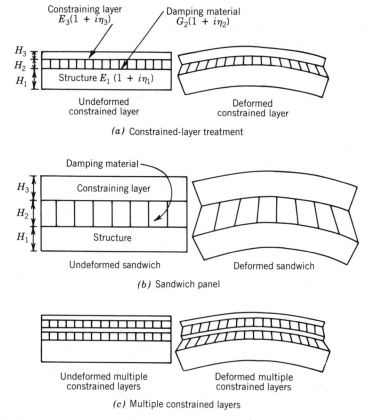

(a) Constrained-layer treatment

(b) Sandwich panel

(c) Multiple constrained layers

**FIGURE 6.16.** Various constrained damping treatments. (a) Constrained-layer treatment. (b) Sandwich panel. (c) Multiple-constrained layers.

these two extremes, the material possesses an optimal modulus value, so that the energy dissipation for the constrained layer goes through a maximum. The maximum shear deformation in the middle layer is a function of the modulus and the thickness of the constraining layer, the thickness of the damping layer, and the wavelength of vibration in addition to the properties of the damping material. The term that contains these variables is the shear parameter $g$, given in equation (6.5).

The variation of the loss factor with $g$ for a structure having a constrained layer treatment is shown in Figures 6.17 through 6.22 for different thickness ratios $h_2 = H_2/H_1$ and $h_3 = H_3/H_1$. The maximum loss factor for each shear parameter can now be plotted in terms of the thickness ratios as shown in Figures 6.23 and 6.24. It is evident from those plots that to fully utilize this type of treatment, the shear parameter must be selected so that the system loss factor is a maximum. At first sight this seems rather difficult to achieve because the shear parameter is a function of many variables. However, this should really be thought of as an advantage because of flexibility in the design process. As an example, for a given vibration problem only the wavelength of vibration is usually fixed. This leaves the type and thickness of both the damping layer and constraining layer as parameters that can be varied until the desired level of system damping is achieved.

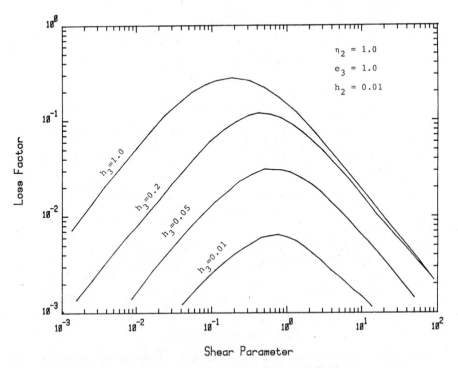

**FIGURE 6.17.**   Variation of the modal loss factor with the shear parameter.

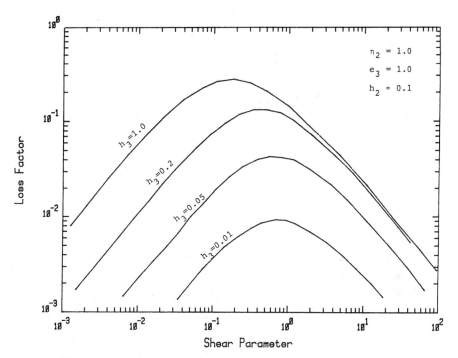

**FIGURE 6.18.** Variation of the modal loss factor with the shear parameter.

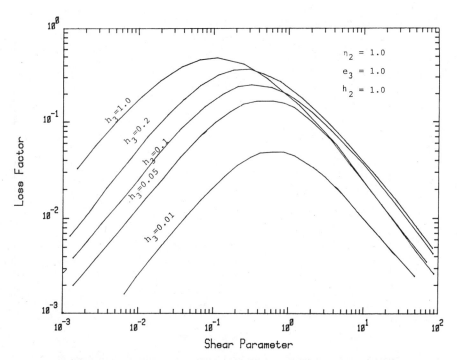

**FIGURE 6.19.** Variation of the modal loss factor with the shear parameter.

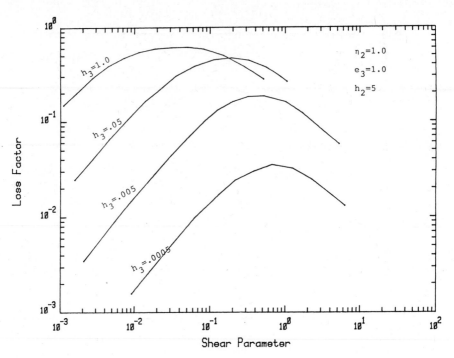

**FIGURE 6.20.** Variation of the modal loss factor with the shear parameter.

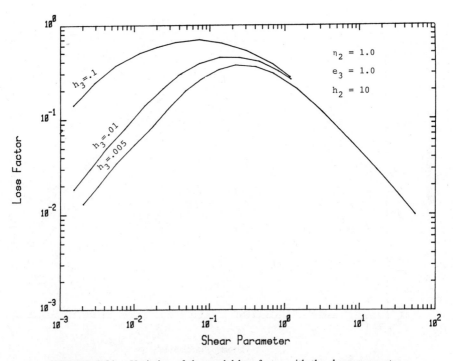

**FIGURE 6.21.** Variation of the modal loss factor with the shear parameter.

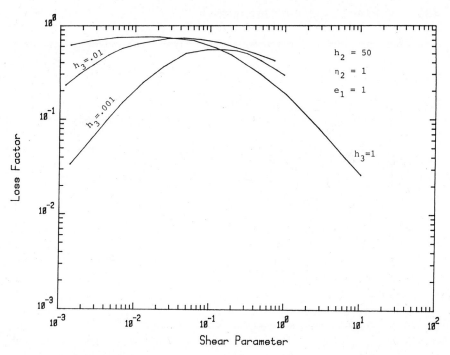

**FIGURE 6.22.** Variation of the loss factor with shear parameter.

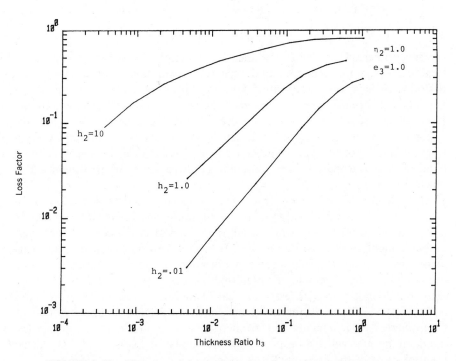

**FIGURE 6.23.** Variation of the maximum loss factor with the thickness ratio.

**283**

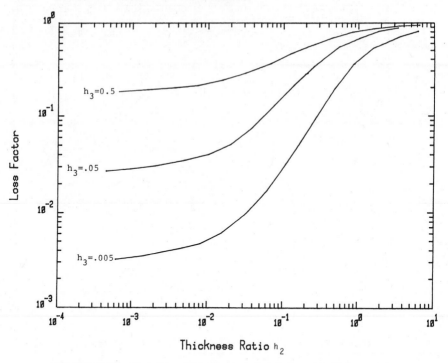

**FIGURE 6.24.**   Variation of the maximum loss factor with thickness ratio.

The performance of the constrained-layer damping treatment depends to a large extent on the geometry and type of constraining layer. Usually, it is desirable to have the constraining layer as stiff as possible to introduce the maximum shear strains into the viscoelastic layer. However, the constraining-layer stiffness should not normally exceed that of the structure. Therefore the maximum amount of shear strain is usually accomplished whenever the constraining layer is of the same type and geometry as that of the structure to be damped. This is usually referred to as the sandwich damping treatment and is illustrated in Figure 6.16*b*.

Multiple constrained-layer treatments are usually used as a means for increasing the damping introduced into a structure. Increasing the number of layers has the same effect as increasing the thickness of the constrained layer, but with slight variations. Figure 6.16*c* illustrates two constrained-layer treatments on a structure. The specific performance of such treatments will be discussed later, but it should be noted that the shear strain introduced into the second viscoelastic layer is less than that of the first layer.

To illustrate the performance of the constrained layer damping treatment with different environments and geometries using actual material properties, the damping material whose properties are shown in Data Sheet 018 will be used for most of the following discussion.

### 6.5.2. Effects of Temperature

As usual, temperature is the first parameter of importance with respect to the damping performance of a constrained-layer treatment. Figure 6.25 illustrates the variation of the composite loss factor with temperature for a selected configuration and semiwavelength of vibration. The specific damping material utilized in this plot is a pressure sensitive adhesive with optimum damping properties around room temperature. It is interesting to note that the variation with temperature for the structure with a constrained-layer treatment is very similar to that of the unconstrained-layer treatment. Again the maximum damping with temperature occurs in the transition region of the material. Therefore it should always be remembered that for a good material to work

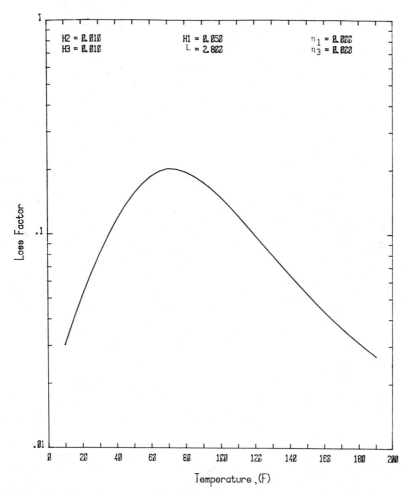

**FIGURE 6.25.** Variation of the damping performance of a constrained-layer treatment with temperature.

well in a constrained-layer configuration, its transition region with temperature must be within the operating temperature range of the structure.

### 6.5.3.  Effects of Thickness

The performance of a constrained-layer damping treatment depends not only on the thickness of the viscoelastic material, as in the case of the unconstrained layer treatment, but also on the thickness and type of the constraining layer. Figure 6.26 illustrates typical variation of the modal loss factor

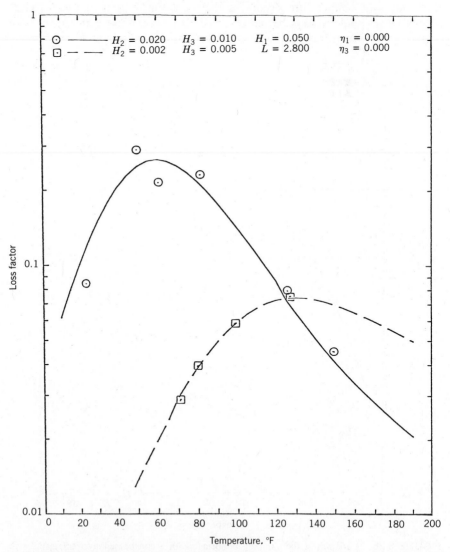

**FIGURE 6.26.** Variation of the damping performance of a constrained-layer treatment with temperature and thickness (⊙, ⊡-experiment).

with temperature for a number of different configurations. It can be seen that the temperature and magnitude of the maximum damping point changes significantly with geometry. This makes it more difficult to design a constrained-layer damping configuration that has its maximum damping occur at the operating temperature of the structure. A useful way of simplifying this task is to utilize what is known as a "Carpet Plot" [6.10]. In this plot only the maximum loss factor and the temperature at which it occurs for each configuration are plotted, as illustrated in Figure 6.27. It is then possible to determine what configuration(s) will give the desired damping at the desired temperature. Having identified the specific configuration, its complete performance with temperature can then be calculated, as illustrated in Figure 6.26.

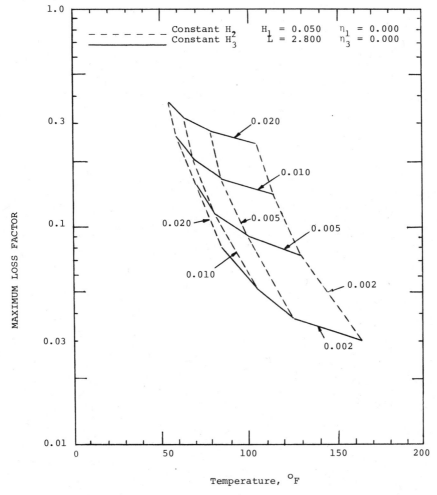

**FIGURE 6.27.**   Carpet plot of maximum composite loss factors versus temperature for different configurations.

## 6.5.4. Effects of Frequency and Wavelength

It is very difficult to separate the effects of frequency and wavelength on the performance of constrained-layer damping treatments. However, the direct effects of frequency are small compared to those of the wavelength of vibration because the properties of the material do not change very much with frequency, but the performance is a function of the square of the semiwavelength of vibration. Figure 6.28 represents the variation of the composite damping

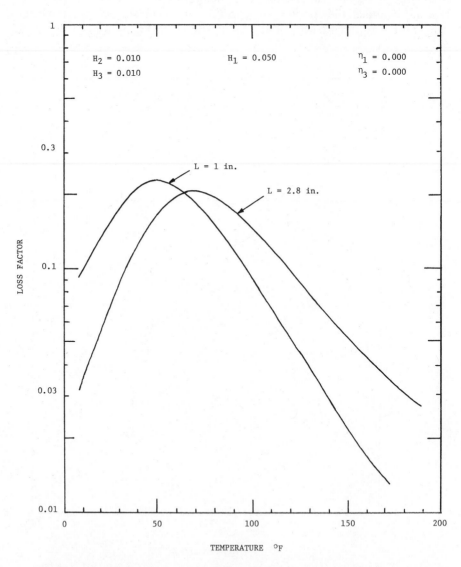

$H_2 = 0.010$   $H_1 = 0.050$   $\eta_1 = 0.000$
$H_3 = 0.010$   $\eta_3 = 0.000$

L = 1 in.

L = 2.8 in.

LOSS FACTOR

TEMPERATURE °F

**FIGURE 6.28.** Effects of wavelength on the performance of a constrained-layer treatment.

loss factor, with temperature, for two different semiwavelengths of vibration of 2.8 in. and 1 in, respectively. It can be seen from this figure that the peak loss factor occurs at a lower temperature for the lower semiwavelength of vibration. The phenomenon is related to the fact that lower wavelengths of vibration have higher modal stiffnesses associated with them, hence the constrained-layer treatment is capable of introducing the same shear strains at lower temperatures. Figure 6.29 illustrates the variation of the resonant frequency with temperature for the two configurations of Figure 6.26.

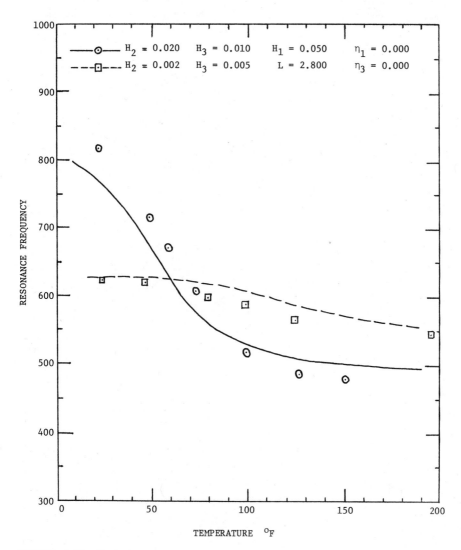

**FIGURE 6.29.**   Variation of the resonant frequency with temperature and thickness (⊙, ⊡-experiment).

### 6.5.5.   Effects of Initial Structural Damping

Typical effects of initial structural damping on the performance of a constrained-layer damping treatment are illustrated in Figure 6.30. This influence is very similar to that for the unconstrained-layer treatment. Compare with Figures 6.27 and 6.7.

### 6.5.6.   Effect of Constraining-Layer Damping

In some cases it is necessary to investigate the performance of the treatment when the constraining layer is itself damped. Figure 6.31 illustrates the effect of

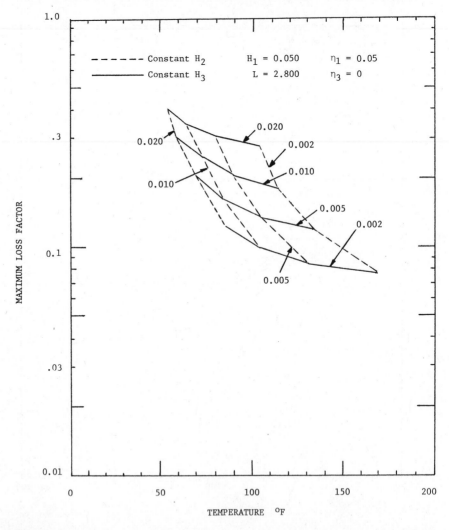

**FIGURE 6.30.** Effects of structural damping on the performance of a constrained-layer treatment (carpet plot).

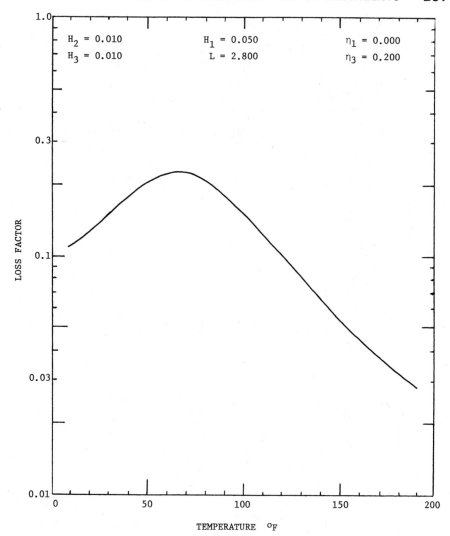

**FIGURE 6.31.** Effects of constraining-layer damping on the performance of the treatment ($\eta_3 = 0.2$).

a constraining-layer loss factor of 0.2, as compared with zero in Figure 6.28 ($L = 2.8$ in.).

## 6.6.  MULTIPLE CONSTRAINED-LAYER TREATMENTS

### 6.6.1.  Introductory Remarks

Multiple constrained-layer treatments are often used to increase the damping of structural applications [6.11, 6.12]. Usually, by increasing the number of layers, more damping can be introduced for a given mode of vibration. However, as a result of many tests on the performance of multiple constrained layers, it has been found that most of the shear deformation occurs in the first damping layer, closest to the structure. In other words, all subsequent layers work mainly to increase the stiffness of the constraining layer to which the first damping layer is subjected. This point is illustrated in Figures 6.32 and 6.33.

These figures illustrate the variation with temperature of the composite loss factor of a cantilever beam with different constrained-layer damping treatments. The results for two constrained-layer pairs, each consisting of a 0.002 in. damping layer and different aluminum constraining layers, are shown in Figures 6.32 and 6.33. It is seen that the different treatments give

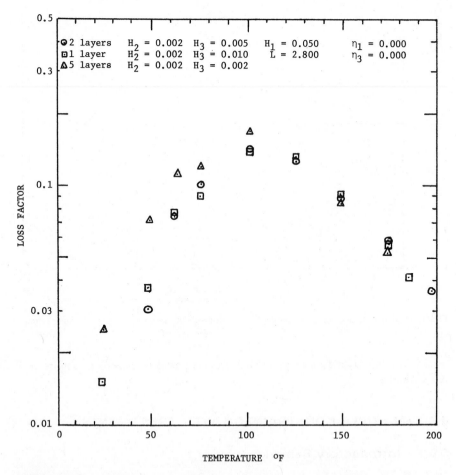

**FIGURE 6.32.** Variation of the performance of multiple constrained-layer treatments with temperature.

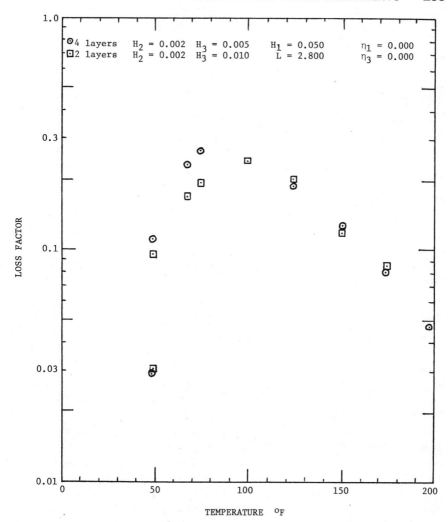

**FIGURE 6.33.** Variation of the performance of a constrained-layer treatment with temperature.

approximately the same damping levels because the net thickness of the aluminum constraining layers was kept the same. This implies that all layers beyond the first tended to work mainly by increasing the net stiffness to which the first layer was subjected. In Figure 6.32 the three different constrained-layer treatment geometries are one layer of 0.002 in. damping layer with 0.01 in. aluminum constraining layer, two layers each consisting of 0.002 in. damping layer and 0.005 in. aluminum constraining layer, and five layers each consisting of 0.002 in. damping layer and 0.002, in. aluminum constraining layers.

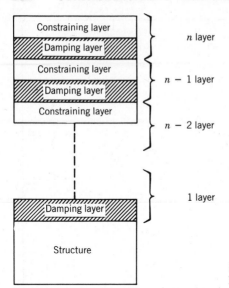

**FIGURE 6.34.** Elements of $n$ constrained layers on a structure.

As a good first approximation for predicting the performance of multiple constrained-layer treatment on structures, one may add the total thickness of all metal constraining layers and use this net thickness for a single constrained layer acting on the first adhesive layer. However, a better approximation can be achieved by modifying the analysis presented in Section 6.2.1. This can be done by using the three-layer equations and working from the last layer (away from the structure) back to the first layer. Figure 6.34 illustrates the case where $n$ constraining-layer pairs are used. Each of the $n$ layer pairs consists of a damping layer and a constraining layer. To start the analysis, one considers the $n$th constraining layer, the $n$th damping layer, and the $n - 1$th constraining layer. The RKU analysis is used to calculate the equivalent stiffness and loss factor of these three layers as if they were a sandwich configuration in themselves. The next step is to take the equivalent modulus and thickness of these three layers, just determined, and use it as a new equivalent single constraining layer acting on the $n - 1$th damping layer and the $n - 2$th constraining layer. By repeating the steps until the first damping layer is reached, the first damping layer will then be subjected to one equivalent constraining layer with a now-known equivalent modulus and thickness. Figure 6.35 represents some results for this analysis and the correlation with experimental measurements. It is seen that the correlation is very good.

### 6.6.2.  Multiple Materials

As in the case of the unconstrained-layer damping treatment, multiple materials can be used for constraining-layer damping treatments to broaden the temperature range. Figure 6.36 illustrates the variation of modal loss

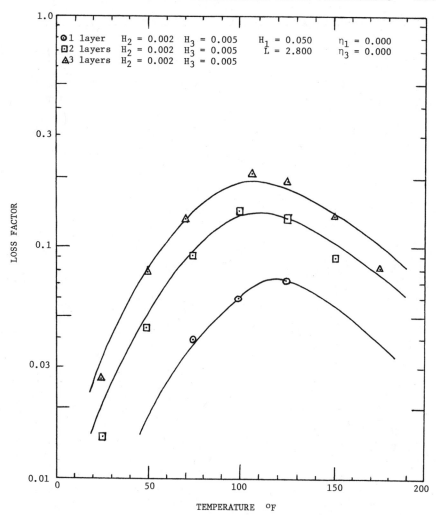

**FIGURE 6.35.** Variation of the performance of multiple constrained-layer treatment with temperature and number of layers.

factor with temperature when two different constrained-layer treatments are applied to alternate sides of a structure. The same effects can be achieved by stacking the layers one on top of the other as with multiple layers, but to observe two distinct peaks, it is necessary to use damping layers that have considerably different properties and in the proper stacking sequence.

### 6.6.3.  Equivalent Complex Modulus Concept

When several alternate layers of viscoelastic adhesive and metal are applied to the surface of a beam or plate, the equations used to define the behavior of a

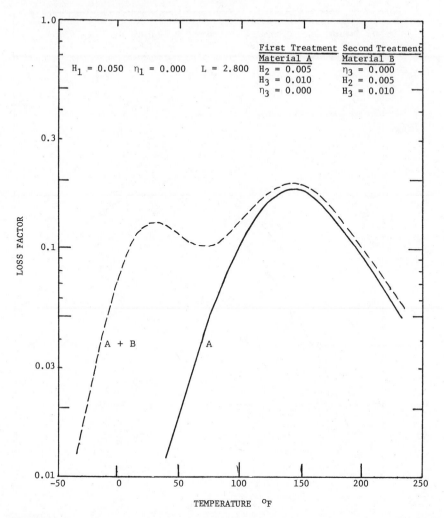

**FIGURE 6.36.** Effects of multiple materials on the performance of a constrained-layer treatment.

sandwich system can be adapted as just shown. However, an alternate approach can be adopted, namely to treat the system as an equivalent homogeneous system whose "average" properties depend on the detailed makeup of the actual treatment. This approach has two advantages: empirical measurements have shown the average complex modulus properties depend in a very simple manner only on a shear parameter $g_N = E_2 \lambda_n^2 / E_C H_C H_D N$ and on the thickness ratio $h = H_C/H_D$, and this equivalent homogeneous treatment can be treated in all respects as if it were a real free-layer treatment, so that all the equations and analyses developed for free-layer systems on simple and stiffened structures can be carried over directly [6.8, 6.12, 6.13].

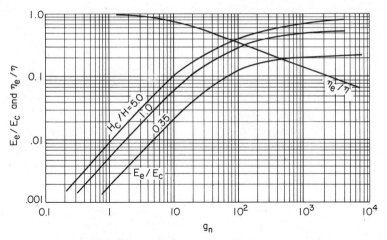

**FIGURE 6.37.** Graphs of $E_e/E_c$ and $\eta_e/\eta_2$ versus $g_N$.

The tests from which the equivalent complex modulus data for multiple layer treatments were derived were conducted on clamped-clamped and clamped-free beams with variations of the mode number $n$, constrained-layer thickness $H_C$, adhesive thickness $H_D$, number of layers $N$, temperature $T$, frequency $\omega$, with acrylic adhesive damping layers. For each test point the known complex modulus properties of the adhesive were used to determine $E_D$ and $\eta_D$ at each specific temperature and resonant frequency, and the shear parameter $g_N$ calculated. $\lambda_n$ was defined as the length of the pinned-pinned beam having the same resonant frequency as for each mode. From the measured modal damping $\eta_n$ in each mode, the resonant frequency $\omega_n$ of the damped beam, and $\omega_{n0}$ of the undamped beam, Oberst's equations were used in reverse to deduce $E_e$ and $\eta_e$ for the treatment. For each value of $h = H_C/H_D$, the graphs of $E_e/E_C$ and $\eta_e/\eta_D$ versus $g_N$ were found to lie along unique curves as shown in Figure 6.37. These curves can be represented empirically [6.13] by the equations

$$\frac{E_e}{E_c} = \left(\frac{2h}{1+2h}\right)\tanh\left(\frac{0.35g'_N}{1+0.3g'_N}\right) \tag{6.25}$$

$$\frac{\eta_e}{\eta_D} = 1 - e^{-2/3\sqrt{g_N}} \tag{6.26}$$

$$g'_N = \frac{g_N}{1+2.5e^{-2.5h}} \tag{6.27}$$

$$h = \frac{H_C}{H_D} \tag{6.28}$$

$$g_N = \frac{3G_2\lambda_n^2}{E_C H_C H_D N} \tag{6.29}$$

### 6.6.4. Illustration for Multiple-Layer Treatment on Clámped-Clamped Beam (Single Material)

The tests were carried out on a beam 7 in. (177.8 mm) long by 0.05 in. (1.27 mm) thick by 1 in. (25.4 mm) wide, of aluminum alloy, bonded to a large rigid block at each end. Two adhesives were used: (1) 3M-467 (Data Sheet 034) and (2) 3M-428 (Data Sheet 035) with $H_C = 0.002$ in. (0.051 mm) and 0.005 in. (0.127 mm), $H_D = 0.002$ in. (0.051 mm), and various values of $N$, the number of layer pairs. From the response spectra at several temperatures, the modal damping $\eta_n$ was determined and plotted as in Figure 6.38 for each material. The calculated values of $\eta_n$ and $\omega_n$ were determined using the equivalent free-layer approach. It is seen that the agreement is good.

### 6.6.5. Application of Equivalent Free-Layer Approach for Multiple Material Treatments

If the physical dimensions of the treatment remain unchanged, but one or more of the adhesive layers comprises a different adhesive, then a very complex dynamical system results, which can be analyzed only with great difficulty. Experimental investigations are as readily conducted as for the single adhesive case, however. The three systems that were investigated are illustrated in Figure 6.39. System A consists of alternate layers of adhesives 1 and 2, separated by aluminum constraining layers. System B consists of several layers of treatment 1 with adhesive 2 added on top. System C is the opposite of system B.

If the treatment depicted as type B in Figure 6.39 is applied to the structure, a fairly simple analysis is applicable, provided that the softer treatment is on the "outside." Otherwise, the "outer" treatment will act itself as a constraining layer for the "inner" treatment, and complex shearing actions, qualitatively comparable to those occuring in a conventional constrained-layer treatment, will take place. This will completely invalidate the assumption that the treatments may be regarded as equivalent free-layer treatments. For the case when the softer treatment is on the "outside," however, no such problem occurs.

### 6.6.6. Illustration of Multiple Material Treatment on Clamped-Clamped Beam

The treatments were applied to the beam in the same way as before, and the modal damping and resonant frequencies were measured in the usual way. The results for configurations A, B, and C are shown in Figure 6.39.

The analytical results for treatment type B are seen to be in good agreement with experiment. No analytical results are available for treatment A; however, it is seen that treatment A behaves in a comparable manner to

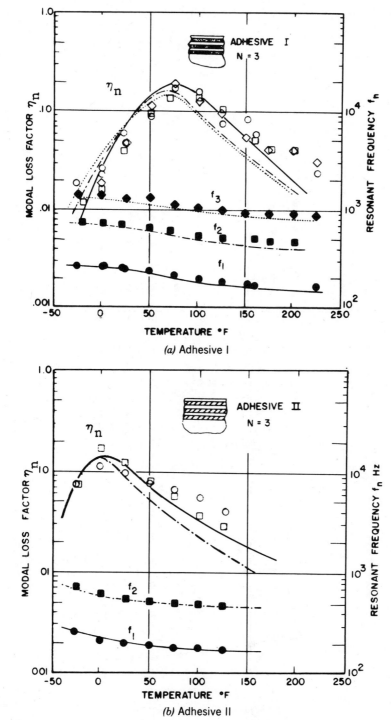

**FIGURE 6.38.** Multiple layer treatment on a clamped-clamped beam ($H_C = 0.005$ in., $H_D = 0.002$ in., (a) Adhesive I, (b) Adhesive II).

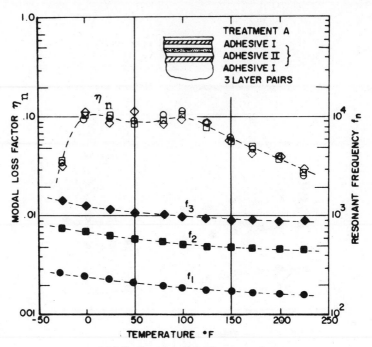

(a) Adhesives I and II, Treatment A

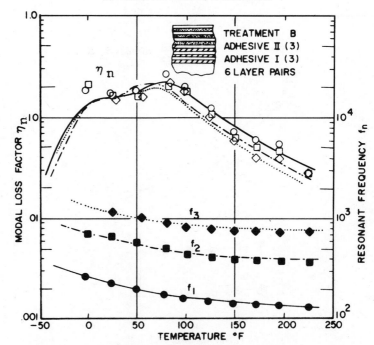

(b) Adhesives I and II, Treatment B

**FIGURE 6.39.** Graphs of $\eta_n$ and $f_n$ versus temperature for multiple materials ($H_C = 0.005$ in., $H_D = 0.002$ in.).

**300**

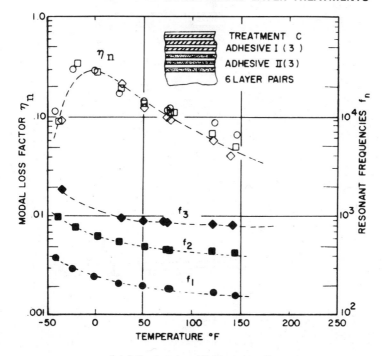

*(c)* Adhesives I and II, Treatment C.

**FIGURE 6.39.** Graphs of $\eta_n$ and $f_n$ versus temperature for multiple materials ($H_C = 0.005$ in., $H_D = 0.002$ in).

treatment B on a per-unit weight basis. Measured results for case C are far poorer, indicating a very ineffective treatment arrangement.

## 6.6.7. Multiple Constrained-Layer Treatments on Stiffened Structure

When a free-layer or equivalent free-layer treatment is applied uniformly to the surface of a stiffened plate structure, the usual equations for a beam are replaced by the more general equations (6.20) to (6.24). It has already been shown that, for the particular structure investigated, $\beta_{nm} = 8.0$. In the same way as before, therefore, we may estimate $\eta_{nm}$ and $\omega_{nm}$ as a function of temperature for the same treatments as used on the beam. Figure 6.40 shows measured and predicted modal damping for the single adhesive type of treatment on the structure, with adhesive system 1 (3M-467) and also the modal damping and resonant frequencies for the structure with a treatment of type B added, adhesive system 1 being in the first three-layer pairs and adhesive system 2 in the outer three-layer pairs. It is seen that analysis and experiment are in good agreement and that the temperature range of useful damping is much broader than for the single adhesive system.

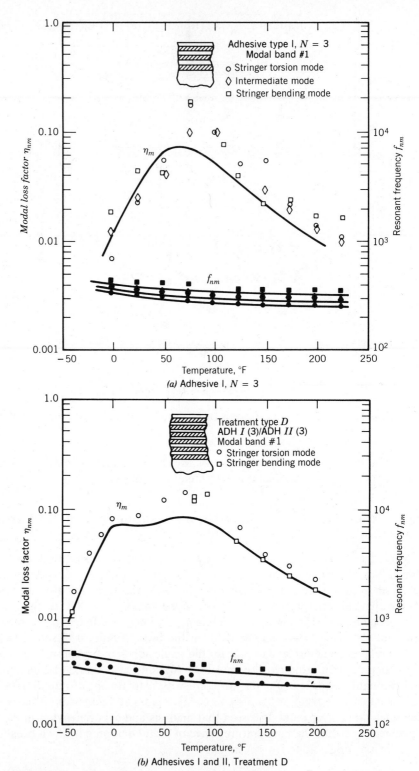

**FIGURE 6.40.** Graphs of $\eta_{nm}$ and $f_{nm}$ versus temperature for single and multiple materials ($H_C = 0.005$ in., $H_D = 0.002$ in.).

## 6.7. APPLICATION OF ANALYSIS TO MEASURE DAMPING PROPERTIES

### 6.7.1. Introductory Remarks

The analysis presented in Section 6.1 can be used to calculate the damping properties of rubberlike materials from measurements made on the composite beam system. To use the analysis in its general form, iterative procedures must be used because the equations are coupled. However, for special cases the equations can be simplified, and only these will be presented here.

The technique is based on utilization of the damping material in conjunction with vibrating metal beams either in shear or extension. If the material is applied to the outside of the metal beam, then the properties can be obtained in extension, whereas if the material is sandwiched between two metal beams, then the properties can be obtained in shear. By studying the various resonances of the damped beam, the effect of frequency on the properties of the material can be established. Also, by placing the system inside an environmental chamber, the effects of temperature can be investigated. More information on this subject can be found in ASTM E75 G-80 Standard Method for "Measuring Vibration-Damping Properties of Materials."

Although the vibrating beam technique can be used to make measurements on damping materials over quite wide temperature and frequency ranges, it does not provide the properties continuously as a function of these two parameters. To surmount this difficulty, the temperature-frequency superposition principle can be used to correlate the properties of the material for all measured temperatures and frequencies.

### 6.7.2. Description of Specimens

Although any type of beam specimen can in principle be used to make measurements on the damping properties of materials, previous experience has concentrated mainly on cantilever beams. This is because of the simplicity of the setups for such beams and also because at least one end can be simulated properly, namely the free end. This implies that careful attention must be applied to ensure that the boundary condition at the other end is fully fixed. The same approach is used for other types of beams, with clamped-clamped, simply supported, or free-free boundary conditions. In making damping measurements on materials, using vibrating beam tests, it is necessary to apply the damping material on the metal beam and make measurements on the composite system. Usually the damping material is combined with a metal beam in different configurations depending on the properties of the material of interest. The main combinations, illustrated in Figure 6.41 are the following:

1. The homogeneous beam. This type is used for measuring the damping properties of metal alloys and composites. Such materials are usually stiff on their own and do not require combination with a metal beam.

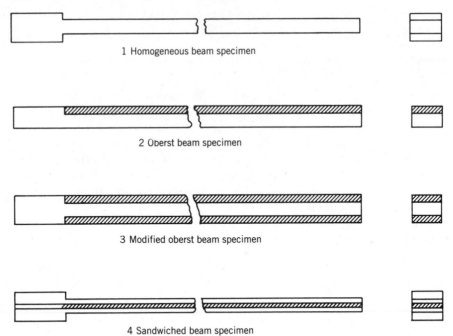

**FIGURE 6.41.**   Beam specimens (1) Homogeneous beam specimen. (2) Oberst beam specimen. (3) Modified Oberst beam specimen. (4) Sandwich beam specimen.

2.  The Oberst or externally coated beam. This type is named for Dr. Oberst [6.2] who developed this technique for making measurements on the damping properties of stiff damping materials under extensional deformation.

3.  The symmetric free-layer beam. This is a modified Oberst beam, where the material is coated on both sides. Again the properties are determined for extensional deformation. This beam allows for a simplification in the equations and data reduction, and it circumvents some experimental difficulties.

4.  The sandwich beam. This technique is used to determine the damping properties of materials under shear deformation and is usually employed for soft damping materials.

5.  The symmetric constrained-layer system. This technique yields the damping properties of soft materials under shear deformation as for the sandwich beams, but with simplified equations (not illustrated).

From Figure 6.41 it can be seen that all beams have massive roots, necessary to simulate properly the fixed boundary condition at the root of the cantilever beam, which is usually clamped in a rigid fixture. These roots can be either integrally machined as part of the beam, welded to the beam, or epoxied to the beam. It is emphasized that for most measurements the roots are essential for generating useful and accurate data.

The damping material is usually bonded to the metal beam by an epoxy adhesive, which must have a modulus higher than that of the viscoelastic material. Also the thickness of the epoxy layer must be kept to a minimum in general and small in comparison with the thickness of the damping material. If these two rules are not followed, deformation will occur in the epoxy layer as well as in the damping layer, and erroneous data will result. In some cases the damping material is, or can be, of the self-adhesive type, and in that case there will be no need of an epoxy layer.

The type of metal used for the beams is usually steel or aluminum. Typical dimensions, used previously and found to be successful, are a width of 0.5 in., a free length of 7 to 10 in., a root length of 1 in., and a thickness of 0.04 to 0.125 in. The thickness of the damping material can vary according to the specific properties of the material and the temperatures and frequencies of interest. These conditions will be discussed later in the text.

### 6.7.3.  Instrumentation and Setup

When using the beam specimens of Figure 6.41 to make measurements of the damping properties of materials, one usually requires two types of transducers. One is needed to apply the excitation force, and the other to measure the response of the beam. Since, when making damping measurements, it is necessary to minimize all extraneous sources of damping apart from the material to be investigated, it is essential to use transducers of the noncontacting type. Such transducers can be of the electromagnetic type and can be used directly with steel beams. For stainless steel or aluminum beams it is necessary to place a small soft iron disc on the beam to permit the transducer excitation.

The beam is mounted in a heavy, rigid, fixture that can provide sufficient clamping force around the root of the beam to simulate a fixed end boundary condition. The transducers are positioned approximately 1 mm away from the beam, as illustrated in Figure 6.42. Either a sinusoidal or random signal can be applied to the excitation transducer via a power amplifier. The response of the beam is measured by the second transducer and is processed through signal conditioning equipment, and the resulting response is stored in a computer or on an $X$-$Y$ recorder plot for later analysis. The setup is usually mounted inside an environmental chamber in order to study such environmental factors as temperature and vacuum. A schematic of this setup is also illustrated in Figure 6.42.

### 6.7.4.  Data Reduction

For all types of beams the response of the specimen is measured as a function of frequency. From such a response spectrum the frequencies and damping values of the various modes of vibration of the specimen are determined at

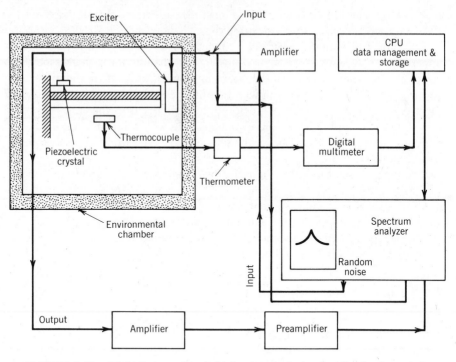

**FIGURE 6.42.** Block diagram of typical laboratory test setup for vibrating beam tests.

each selected temperature. By varying the temperature of the environmental chamber, these tests are repeated at various temperatures to investigate the effects of temperature. The response of the undamped beam is also needed. Before these data can be processed, it is necessary to have three sets of information, as follows:

1. The frequency and the loss factor of the damped beam for each mode of vibration.
2. The frequencies of the undamped beam for the same modes of vibration as for the damped beam.
3. The geometrical properties of the damped and undamped beams, along with the densities of the materials.

The following data-reduction process is used to calculate the damping properties of materials from measurements conducted on the four types of beams outlined previously.

**Homogeneous beam technique.** The loss factor $\eta$ and Young's modulus $E$ of the damping material can be calculated from

$$\eta = \frac{\Delta f_n}{f_n} \tag{6.30}$$

and

$$E = \frac{\rho b H L^4 \omega_n^2}{(a_n^2 I)} \tag{6.31}$$

where $f_n$ is the $n$th natural frequency, $\omega_n$ is the $n$th circular frequency $(2\pi f_n)$, $\rho$ is the mass density, $L$ is the length of beam, $a_n$ is the characteristic number of the $n$th mode $(= \xi_n^2)$, $I$ is the second moment of area, and $\Delta f_n$ is the half-power bandwidth.

**Oberst beam technique.**    The properties of the damping material are computed from the measured values of the composite beam using equations (6.17) and (6.18) and rewriting them to give

$$\left(\frac{\omega_n}{\omega_{1n}}\right)^2 (1 + h_2 \rho_r) = \frac{1 + 2e_2 h_2(2 + 3h_2 + 2h_2^2) + e_2^2 h_2^4}{1 + e_2 h_2} \tag{6.32}$$

$$\frac{\eta_n}{\eta_2} = \frac{e_2 h_2(3 + 6h_2 + 4h_2^2 + 2e_2 h_2^3 + e_2^2 h_2^4)}{(1 + e_2 h_2)(1 + 4e_2 h_2 + 6e_2 h_2^2 + 4e_2 h_2^3 + e_2^2 h_2^4)} \tag{6.33}$$

from which $e_2$ and $\eta_2$ can be calculated, where $e_2 = E_2/E_1$, $E_1$ is the Young's modulus of the metal beam, $h_2 = H_2/H_1$, $H_2$ is the thickness of the damping material, $H_1$ is the thickness of the metal beam, $\rho_r = \rho_2/\rho_1$, $\rho_2$ is the density of the damping material, $\rho_1$ is the density of metal beam, $\eta_n$ is the loss factor of the composite beam in the $n$th mode, $\eta_2$ is the loss factor of the damping material, $\omega_{1n}$ is the $n$th circular frequency of the metal beam, and $\omega_n$ is the $n$th circular frequency of composite beam.

**Symmetric free-layer beam technique.**    Although the analysis of Section 6.1 can be modified for the symmetric beam case, it is considerably easier to use the relationship

$$EI(1 + i\eta) = 2E_2 I_2(1 + i\eta_2) + E_1 I_1 \tag{6.34}$$

Using equation (6.34), the resulting equations for such symmetric beams become, after some simplification,

$$E_2 = \frac{[(\omega_n/\omega_{1n})^2(1 + 2h_2 \rho_r) - 1]E_1}{8h_2^3 + 12h_2^2 + 6h_2} \tag{6.35}$$

$$\frac{\eta_2}{\eta_n} = \frac{E_1}{(8h_2^3 + 12h_2^2 + 6h_2)E_2} + 1 \tag{6.36}$$

**Sandwich beam technique.** For the sandwich beam we note that $E_3 = E_1$ and $H_3 = H_1$, so that

$$H_{21} = \frac{(H_1 + H_2)}{2} \tag{6.37}$$

$$H_{31} = H_1 + H_2 \tag{6.38}$$

$$D = (H_1 + H_2)\frac{g}{1 + 2g} \tag{6.39}$$

At first sight the relation for $D$ does not seem right because $D$ is the distance between original location of the neutral axis of the structure and the new location of the neutral axis of the structure with the constrained layer. For sandwich beams $D$ should be equal to $(H_1 + H_2)/2$, which is different from the relation given by equation (6.39). However, the relation is consistent with the analysis for the following two reasons.

1.  For large values of $g$, when there is rigid coupling between the structure and the constrained layer, $D$ does approach $(H_1 + H_2)/2$.
2.  For small values of $g$, where there is little or no coupling between the two outer layers, there is minimal shift of the neutral axis location so that as $g \to 0$, $D \to 0$, as it should.

Therefore the value of $D$ is not only complex but also variable and is dependent on the geometry and, more important, the material properties. Using the foregoing relations and noting that for most rubberlike materials $E_2 \ll E_1$, equation (6.1) can be written:

$$EI(1 + i\eta) = \frac{E_1 I_1}{6} + E_1 H_1 (H_1 + H_2)^2 \frac{g(1 + i\eta_2)}{1 + 2g(1 + i\eta_2)} \tag{6.40}$$

Equating the real and imaginary parts of equations (6.40) yields the following relations for the shear modulus and loss factor of the damping material, along with the following values of $a_n = \xi_n^2$;

| $n$ | Clamped-Free | Clamped-Clamped | Free-Free |
|---|---|---|---|
| 1 | 3.516 | 22.373 | 0 |
| 2 | 22.035 | 61.673 | 0 |
| 3 | 61.697 | 120.90 | 22.373 |
| 4 | 120.90 | 199.86 | 61.673 |
| 5 | 199.86 | 298.56 | 120.90 |
| 6 | 298.56 | — | 199.86 |
| ⋮ | ⋮ | ⋮ | ⋮ |
| $n$ | $(2n - 1)^2 \pi^2/4$ | $(2n + 1)^2 \pi^2/4$ | $(2n - 3)^2 \pi^2/4$ |

Therefore

$$G_2 = \frac{(A - B) - 2(A - B)^2 - 2(A\eta)^2}{(1 - 2A + 2B)^2 + 4(A\eta)^2} \times \frac{E_1 H_1 H_2 a_n}{L^2} \tag{6.41}$$

$$\eta_2 = \frac{A\eta}{A - B - 2(A - B)^2 - 2(A\eta)^2} \tag{6.42}$$

where

$$A = \left(\frac{\omega_n}{\omega_{1n}}\right)^2 (2 + \rho_r h_2)\left(\frac{B}{2}\right) \tag{6.43}$$

$$B = \frac{1}{6(1 + h_2)^2} \tag{6.44}$$

$G_2 = $ the shear modulus of the damping material

**Symmetric constrained layer.**  By applying a constrained layer on each side of a cantilever beam, the neutral axis position remains truly unchanged, and therefore $D$ becomes exactly zero. Making this simplification, along with the fact that $E_2 \ll E_1$, equation (6.1) becomes

$$EI(1 + i\eta) = E_1 I_1 + 2E_3 I_3 + 2E_3 H_3 H_{31}^2 \frac{g(1 + i\eta_2)}{1 + g(1 + i\eta_2)} \tag{6.45}$$

Equating the real and imaginary parts, we get

$$g = \frac{+\alpha}{\alpha^2 + \beta^2} - 1 \tag{6.46}$$

$$\eta_2 = \frac{\beta}{g(\alpha^2 + \beta^2)} \tag{6.47}$$

where

$$\alpha = \frac{1}{B}\left[\frac{EI}{E_1 I_1} - A\right] \tag{6.48}$$

$$\beta = \frac{\eta}{B}\frac{EI}{E_1 I_1} \tag{6.49}$$

$$A = 1 + 2e_3 h_3^3 + B \tag{6.50}$$

$$B = 6e_3 h_3 (1 + 2h_2 + h_3)^2 \tag{6.51}$$

$$g = \frac{G_2 L^2}{E_3 H_3 H_2 a_n} \tag{6.52}$$

Also

$$\frac{EI}{E_1 I_1} = \left(\frac{\omega_n}{\omega_{1n}}\right)^2 \left(1 + 2\frac{\rho_2 h_2}{\rho_1} + 2\frac{\rho_3 h_3}{\rho_1}\right) \qquad (5.53)$$

Since the properties of damping materials can be computed for different modes of vibration, and at different temperatures, the effects of both frequency and temperature can be established.

Usually the loss factor of the composite beam is measured by the half-power bandwidth method. This is not the only method to measure the damping for a given resonant condition of the beam, and other techniques can be used equally well, within their limitations, such as decay, modal curve fitting, and Nyquist diagrams.

### 6.7.5.  Assumptions and precautions

**Assumptions.**   As with all techniques and analyses certain important assumptions have to be made and are listed here. They should be kept in mind when making damping measurements on materials.

1.   All damping measurements made by beam tests must be made in the linear range. It is important to select a force level that will be meaningful, because the analysis for the data reduction will otherwise not be applicable.

2.   It is important to keep the amplitude of the force signal, applied to the excitation transducer, constant with frequency. This is particularly important when making bandwidth measurements on modes of vibration with high damping. If the force cannot be kept constant, then the response of the beam must be divided by the input force in order to obtain the normalized transfer functions. The force can be measured by using the current signal applied to the force transducer, since the two are proportional.

3.   For the Oberst and symmetric free-layer specimens the analysis is the classical analysis for beams. It does not include the effects of rotatory inertia or shear deformation. The analysis assumes that plane sections remain plane, and therefore care must be taken not to use specimens with a damping material thickness that is too much higher than that of the metal beam.

4.   The equations presented for computing the properties of damping materials in shear do not include the extensional terms of the damping layer. This is a good assumption as long as the stiffness of the damping layer is considerably lower than that of the metal beam. Also these equations were developed and solved using sinusoidal expansions for the mode shapes. For cantilever beams this approximation is good only for the higher modes. For the first mode it does not apply, and an equivalent wavelength of vibration must be assumed empirically to generate useful data. It has been the common practice to ignore the first mode results for sandwich cantilever beams.

5.  The loss factor of the metal beams is assumed to be zero. This is usually a good assumption because steel and aluminum have loss factors of the order of 0.001 or less, which is significantly lower than those of the composite damped beams. This point should be kept in mind if other beam materials are used, such as plastic or epoxy.

**Precautions.**    The following is a list of precautions that must be observed concerning the geometry of the specimen, instrumentation, and assumptions mentioned before. It is very important to pay close attention to these precautions, in order to obtain good measurement at all times and for all conditions.

1.  The first point is rather general in nature. Making measurements on the damping properties of materials is usually a very complex and involved process because it involves materials, vibration, instrumentation, computers, and so on. It is necessary to pay close attention to all details of the experiment from the fabrication of the specimen to the actual data reduction and analysis. Good and careful experimentalists are needed to work successfully in this area.

2.  The first mode of vibration of the cantilever beam should not be used to calculate the properties of the material being investigated. This precaution is necessary because in the first mode the high amplitudes that usually result may introduce nonlinear effects into the measurements. Also the assumptions made in the analysis of the sandwich beams are not well duplicated for this mode.

3.  The damping of the metal beam must be very small. This is usually true because metal damping is considerably lower than that of good damping materials.

4.  When applying a good damping material on a metal beam, the resulting response is usually well damped and the signal-to-noise ratio is not very high. Therefore it is important to select an appropriate ratio of the thickness of the damping material to that of the metal beam so as to end up with moderate amounts of modal damping. Also extremely low damping in the system should be avoided because the differences between the damped and undamped systems will then be too small.

5.  With the exception of the homogeneous beam technique, all other techniques are based on measurements on the damped and undamped systems and making use of the differences, which are often small. The expressions for such differences are in the denominator of the equations used for data reduction, so if small errors are made during the measurements, or in the geometry of the beam, then those small errors will be magnified considerably during the data reduction processes. This will lead to erroneous results. As a general guideline to prevent such conditions from occurring, the following precautions are usually recommended:

   a.  For the Oberst beam the term $(\omega_n/\omega_{1n})^2(1 + \rho_r h_2)$ must be at least equal to or greater than 1.1.

b.  For the symmetric free-layer beam the same precaution as that of the Oberst beam applies for the expression $(\omega_n/\omega_{1n})^2(1 + 2\rho_r h_2)$ $\geq 1.10$.

c.  For the sandwich beam technique the same precaution applies for the expression $(\omega_n/\omega_{1n})^2(2 + \rho_r h_2)$.

d.  For the symmetric constrained-layer beam the same precautions apply for the term $(\omega_n/\omega_{1n})^2[1 + 2(\rho_2/\rho_1)h_2 + 2(\rho_3/\rho_1)h_3]$.

6.  The Oberst and symmetric free-layer beam techniques are usually used for stiff materials (Young's modulus of $10^4$ psi or higher), where the properties are measured in the glassy and transition regions. These materials are usually applied as free-layer treatments and include enamels and loaded vinyls. The sandwich and symmetric constrained-layer beam techniques are usually used for softer viscoelastic materials, with moduli of the order of $10^5$ psi or less.

## 6.8.  APPLICATIONS AND EXAMPLES

### 6.8.1.  Development of a Constrained-Layer Treatment for Vibration Control in an Aircraft Weapons Dispenser [6.8, 6.9]

This vibration induced failure problem was encountered on a weapons dispenser used on a high performance aircraft (the F-4). The dispenser, illustrated in Figure 6.43, was slung under the aircraft, with the munitions stored in the rectangular cavities. After a mission the aircraft returned to base with the cavities open, and acoustically induced resonances within the cavities generated high levels of unsteady dynamic pressure which caused the center webs, illustrated more clearly in Figure 6.44, to resonate and fail quickly. The problem became important because the failures were rapid enough to damage irreparably a dispenser during the course of a single mission, which was a very inefficient way to use the dispensers. The authors were assigned the task of developing an easily applied, low cost, damped treatment to increase the service life of the dispensers in the field. A typical cracked center web is illustrated in Figure 6.45.

**Mode shape determinations.**   In order to obtain some idea of the scope of the problem, and of the possible approaches toward a solution, a single center web was set in a stiff fixture as illustrated in Figure 6.46 and excited inertially on a shaker table. The input acceleration was controlled at selected levels (0.1 to 1.0 g), and the response at a large number of points on the web was measured by moving a small accelerometer to each point in turn and conducting a sine-sweep test. A typical measured response spectrum is illustrated in Figure 6.47. A large number of modes are evident, their shapes being determined from peak response measurements at all the selected points on the web. Figure 6.48 shows some of the measured mode shapes. They are

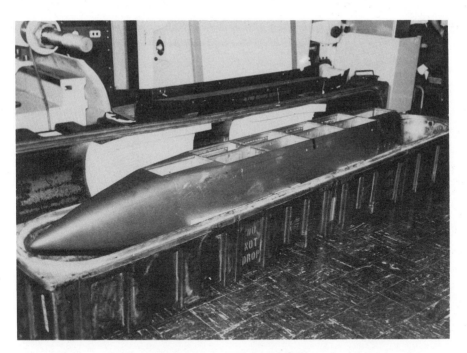

**FIGURE 6.43.** Weapons dispenser.

**FIGURE 6.44.** Center web of dispenser.

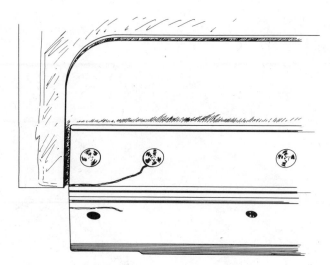

**FIGURE 6.45.** Crack in center web.

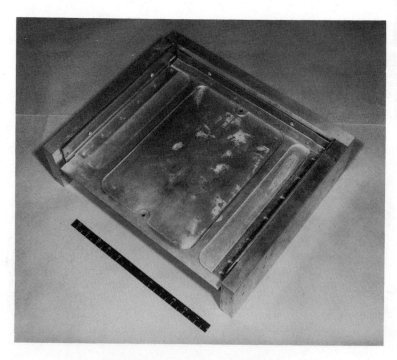

**FIGURE 6.46.** Center web in fixture.

314

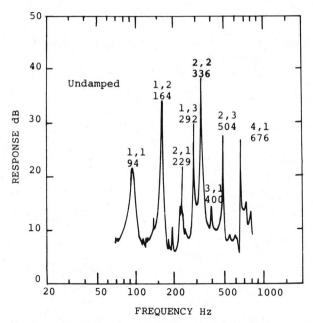

**FIGURE 6.47.** Typical undamped response spectrum.

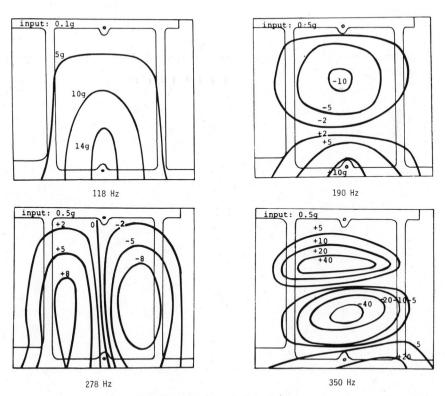

**FIGURE 6.48.** Typical mode shapes.

seen to be platelike. This, together with the limited clearances involved in the dispenser design, and the wide frequency range over which the modes were distributed, ruled out the use of any damping treatment except the layered type. The problem was to determine the best layered treatment.

**Free-layer treatment tests.**  Since the center webs were of complex geometry with integral stiffners, and the space available for applying the treatment only partially covered the web surface, as Figure 6.46 shows, it was decided to first study the effect of a free-layer treatment on the structural response in order to demonstrate whether or not such a treatment could give adequate damping in all the modes of interest. The center area of the web was covered by a layer of LD-400 damping material (a filled PVC tile) (Data Sheet 010). The tile was 0.055 in. thick (1.40 mm) and was bonded to the surface, of local thickness 0.063 in. (1.60 mm), by means of 3M-410 adhesive tape (3M Company) for convenience in application and removal. This adhesive was fairly thin and stiff so as not to affect seriously the behavior of the treatment. The damping in each mode was measured by the half-power bandwidth method, since the modes were sufficiently well separated. Figure 6.49 shows the measured variation of damping, in several modes, as a function of temperature.

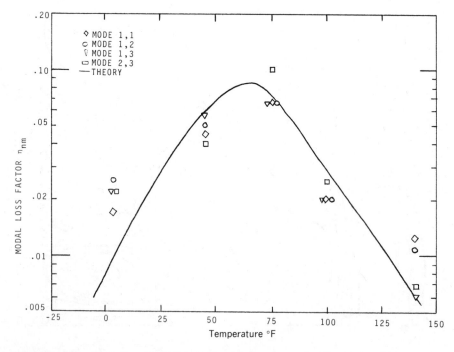

**FIGURE 6.49.**  Variation of modal damping with temperature for LD-400 damping treatment.

**Analysis of free-layer test results.** Since the plate was partially covered, equations (6.20) to (6.23) were used to determine the corresponding value of $\beta_{nm}$ for each mode. The damping properties of LD-400 were taken from Data Sheet 010, at an average frequency of about 250 Hz. The thickness ratio $h$ was given by

$$h = \frac{0.055}{0.063} = 0.873$$

The value of $e = E_D/E$ was determined at each temperature by reading $E_D$ off Data Sheet 010, dividing by $10^7$ lb/in.$^2$ (for aluminum) and inserting into equation (6.20) for several values of $\beta_{nm}$. The solid line in Figure 6.49 corresponds to $\beta_{nm} = 8.0$, an average value for all the modes of interest. This information is needed for design of the constrained-layer treatment actually used, since the room temperature treatment used in the free-layer tests did not give high damping down to $-30°$ F $(-34.4° \text{C})$ as required in service.

To illustrate a typical calculation, consider a temperature of 75° F. Then $\eta_{nm} \simeq 0.060$, $h = 0.873$, $e = E_D/E \simeq 4.5 \times 10^5/10^7 = 0.045$ and $\eta_D \simeq 0.60$. Then from equations (6.22) and (6.23)

$$A = \frac{[1 - (0.873)^2 \times 0.045]^3 + [1 + 2.508 \times 0.045]^3}{(1 + 0.873 \times 0.045)^3} = 2.0301$$

$$B = \frac{[2.76 + (0.873)^2 \times 0.045]^3 - [1 - (0.873)^2 \times 0.045]^3}{[1 + 0.873 \times 0.045]^3} = 18.647$$

Therefore from equation (6.20)

$$0.055 = \frac{0.060}{1 + \dfrac{2.0301 - 2 + \beta_{nm}}{18.6347 \times 0.045}}$$

$$\therefore \quad \beta_{nm} = 8.01$$

The average value of $\beta_{nm}$ is about 8.0. Figure 6.49 compares predicted and measured modal damping versus temperature for this value of $\beta_{nm}$. The agreement is sufficiently good for this value to be used in subsequent predictions for constrained layer treatments.

**Multiple constrained-layer treatments.** Since a lower temperature peak damping was required, a material having suitable properties at 0° F was sought. The treatment selected was 3M-428A damping tape, consisting of a 0.002 in. adhesive layer of viscoelastic adhesive on a 0.005 in. soft aluminum sheet. The two-layer pair treatment was then built up, one upon the other, to make a multiple-layer constrained damping treatment in the center area of the panel.

The design procedure is, for each mode, to make an initial estimate of the frequency. Then $E_D$ and $\eta_D$ are read off the complex modulus data for 3M-428A, given in Data Sheet 035, using an estimate of $\lambda_{nm}$ for the mode, which of course involves some uncertainty (the effects can be assessed by trying several estimates for $\lambda_{nm}$), and knowing $E_C$, $h_C$, $h_D$, and $\beta_{nm}$, the values of $E_e$ and $\eta_e$ can be obtained for the equivalent free layer from Figure 6.37. These values can then be used in conjunction with equations (6.20) to (6.22) to estimate $f_{nm}$ and $\eta_{nm}$. If the calculated value of $f_{nm}$ is markedly different from the initial estimate, the procedure should be repeated. Equations (6.25) through (6.29) may also be used.

Calculated graphs of $\eta_{nm}$ versus temperature for five-layer pairs of 3M-428A are shown in Figure 6.50, for an assumed value $\beta_{nm} = 8.0$, $\lambda_{nm}$ being assumed to be about 8.0 in. for all the modes examined and $r \simeq \frac{2}{3}$. It is seen that the results of the design procedure outlined are in reasonable agreement with the experimental data. It is clear therefore that this treatment was acceptable for introducing significant amounts of damping from $-30°$ F to $+50°$ F, or possibly to $+75°$ F. At higher temperatures the damping would probably prove to be inadequate. Subsequently it was found that higher temperature damping was not needed to solve the service problem, but tests on multiple-layer treatments were carried out using 3M-467 adhesive (Data Sheet 034) instead of 3M-428). The results of analysis and experiment are shown in Figure 6.51. A typical measured damped response spectrum is illustrated in Figure 6.52.

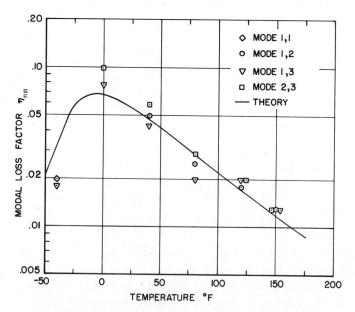

**FIGURE 6.50.**   Effect of 3M-428 treatment on modal damping.

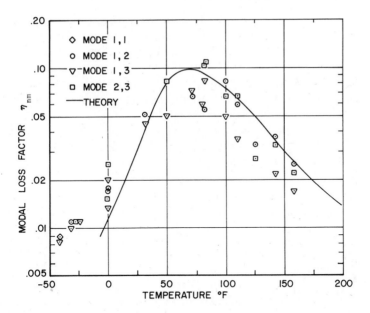

**FIGURE 6.51** Effect of 3M-467 treatment on modal damping.

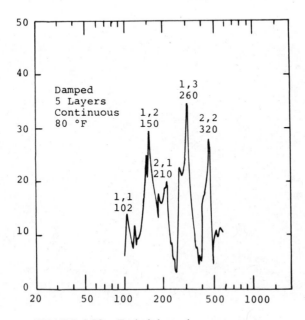

**FIGURE 6.52.** Typical damped response spectrum.

**Summary.** The multiple-layer treatment using 3M-428A adhesive and aluminum constrained layers was manufactured in substantial quantities and applied to the center webs of all weapons dispensers of this type. Crack initiation rates in service were drastically reduced, with the result that each dispenser survived many missions, instead of less than one mission before application of the treatment. The cost savings were very great, since the cost of treating each dispenser was less than 2% of the cost of each dispenser, and this expenditure in turn saved the cost of the several replacement dispensers which would have been required in the absence of the treatment. The improvement of operational and supply system effectiveness was also great.

### 6.8.2. Development of a Damping Wrap for Vibration Control in a Jet Engine Inlet Guide Vane [6.16]

The TF-30-P100 Jet Engine is employed in the U.S. Air Force F-111F Fighter Aircraft. It is a modern, highly sophisticated, high thrust-to-weight ratio jet engine. Inlet air first encounters the inlet guide vanes (IGVs). These stationary airfoils slightly turn the air for entry into the first stage fan blades, which pass near the trailing edge of the IGVs, which are titanium weldments that in turn are welded to inner and outer shrouds (cylindrical sections) to form the IGV assembly. Hot bleed air passes through the IGVs during climatic icing conditions. In the past few years cracks were discovered in the IGVs prior to completion of the required lifetime. The cracks were of such size and number that maintenance/refurbishment became a high cost item. Figure 6.53 illustrates the IGV case and vanes, along with the fully developed damping treatment.

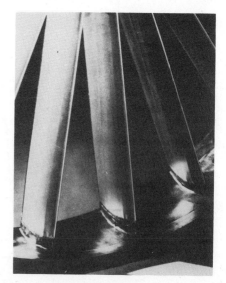

**FIGURE 6.53**  Inlet guide vanes with damping treatment.

An examination of cracked surfaces proved that they were due to high cycle fatigue. Test cell and flight test measurements showed that vibratory stresses were very high, whereas static stresses from thermal gradients and residual manufacturing strains were contributory. Spectrum analysis showed that excitation was predominantly discrete at engine-order and blade-passage frequencies with some wide-band excitation from inlet air turbulence. Some results from one-third octave band analysis are presented as a Campbell diagram in Figure 6.54 where the vibratory frequencies of peak stresses in test cell operations are plotted versus the corresponding engine speed. Secondary peaks within 15% of the maximum amplitude are also shown. Individual blade passage frequency (at 28 times engine speed, represented by the 28E line

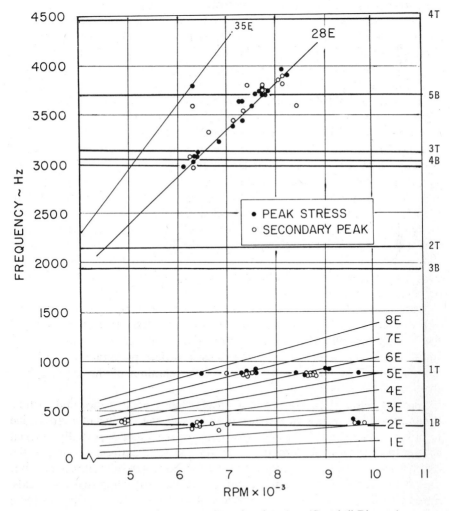

**FIGURE 6.54.**  Frequency and RPM of peak stresses (Campbell Diagram).

in Figure 6.54) excited the fourth torsion mode at 4000 Hz, fifth bending at 3600 Hz, and third torsion at 3000 Hz. These are nominal frequencies as they vary with temperature and other engine conditions. The first torsion mode at 850 Hz and first bending mode at 350 Hz were excited at lower engine orders. Analysis revealed that the highest vibratory stresses were occurring in the 3000 and 4000 Hz modes. An analysis of service operations indicates engine speeds that excite these modes occur throughout the flight envelope. An attractive approach to vibratory stress reduction and consequent crack abatement and life extension was judged to be an additive, multiple constrained-layer, broad temperature range, viscoelastic damping treatment.

**Requirements for treatment design.**   In the application of viscoelastic damping materials, an essential requirement is a knowledge of the temperature at which vibratory damage occurs in service. Figure 6.55 shows cumulative percent of time versus total outside air temperature for service operation of the F-111 aircraft. On the basis of this data it was decided to design the damping treatment for an average temperature of 62° F. This compares closely with the 59° F standard day temperature. Figure 6.56 shows a summary of the temperature requirements of the IGV damping treatment. The operating range is between 0° and 125° F and accounts for 98 % of engine operation and also, presumably, damage. The values in the individual bars are the percent of operating time for that particular temperature increment. In addition to the operating temperature, the survival temperature is an overriding consideration. During operation of the F-111 aircraft the ram temperature is not to exceed 307° F for more than five minutes during any excursion, and is not to exceed 417° F at any time. The autoclave temperature for curing the structural adhesive used to bond the damping treatment to the vane was 350° F. Heat transfer analysis for normal anti-icing operation showed that the maximum bond line temperature is 420° F. It follows that 420° F is the maximum temperature for survival.

Other requirements are crucial in the design of the damping treatment. One of the other major issues is adhesion. The damping treatment must adhere to the existing structure and remain in place during service operation. A major factor in adhesion and bonding technology is surface preparation of the aluminum foil in the damping wrap and of the titanium of the existing IGV structure.

Another factor is airflow, since it affects engine performance and turbine inlet temperature, where an increase of even 5° F will significantly reduce life of turbine components. The inlet area between the vanes is partially blocked by the thickness of the damping wrap on the vane surface. Obviously the blocked area consideration alone would result in degraded performance, but the smoothing effect of the damping wrap over the protruding spanwise welds, introduced during fabrication, completely offset the blockage effect, and there was no measurable net effect on performance. Another related factor was

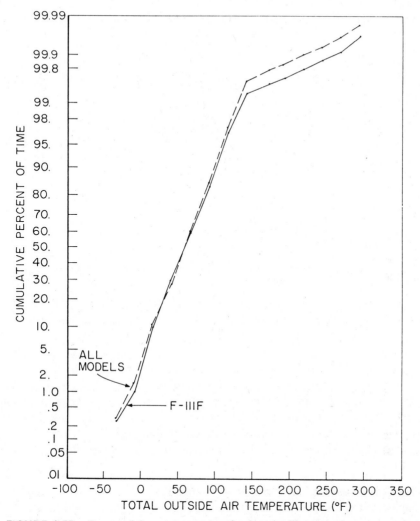

**FIGURE 6.55.**   Percent of time at temperature from service life monitoring program data.

distortion tolerance of the engine, which the damping wrap influenced beneficially.

Formability of the damping wrap to the vane surface was also a major consideration. Other factors were foreign object damage (FOD), to which the aluminum foil, backed by the soft viscoelastic adhesive, was susceptible, erosion, and the effectiveness of the anti-icing. The channels between the interior stiffeners of the vane carry hot anti-icing air under certain climatic conditions. Thus the damping wrap must provide not an insulating effect but efficient heat transfer to the vane leading edge surface to prevent the formation of ice. The survival temperature of 420° F, which results from the most severe anti-icing

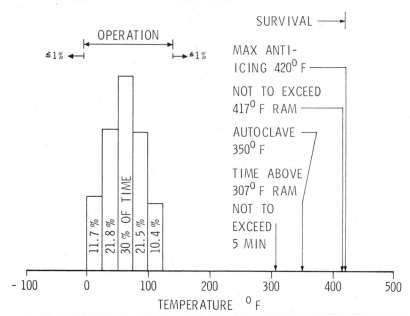

**FIGURE 6.56.**   Temperature requirements of IGV damping treatment.

conditions, has already been mentioned. This 420° F temperature approaches the upper limits of the material used. The handling of anti-icing effectiveness and of maximum bondline temperature required extensive heat transfer analyses. Overhaul was another major item. When service time accumulates on the engine such that it needs to be overhauled, the IGV case itself must be overhauled. If the damping wrap is sufficiently damaged from FOD, and so forth, it must be replaced. This must be a relatively easily performed operation.

**Laboratory tests of prototype treatments.**   Because of the high cost of engine test cell experimental programs, a laboratory test program was undertaken to design an additive damping treatment and investigate the effect it could have on the vibratory stress levels of the IGVs.

The initial phase of the work focused on selecting a suitable experimental setup that would give sufficiently accurate modal damping measurements of additive damping treatments. A number of support configurations for the IGVs were devised and analyzed. Finally, a vane welded between two large titanium blocks was chosen. This arrangement adequately simulated the actual boundary conditions to which the vanes were subjected in the engine, and the low inherent, baseline specimen, modal damping ensured a reliable comparison between the vane damping levels before and after damping treatment.

A modal survey of the specimen, using holographic techniques, revealed several vibration modes. These included the first through fifth bending modes and the first through fourth torsional modes. Figure 6.57 illustrates some of

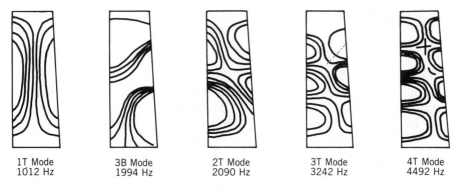

| 1T Mode<br>1012 Hz | 3B Mode<br>1994 Hz | 2T Mode<br>2090 Hz | 3T Mode<br>3242 Hz | 4T Mode<br>4492 Hz |

**FIGURE 6.57.** Mode shapes.

the most damaging of these modes along with their respective frequencies. These mode shapes and frequencies compare quite favorably with those identified by the engine manufacturer in independent tests.

Tests were conducted on the single vane specimen over a wide temperature range to determine the effect of additive damping wraps on the modal damping of the vane as a function of temperature. The vane was excited using a magnetic transducer driving an iron disc bonded to the vane. The response was picked up by an accelerometer. Both the transducer and accelerometer were located to ensure excitation and response of the third and fourth torsional modes. Figure 6.58 is a schematic of the test system used.

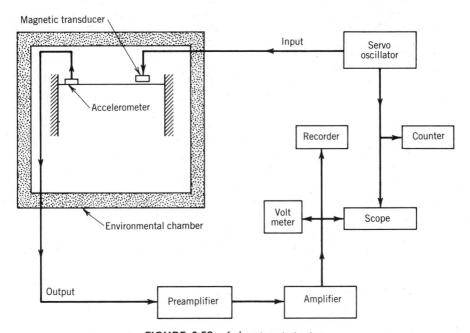

**FIGURE 6.58.** Laboratory test setup.

Four damping wrap designs were evaluated. Each had the same geometry but varied as to viscoelastic and constraining layer materials. By varying the material composition of the wrap, both the level of damping and the effective temperature range of the wrap could be changed as desired. The basic wrap design is illustrated in Figure 6.59.

Figure 6.60 is a plot of modal loss factor versus temperature for the four damping wrap designs. Each curve is designated as to its inner constraining layer and damping material used. As can be seen, the wrap can be designed to be effective over different temperature ranges by changing its composition. Figure 6.60 includes a bar graph depicting percent time at temperature increments which the vanes experience during their service life. This bar graph, in conjunction with the loss factor curves, indicated that the ISD-830-110-aluminum configuration is less suitable for the desired temperature range. The other three configurations all provide acceptable damping levels over the temperature range of interest.

The final choice of wrap configuration was made on the basis of factors other than composite loss factor versus temperature. Difficulties with tight process control requirements for the viscoelastic materials adhering to the Ultra-High Modulus (UHM) Graphite eliminated the ISD-112-113-UHM graphite design. The ISD-113-830-aluminum configuration provided damping over the temperature range. In the final analysis, however, a wider margin in the effective damping level for the temperature region above 125° F dictated the use of the ISD-112-830-aluminum wrap design. This design also provided significant damping of the lower modes at the service operating temperatures.

The complex modulus properties of the damping materials were measured using a vibrating beam technique. Curves of shear modulus and loss factor versus temperature for various frequencies are shown in Data Sheets 017, 018, 019, 021.

The final damping treatment configuration, illustrated in Figure 6.59, consisted of aluminum foil constraining layers and viscoelastic damping material.

.002 in. ALUMINUM FOIL
.002 in. 3M ISD MATERIAL
.005 in. CONSTRAINING LAYER
.002 in. 3M ISD MATERIAL
.005 in. ALUMINUM FOIL

T.E  CONCAVE  L.E.  CONVEX  T.E.

.002 in. ALUMINUM FOIL
.002 in. 3M ISD MATERIAL
.005 in. CONSTRAINING LAYER
.002 in. 3M ISD MATERIAL
.005 in. ALUMINUM FOIL

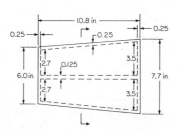

**FIGURE 6.59.** Damper wrap kit.

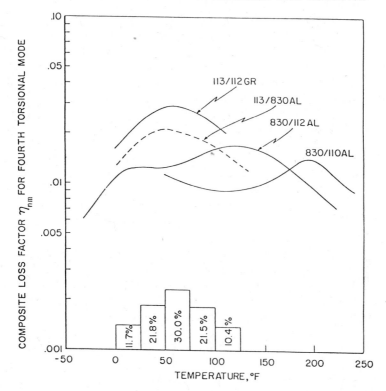

**FIGURE 6.60.** Composite loss factor versus temperature for four damping treatments.

The viscoelastic material used was the 3M Company ISD-830 on the concave surface (two layers) and ISD-112 on the convex surface (again, two layers). Each constraining layer was 5 mils of aluminum foil. Under the viscoelastic layer was 2 mils of aluminum foil, necessary to prevent air entrapment which would severely degrade the durability. The 2 mils of aluminum foil were bonded directly to the titanium surface of the vane with a structural adhesive (AF 126 epoxy, nominally 5 mils thick). The total thickness was about 20 mils. The circumference (i.e., picture frame) of the damping treatment, including the leading edge, was bonded to the titanium surface. An autoclave operation was used to cure the wraps in place on the IGVs.

**Engine test cell evaluations.** During a series of engine operations in a test cell, the vibratory stresses were investigated for the IGV case with and without damping treatment. Numerous strain gages and thermocouples were located to give data on temperature distribution and stress reduction. Performance tests were also made. Parameters were measured that would permit calculation of turbine inlet temperature, which significantly affects the life of turbine components. Distortion tolerance was also measured, and it was found that the additive damping treatment enhanced the distortion tolerance. Durability was also investigated, that is, the tolerance of the damping wrap to the anti-icing temperature cycles, and steady state temperature distributions were investigated. The main durability test consisted of 50 cycles of anti-icing air. This is representative of 1200 hours of engine service operation. In addition modal damping of the significant modes was measured in the test cell with and without damping treatment. In all instances modal damping was significantly increased with the damping treatment in place.

**Field evaluation tests.**  Prior to fully qualifying the wrap design, but at such a time that its development was considered adequate for service confirmation, a set of damping wraps was installed on several IGV cases mounted on engines in service. These wraps were instrumented with temperature indicating paints and patches for the purpose of acquiring additional information on the in-service thermal environment. Aircraft on which these cases were installed were equipped with service life monitoring program recorders which allowed a review of the aircraft flight usage in the event any change in the wrap was detected. Pilot debriefings included the number of times the anti-icing system was activated during flight and the time duration. Preflight inspection of the wraps was requested. In the event damage to a wrap was found, a quick and simple repair procedure was outlined, which consisted of trimming away any disbonded region of foil and applying a quick-setting, two-part epoxy to the exposed edges. This procedure circumvented any need to perform lengthy maintenance on the aircraft because of the wrap.

As of 1984 the field service evaluation wraps had accumulated over 1000 hours of experience on engines in the field. During this time no vibration induced cracks were detected. The wraps successfully withstood the thermal environment with no loss of damping effectiveness. Minor foreign object damage (FOD) occurred on a few wraps, but these wraps were repaired using the procedure outlined earlier without impact on the aircraft flight schedule or operational readiness.

Visual inspection and coin tap tests demonstrated the excellent condition of the field service wraps, and the absence of vane cracks indicate the wraps effectively reduced dynamic vane stresses. Temperature indicating paints and patches confirmed the upper temperature bound used for the wrap design.

The field evaluation tests showed that wrapped inlet guide vanes do abate in-service cracking and that the wraps can withstand the severe service environment. They represent a practical, low cost, easily implemented remedy to a high maintenance cost, vibration-induced, cracking problem.

### 6.8.3.  Constrained-Layer Damping for Noise Reduction in a Helicopter Cabin

**Problem definition.**  The particular system investigated as an Air Force HH-53C helicopter manufactured by Sikorsky Aircraft Company. The HH-53C is powered by 2 GE T-64-7 engines, with a 6-bladed main rotor. Its

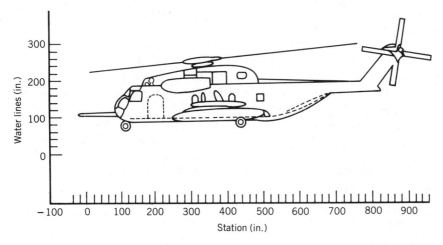

**FIGURE 6.61.** Side view of HH-53 helicopter.

primary mission in the Air Force is Air Rescue, with transportation of personnel and equipment as a secondary function. Figure 6.61 gives a side view of this system with fuselage station numbers and waterline references. Acoustic data measured, prior to this investigation, in the middle of the cabin, near station 342, showed high sound pressure levels at three frequency peaks as illustrated in Figure 6.62.

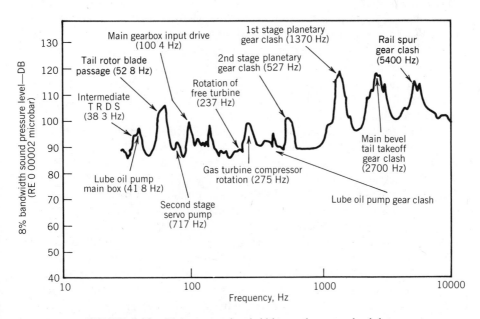

**FIGURE 6.62.** Eight percent bandwidth sound pressure level data.

These frequencies were nominally 1370 Hz, corresponding to the first-stage planetary gear clash, 2700 Hz corresponding to the main bevel and tail takeoff gear clash, and 5400 Hz corresponding to the rail spur gear clash. It was estimated that a major noise source in the cabin was the resonant vibration of the fuselage skin which was excited by these gear clash vibrations of the transmission. If this was indeed the case, then damping of the skin would be an effective means of reducing cabin noise at these frequencies.

Standard-equipment acoustic blankets, consisting of a layer of fiberglass sandwiched between layers of heavy vinyl cloth can easily be attached to the inside of fuselage frames for the reduction of cabin noise, but are seldom used in the field. The objections to these standard acoustic treatments are that they interfere with maintenance and make rapid inspection and repair of battle damage in combat situations impossible. Therefore the using command was interested in the layered damping treatments, which have the advantage of being attached directly to the skin out of the way of maintenance and inspection operations.

The purpose of this investigation was, first of all, to obtain additional data on the nature of skin panel resonant vibrations and their contribution to the noise problem in the cabin. Second, we desired to evaluate the reduction of these resonant responses through the application of a layered damping treatment known to be effective over the temperature range expected in these tests. No attempt was made to treat the entire aircraft, or test a material that would be suitable for the entire range of temperatures of environments expected in worldwide USAF operations.

**Data acquisition.**   A small portion of the cabin structure, near the transmission between stations 322 and 362 and above waterline 140, was chosen for detailed measurements and the application of the damping material. Seventeen accelerometers were mounted at various locations on the transmission housing, on the heavy frames supporting the transmission, and on several skin panels in the area. Locations of some of the accelerometers are shown in Figure 6.63. Care was taken when choosing accelerometer sizes to avoid undue mass loading of the structure being measured.

Five piezoelectric microphones were also located within this section of the cabin, as well as six thermocouples for monitoring structural temperatures. In-flight data recording equipment, calibration, and data reduction were provided by the Vehicle Dynamics Division of the Flight Dynamics Laboratory. Since only a small portion of the aircraft was treated with damping materials, flanking paths for noise radiated from other sources were reduced by installing fiberglass blankets over the untreated panels and reducing the noise radiated through the transmission drip pan with alternate layers of fiberglass and lead-vinyl cloth. Photos of the treated area of fuselage skin are shown in Figure 6.64. Vibration and acoustic measurements were recorded during four flights of the helicopter to assess the effectiveness of the damping treatment. Several maneuvers were performed in each flight to show the effects of the flight variables. Vibration measurements were also conducted on the ground, utiliz-

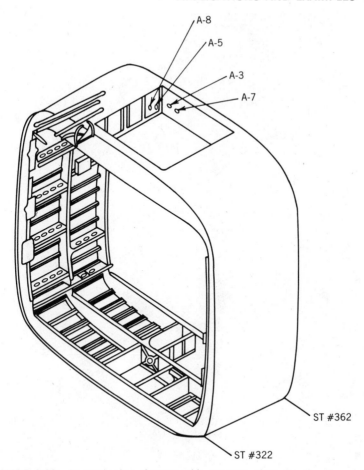

**FIGURE 6.63.**   Isometric view of center cabin structure where damping was applied.

ing harmonic excitation to compare dynamic response before and after adding the layered damping treatments.

**Development of damping treatment.**   Elastomeric damping materials can be very effective if applied in an appropriate geometry and utilized within their transition temperature region to achieve optimum performance. Such materials reach their peak loss factor and optimum shear modulus in the transition temperature region between the glassy state at low temperature and the rubbery state at higher temperatures. The damping materials applied in this investigation were pressure-sensitive adhesives that had optimum damping near room temperature. The complex shear modulus properties of the materials as a function of temperature are shown in Data Sheets 034 (Material A) and 035 (Material B).

The geometry of the layered damping treatment consisted of three (3) layers (0.002 in. thick each) of damping adhesive and three (3) alternate constraining

**FIGURE 6.64.** Photograph of treated area.

layers (0.005 in. thick each). This treatment was selected because it would give high damping in the panels over the anticipated temperature range of the tests. However, for in-service applications it may be necessary to cover a wider temperature range. This can be accomplished by combining different damping adhesives with different transition regions.

Figure 6.65 illustrates the effect of this procedure for a simple clamped-clamped beam with the same thickness and frequency as the panels considered in the helicopter tests. It can be seen from the figure that high damping can be achieved, by the proper selection of damping materials, over a wide temperature range from $-25°$ F to $+200°$ F.

**Ground test evaluation of damping treatment.** Artificial excitation tests were conducted, with the helicopter on the ground, to establish the effectiveness of the damping treatment in reducing the skin vibrational amplitudes.

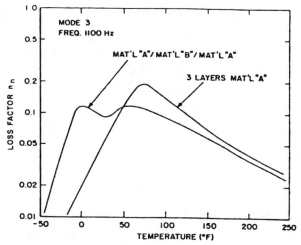

**FIGURE 6.65.**  System loss factor of a clamped-clamped beam with multiple-layer damping treatments.

A harmonic force with constant amplitude was applied to the transmission supporting frame by means of an electromagnetic shaker through an impedance head, as illustrated in Figure 6.66. Frequency response spectra were obtained at the center of a panel by means of an accelerometer (A-14). Figure 6.67a, b, and c represents the measured response spectra without damping treatment, with damping treatment on the panel of interest only, and with

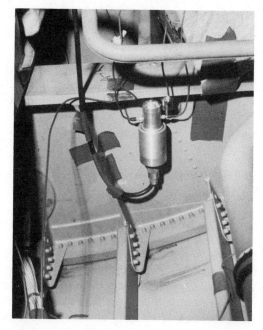

**FIGURE 6.66.**  Photograph of impedance head and exciter.

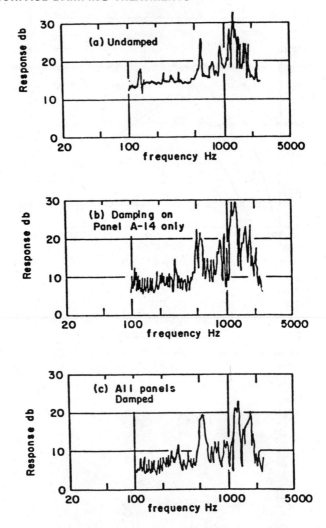

**FIGURE 6.67.** Typical ground vibration test spectra.

damping treatment on all panels, respectively. It can be seen from this figure that a considerable reduction, approximately 10 db, in amplitude was achieved in the peak near 1370 Hz, as anticipated.

Mode shapes were measured by moving a very light accelerometer along the panel while exciting a panel resonance. In general, for modes near 1370 Hz it was found that node lines parallel to the stringers were spaced at intervals of about 1.5 in. and that node lines parallel to the frames were spaced about 5 in. apart. Segments of panels on opposite sides of node lines were generally out of phase. There were several resonant frequencies of the skin near 1370 Hz so that slight changes in excitation frequency excited different modes in the skin-stringer structure. The higher frequency peaks near 2700 Hz and 5400 Hz correspond to resonances in the heavy supporting frame as well as in the skin-stringer structure.

**Flight test evaluation of damping treatment.** The narrowband (50 Hz) analysis of a small portion of the flight test data is presented in Figures 6.68 through 6.71 and summarized in Table 6.1, for purposes of illustrating the effects of the damping treatment. Accelerometer A-14 was located on a skin panel at fuselage station 352, and waterline 172. Microphone M-3 was located 5 in. below the transmission drip pan. Flight 1 was with an untreated aircraft, and flight 3 was with damping treatment installed.

It can be seen from these plots that the three-gear clash frequencies correspond to peaks in the accelerations measured on the skin and in sound pressure levels measured in the cabin. Accelerations of the damped skin of the structure in the frequency band around 1370 Hz were generally reduced as much as 12 dB in the one-third octave band analysis and 14 dB in the narrowband analysis, as compared with the undamped structure. There are indications that damping reduced the sound pressure levels at microphone M-3 near the skin and microphone M-5 under the transmission drip pan. Reductions in narrowband acceleration, and acoustic levels for the three frequency bands of interest, are shown in Table 6.1.

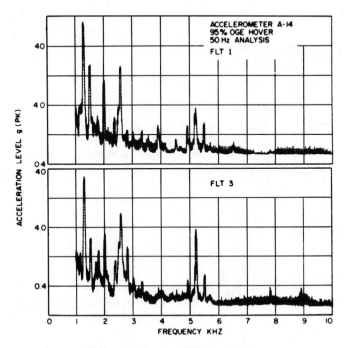

**FIGURE 6.68.**   Acceleration, 95% OGE hover.

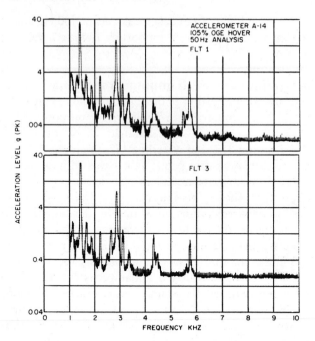

**FIGURE 6.69.** Acceleration, 105% OGE hover.

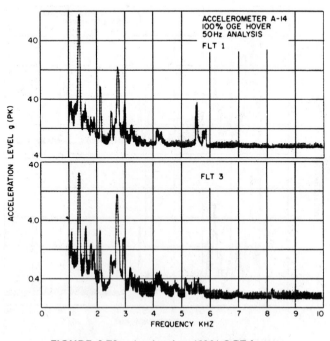

**FIGURE 6.70.** Acceleration, 100% OGE hover.

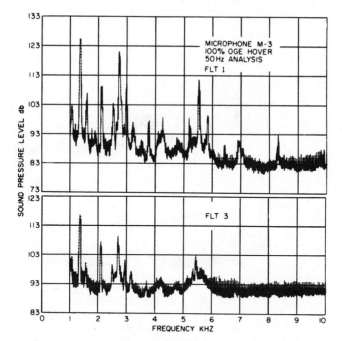

**FIGURE 6.71.** Microphone M-3 data.

It must be remembered that large changes in acoustic levels due to damping were not anticipated because only a small portion of the structure was damped, noise radiating from other sources such as undamped skin and the transmission drip pan was not completely controlled, and the fact that cabin noise at these high frequencies (and hence short wavelengths) is not fully defined with measurements at few microphone locations. Despite these difficulties there does appear to be a general trend in the data indicating a reduction of high frequency noise levels at microphones M-3 and M-5.

**TABLE 6.1. Narrowband Acceleration and Acoustic Levels of Flight 3 Compared with Flight 1**

| Acceleration (A-14): dB ref. flight 1 | 1370 Hz | 2700 Hz | 5400 Hz |
|---|---|---|---|
| O.G.E. Hover 95% | −11 dB | −9 dB | 0 dB |
| O.G.E. Hover 100% | −13 | −2 | −10 |
| O.G.E. Hover 105% | −2 | −5 | −9 |
| Sound Pressure Level—O.G.E. Hover 100% | | | |
| M-3 | −10 | −12 | −9 |
| M-5 | −11 | −17 | −7 |

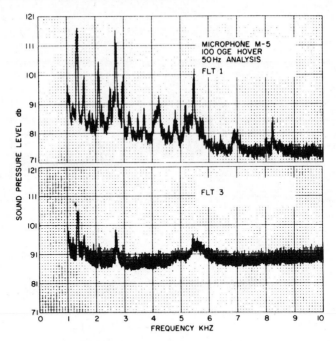

**FIGURE 6.72.** Microphone M-5 data.

Figures 6.68 to 6.71 illustrate some measured acceleration and noise spectra for flights 1 and 3. Figure 6.72 illustrates the noise spectra for microphone M-5 under the transmission drip pan.

**Summary.** Four flight tests plus additional ground tests were performed on an HH-53C helicopter for the purposes of evaluating the effectiveness of multiple-layer damping treatments, applied to the skin, as a means of reducing high frequency cabin noise. Acceleration, acoustic, and temperature measurements were made on the aircraft at several locations and for several flight conditions. Although only a small portion of the structure was covered with damping treatments, significant reductions (up to 12 dB) in skin accelerations were noted at frequencies where the skin is a major noise source. Acoustic measurements indicated a reduction in high frequency noise levels in the cabin, of approximately 5–11 dB, although quantitative measurements were difficult to make because of the small area treated, the multiplicity of noise sources, and other variations in flight conditions.

The following conclusions were drawn from the results of this investigation:

1. Cabin noise peaks that occur near 1370 Hz, 2600 Hz, and 5400 Hz correspond to specific gear clash frequencies. At these frequencies the transmission causes the heavy supporting frame to vibrate. This in turn excites resonant frequencies in the skin-stringer structure. Much of the

noise in the cabin is a result of the resonant response of the cabin structure.

2.  The damping treatment applied in this test was shown to be effective, at the test temperatures, in reducing skin vibrations, particularly near the 1370 Hz peak, and thus reducing radiated noise in the cabin.

3.  Although the feasibility of using damping treatments to control cabin noise has been demonstrated for one limited temperature range and for partial coverage of the cabin skin, different materials for broader temperature ranges and additional coverage would be required in a practical HH-53 modification.

4.  The transmission drip pan is a major cabin noise source and should be modified to increase acoustic transmission loss in any future noise control efforts.

Subsequent to the completion of this investigation, damping treatments, similar to these, have been used as part of noise control treatments on some HH53, VH-3, and SH-3H helicopters. In the Air force HH-53 helicopters, approximately 150 ft$^2$ of coverage was used, with a total weight of about 37 lb per aircraft. This treatment could be installed in the aircraft with about 24 worker-hours of labor. In each case the damping materials were chosen to provide maximum damping over the expected environmental temperature range, and additional acoustic absorption material was used to reduce cabin reverberation time.

### 6.8.4.  Development of a Free-Layer Treatment for an Engine Exhaust Stack

Increasing the service life of helicopter engine exhaust extensions has been of great concern because of fatigue failures that have occurred within a few hours of operation [6.14]. Previous attempts to resolve this problem by introducing structural stiffness failed because of the broadband excitation generated by the engines, which resulted in small reduction in structural resonant response.

The most promising way of increasing the fatigue life of the exhaust extension is to reduce the vibrational amplitudes at resonance. This can be achieved by additive damping. However, most damping materials, being of the elastomeric type, will not survive the high temperatures associated with the subject tailpipe. Vitreous enamels have been used as high temperature damping material on exhaust systems and have successfully reduced their associated vibrational amplitudes and noise levels. The vibration failure problem on this component represented an opportunity for application of this material.

The major objectives of the effort were as follows:

1.  An investigation to define the mode shapes and natural frequencies of the engine exhaust extension.

2. The selection of a proper damping treatment optimized at the operating temperature.

3. A thickness ratio to yield at least a 10:1 increase in damping throughout the desired frequency range.

4. A 100 hour flight test program to evaluate the damping properties.

As will be further explained, each of these objectives was achieved.

**Definition of the problem.**   Introducing damping into structures at high temperatures has been restricted in the past because most damping materials are of the elastomeric type and do not survive beyond 400° F. However, vitreous enamels are capable of providing good damping values at high temperatures if properly selected and designed. The main problem was in selecting the proper material that would maximize damping at the operational temperature of the tailpipe extension.

A flight test was performed to determine the operational temperature of the tailpipe extension. Four thermocouples were installed on the tailpipe flange, 90 degrees apart at the four clock positions (12, 3, 6, 9). The test aircraft (CH-54; S/N 67-18417) maintained a stable hover condition for 30 minutes at a gross weight of 42,000 lb and a neutral center of gravity. Temperature data were obtained at 5 minute intervals and are presented in Table 6.2. The critical temperature was obtained by finding the highest temperature of the exhaust extension during the hover survey. Since 742° F was the highest temperature value, it was decided to optimize the damping treatment at 800° F.

Testing was then initiated to define the vibrational characteristics of the exhaust extension for proper optimization of the damping treatment. Both analog and digital techniques were used to generate the transfer functions and mode shapes necessary for this investigation. These were the Transfer Function Analyzer (TFA) and the Digital Fourier Analyzer (DFA), respectively. The engine exhaust stack was bolted through the existing mounting holes to a rigid plate which simulated actual boundary conditions.

The impulse technique was used to generate the data with the DFA setup. The input force impulse was applied to the exhaust extension, and this signal was measured simultaneously with a load cell, while the output response was measured with an accelerometer at the same location. The input and output analog signals were then filtered, digitized, and Fourier transformed to yield magnitude and phase information in the frequency domain. The complex ratio of response to force was then computed, and the resultant transfer function was plotted as compliance versus frequency. Typical driving point frequency response obtained by this technique is shown in Figure 6.73. Due to the large number of observed modes, only typical resonances were further investigated to determine their associated mode shape of vibration. These were 52.7, 84.0, 207.0, and 339.8 Hz. Animated mode shapes of vibration were obtained by the impulse technique by generating transfer function plots on various locations

**TABLE 6.2. Flight Test Temperature Evaluation**

Flight Condition, Hover 100% $N_R$—Nose into Wind
Engine Parameters, #1 Engine 89% $N_1$—Engine Speed
                              1900 lb/HR—Fuel Flow
                              1.75 Engine Pressure Ratio

              #2 Engine 915 $N_1$—Engine Speed
                        1920 lb/NR—Fuel Flow
                        1.79 Engine Pressure Ratio
                        $T_5$ = Turbine Exhaust Temperature

5 *Minute Hover*—Nose into Wind

$T_5$—480° C(896° F)   $T_5$—460° C(860° F)

| #1 Engine | #2 Engine |
|---|---|
| 1—330° C(626° F) | 5—300° C(572° F) |
| 2—290° C(554° F) | 6—220° C(428° F) |
| 3—190° C(374° F) | 7—250° C(482° F) |
| 4—230° C(446° F) | 8—300° C(572° F) |

10 *Minute Hover*—Nose into Wind

$T_5$—500° C        $T_5$—490° C

| #1 Engine | #2 Engine |
|---|---|
| 1—340° C(644° F) | 5—300° C(527° F) |
| 2—290° C(554° F) | 6—250° C(482° F) |
| 3—190° C(374° F) | 7—260° C(500° F) |
| 4—220° C(428° F) | 8—310° C(590° F) |

15 *Minute Hover*—Nose into Wind

$T_5$—495° C        $T_5$—490° C

| #1 Engine | #2 Engine |
|---|---|
| 1—340° C(644° F) | 5—300° C(572° F) |
| 2—295° C(563° F) | 6—240° C(464° F) |
| 3—190° C(374° F) | 7—250° C(482° F) |
| 4—210° C(410° F) | 8—310° C(590° F) |

20 *Minute Hover*—Nose into Wind

$T_5$—490° C        $T_5$—460° C

| #1 Engine | #2 Engine |
|---|---|
| 1—330° C(626° F) | 5—300° C(572° F) |
| 2—270° C(618° F) | 6—220° C(428° F) |
| 3—180° C(356° F) | 7—260° C(500° F) |
| 4—190° C(374° F) | 8—314° C(599° F) |

25 *Minute Hover*—Nose into Wind

$T_5$—490° C        $T_5$—480° C

| 1—330° C(626° F) | 5—300° C(572° F) |
|---|---|
| 2—265° C(509° F) | 6—230° C(446° F) |
| 3—180° C(356° F) | 7—240° C(464° F) |
| 4—200° C(392° F) | 8—310° C(590° F) |

30 *Minute Hover*—Nose into Wind

$T_5$—490° C        $T_5$—480° C

| 1—330° C(626° F) | 5—310° C(590° F) |
|---|---|
| 2—295° C(563° F) | 6—220° C(428° F) |
| 3—200° C(392° F) | 7—260° C(500° F) |
| 4—220° C(428° F) | 8—310° C(590° F) |

*Right Side into Wind*

$T_5$—490° C        $T_5$—480° C

| #1 *Engine* | #2 Engine |
|---|---|
| 1—350° C(662° F) | 5—310° C(590° F) |
| 2—395° C(742° F) | 6—220° C(428° F) |
| 3—200° C(392° F) | 7—260° C(500° F) |
| 4—220° C(428° F) | 8—315° C(599° F) |

*Left Side into Wind*

$T_5$—490° C        $T_5$—490° C

| #1 Engine | #2 Engine |
|---|---|
| 1—350° C(662° F) | 5—310° C(590° F) |
| 2—320° C(608° F) | 6—250° C(482° F) |
| 3—200° C(392° F) | 7—270° C(518° F) |
| 4—235° C(455° F) | 8—315° C(599° F) |

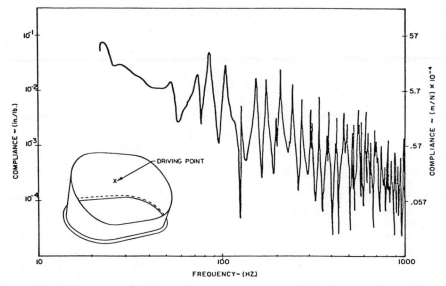

**FIGURE 6.73.**   Driving point frequency response of undamped exhaust extension.

on the engine exhaust extension. The animated mode shapes of vibration proved to be a useful aid in determining the general vibratory characteristics of the tailpipe extension.

After studying the driving point frequency response plots and the animated mode shapes of vibration, the following was found:

1.  A large number of resonances occurred over a large frequency range.
2.  Damping values at the various resonances were small. Typical loss factors ($\eta_n$) ranged between 0.001 and 0.01.
3.  Deformation at resonance was associated with all surfaces of the engine exhaust extension, over a wide frequency range.
4.  Typical semiwavelengths of vibration ranged from 5 in. to 15 in. for most observed modes.

The transfer function tests were performed to aid in determining the damping loss factors of the various modes of vibration. Transfer function plots were obtained by placing an impedance head between the shaker and the exhaust extension. Sinusoidal excitation was used to drive the structure. Both the input force and the output acceleration were measured by the impedance head and later used to obtain the compliance versus frequency plots. Data were generated at a number of locations close to the mounting holes in the flange in order to minimize mass loading effects. Some data are presented in Figures 6.74 through 6.76. The half power bandwidth method was used to obtain the damping for the various modes. The sharpness of the peaks indicates that

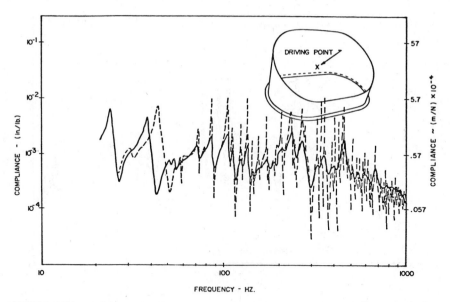

**FIGURE 6.74.** Driving point frequency response comparison of damped and undamped exhaust extension (room temperature).

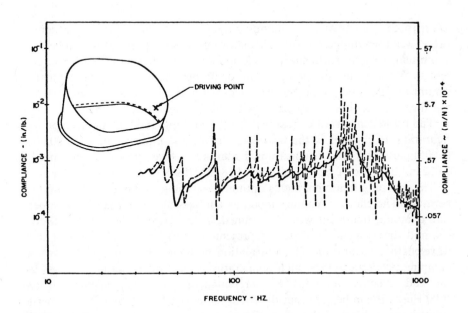

**FIGURE 6.75.** Driving point frequency response comparison of damped and undamped exhaust extension.

**343**

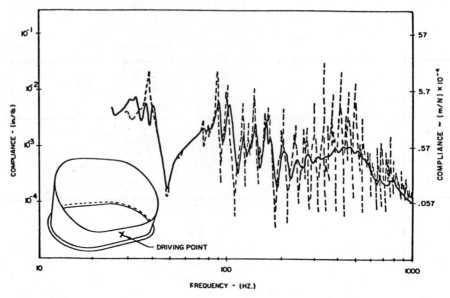

**FIGURE 6.76.** Driving point frequency response comparison of damped and undamped exhaust extension.

damping loss factor values are very low (between 0.001 and 0.01) for most modes.

**Design of treatment.** A damping treatment, optimized for room temperature, was then applied to the exhaust extension for optimization studies. The room temperature treatment was selected to have damping properties equivalent to those of the high temperature damping treatment. In this manner all optimization studies could be evaluated at room temperature, thereby avoiding the usual problems associated with high temperature testing.

The room temperature treatment was a multiple-layer damping tape configuration. Each layer of the tape consisted of a damping adhesive 0.002 inch (0.0051 cm) thick and a constrained aluminum layer having a thickness of 0.002 in. (0.0051 cm). See Data Sheet 018 (3M-ISD-112).

The damping treatment yields optimum damping capabilities at room temperature. The effective Young's modulus and loss factor of the room temperature damping treatment were determined, in the manner discussed in Section 6.6, and these calculated data are presented in Figure 6.77 for typical semi-wavelengths of vibration. The information presented in Figure 6.78 shows the damping properties of the high temperature treatment material. The loss factor and effective modulus are presented at two frequencies, 100 and 1000 Hertz. It can be seen that the room temperature treatment has properties equivalent to those of the selected high temperature vitreous enamel at the operational temperature of the exhaust extension. Therefore any optimization

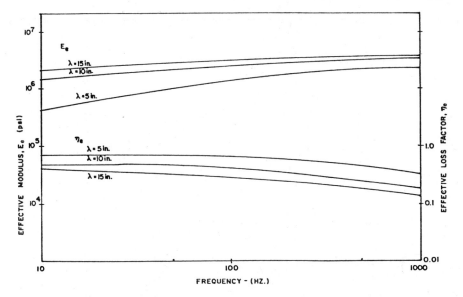

**FIGURE 6.77.**    Equivalent damping properties of the room temperature treatment.

for the room temperature damping treatment will be directly applicable to the vitreous enamel. Several layers of the room temperature damping treatment were applied until adequate reductions of the resonant vibrational amplitudes were achieved. This was accomplished with five layers (total thickness of 0.020 in., 0.051 cm) of the damping material. Testing was performed exactly as for the untreated tailpipe extension. The results of the room temperature damping evaluations are shown in Figures 6.74 through 6.76. Comparing the results of the damped and undamped cases, it can be seen that the damping values have been increased by a factor of 10 or more for most modes of vibration.

Additional tests were performed which revealed that 15% of the damping treatment could be removed with no significant change in spectrum shape. This is clearly shown in Figure 6.79. These tests were conducted to investigate the possibility that some of the damping treatment may chip during installation. The flaking problem which was encountered did not result in any damping degradation. Therefore it was decided that for the high temperature enamel to be as effective as the room temperature treatment, and to completely alleviate any possible decrease in damping characteristics, the treatment should be applied to the entire surface in the same thickness (0.020 in., 0.051 cm).

Since the engine exhaust temperature ranged between 700° F (370° C) and 800° F (428° C), it was decided to select a damping material that exhibited optimum capabilities in that range. The enamel that conformed to these requirements was found to be Corning Glass No. 8363. The damping properties

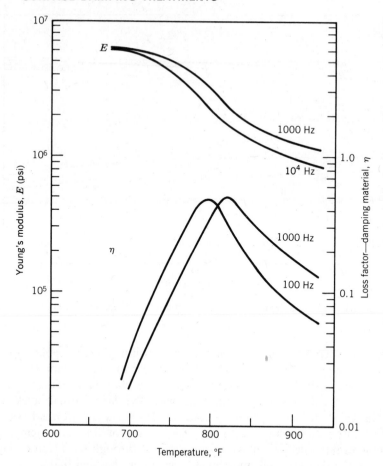

**FIGURE 6.78.**   Damping properties of Corning Glass No. 8363.

of this enamel have been measured and were previously presented in Figure 6.78 in terms of temperature and frequency. It can be seen from the data of Figure 6.78 that maximum damping for the enamel occurs at a somewhat lower temperature than that of the engine exhaust. This shift is in the appropriate direction since the outside surfaces of the exhaust extension were subjected to lower temperatures than those of the exhaust.

To verify the high temperature damping properties shown in Figure 6.78, a 321 stainless steel cantilever beam was coated with the vitreous enamel on both sides. This beam, having properties similar to the engine exhaust extension, was used to simplify further testing and verification of the high temperature damping treatment. The beam dimensions were 8.0 in. (20.32 cm) long, 0.75 in. (1.91 cm) wide, and 0.125 in. (0.317 cm) thick. The coating thickness was 0.008 in. (0.02 cm) on each side. The damping of the beam was measured in terms of frequency and temperature, and results are presented in Figure

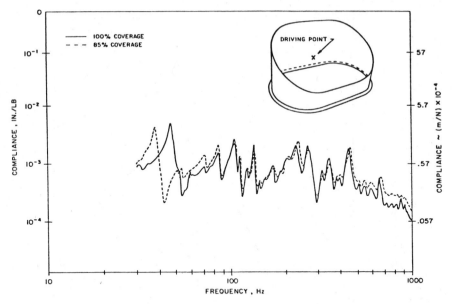

**FIGURE 6.79.**   Driving point frequency response of the damped exhaust extension showing 100% and 85% coverage.

6.80 for the third mode of vibration. Superimposed on these data are the predicted performances of the beam using the characterized properties obtained from Figure 6.76. The predicted values were computed using the equations of Section 6.7 for symmetric free-layer treatment. It can be seen from Figure 6.80 that there is good correlation between the two. Therefore the measured properties of the enamel can be used for optimization purposes with confidence.

The thickness of the enamel coating for the tailpipe extension was chosen to be 0.020 in. (0.051 cm) in order to achieve the same performance as that achieved with the room temperature damping tape. This selection was also confirmed by using the characterized material properties to predict the damping of the tailpipe extension for various thicknesses of enamel. It was found that high damping could be achieved over the operational temperature range with a 0.010 in. (0.025 cm) coating on each side of the structure. This is illustrated in Figures 6.81 and 6.82.

The entire tailpipe extension was coated with the exception of the flange. The flange was not coated so as to alleviate any potential installation problems. The applied damping treatment resulted in a total tailpipe extension weight of 15.28 lb (6.95 kg), an increase of 4.25 lb (1.93 kg).

**Verification of treatment performance.**   The coated tailpipe was inspected and installed on a CH-54 Helicopter (A/C 18470). It was observed during the first 30 hours that some flaking of the damping material occurred

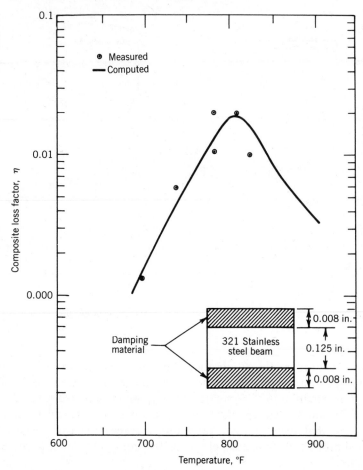

**FIGURE 6.80.** Comparison between computed and measured damping values for a 321 stainless steel coated beam.

on the inside section of the tailpipe. It was determined that the exhaust flow ablated the damping treatment from the inner tailpipe surface. Further investigations showed that the damping material could be applied to the outside surface only, maintaining the 10:1 increase in damping achieved previously. This would eliminate the flaking problem and tend to increase further the service life of the tailpipe extension.

Upon completion of the 100 hour flight test evaluation, the treated tailpipe extension was removed, and a detailed inspection revealed no indications of metal fatigue. The flaking on the inner surface, discussed previously, was again observed. This degradation did not significantly increase, subsequent to its discovery during the initial 30 hours. The outer surface of the extension showed no evidence of flaking, and substantiates the recommendation to apply the material to the outer surface only.

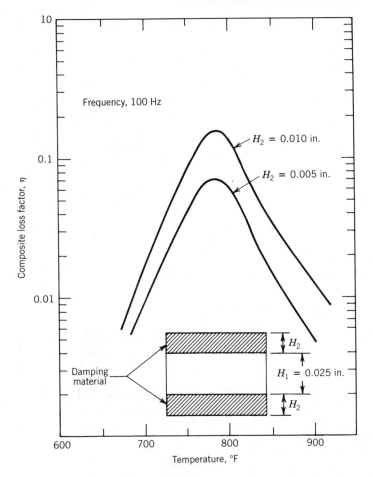

**FIGURE 6.81.**   Composite loss factor of the damped exhaust extension at 100 Hz.

## 6.8.5.  Application of Damping to Noise Control in a Diesel Engine

**Introductory remarks.**   Although noise and vibration problems can vary in their nature and severity, there exist only a limited number of ways to solve them. The most frequently used approaches to solve such problems are source modification, structural modification, damping, isolation, enclosures, and barriers. Considering the various options, it is essential that the proper one(s) be used for each application to achieve the most cost- and time-effective solution.

To illustrate the role of damping in the overall design procedure for solving a given problem, let us consider a typical structure-borne noise problem. The approach to vibration problems is similar because structure-borne noise problems are concerned with the noise resulting from vibration of the structure.

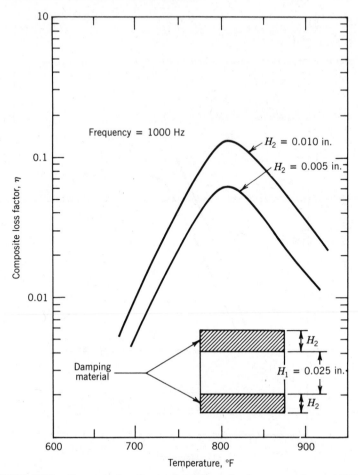

**FIGURE 6.82.**   Composite loss factor of the damped exhaust extension at 1000 Hz.

This discussion will concentrate mainly on experimental approaches and procedures for defining the problem. The reason for such emphasis is that most noise and vibration problems are quantified after the structure is built because most designers and manufacturers do not like to admit that they are going to have any noise and or vibration problems with their designs and/or are unaware of potential problems until they arise. However, after the structure is built, unforeseen problems frequently arise and/or the use and specifications for the structure may change, thus causing new problems. In such cases, where the structure is available for test, it is more cost- and time-effective to follow an experimental approach to define the problem. For such cases the structure is usually instrumented with a number of measurement transducers to define the problem better and develop the appropriate treatment. Such an approach can be illustrated by considering the overall noise of an internal combustion engine.

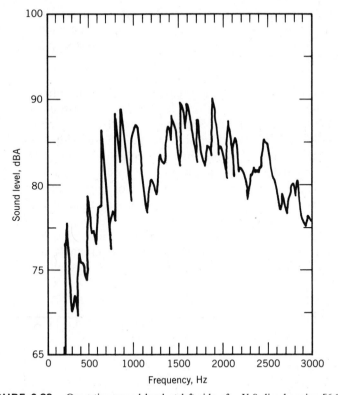

**FIGURE 6.83.**   Operating sound level at left side of a V-8 diesel engine [6.15].

Figure 6.83 represents the overall noise level of a V-8 diesel engine measured 1 m away from its left side. This spectrum represents the structure-borne
noise of the engine only. Both the intake and exhaust noise were ducted away,
and the fan was removed from the engine. Thus the noise shown in the figure
is caused by the vibration of the external surfaces of the engine only. The
important question now is to determine which of the many surfaces of the
engine can be treated by damping and how to predict its effect on the engine
overall noise. As part of an overall noise reduction program, the engine is
usually instrumented with a number of internal and external transducers to
define its internal forces, transmission paths and the radiation characteristics
of its external surfaces. The overall engine noise can be reduced by working on
these three items. Although the design of the damping treatment, discussed in
this chapter will be limited to the external surfaces, a brief discussion of the
other options is given.

**Excitation forces.**   In the internal combustion engine there are two primary excitaton forces—the combustion and mechanical forces. Changes to the
combustion forces, such as timing, could affect the overall engine noise. However, such changes also affect the engine performance, emission, and economy,

and therefore only limited changes can be made. The mechanical forces are governed by piston slap, and a number of parameters are available to reduce them, such as piston and cylinder wall dynamics, piston mass, piston clearance, and piston offset.

**Transmissibility paths.**   The engine forces are transmitted to the external surfaces along two major paths. The first is the piston–connecting rod–crankshaft–crankcase path and is primarily excited by the combustion forces. The second is the impact of the piston on the cylinder wall or the liner on the head and then to the rest of the engine. Changes can be made to either path, in isolation, to reduce the transmission of forces to the external surfaces. Damping usually is not an effective option for engine internal components because of their high stiffness and high initial structural damping.

**Radiating surfaces.**   The vibration of the various external surfaces of the engine affect its overall noise in different ways. Figure 6.84 shows the level of noise contributed by several major components to overall engine noise. To

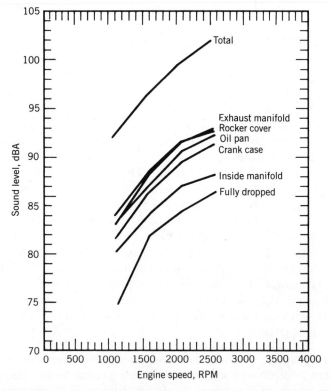

**FIGURE 6.84.**   Noise contribution of various components and surfaces of a diesel engine [6.15].

reduce the engine noise level, it is essential to reduce the level of each one of these major contributing components. To identify which approach, such as damping, isolation, or stiffening, is more suited for a given component, it is very important to identify the means by which particular radiation characteristics are affecting the overall noise problem within the frequency range where the problem exists. This depends on the dynamics of the structure that controls its response and therefore on whether the noise associated with the structure is forced or resonant. If the response is forced, damping is not likely to introduce any effect into the system, and other approaches are necessary to solve the problem, such as stiffening, mass, or isolation treatments. If the structure is responding in a so-called resonant response, where there are many resonances in the frequency range of the noise or vibration problem, then there are a number of available approaches, one of which is damping. At this point it is very important to consider which of the various approaches will result in the most cost- and time-effective solution to the problem. Consider, for example, that the valve cover is made of sheet metal construction. The various approaches that can be undertaken to solve the problem include isolation to reduce the forces coming from the head to the valve cover, damping to control the resonant response of the valve cover, or a new material and structural configuration of different shapes, such as cast plastic components with high damping–high stiffness properties. For each of these approaches the production costs, the material feasibility, the material manufacturing availability, and the noise reductions that can be achieved are considered. Unless these steps are followed in sequence, it will not be feasible to say whether damping is the best approach to the problem or not. In either case, in the following discussion, it will be assumed that the damping approach is adaptable to a given problem, and we will introduce the expected improvements. Under such conditions the following procedures for the design of the damping material could be adopted.

**Definition of the problem.** The first step in controlling the noise generated by any specific component on an engine structure is to rank and quantify its contribution to the overall engine noise. Customarily noise measurements are performed for a completely wrapped engine, with each one of its major noise-contributing components independently exposed in sequence. However, recently developed techniques are now being used to measure the noise contribution of various components on surfaces of an engine structure without the need for wrapping. Of these the acoustic intensity technique [6.15] has been most widely used. This technique utilizes the sound pressure levels measured by two microphones located near the surface of a given component. From these two sound measurements the velocity of the sound particles can be determined, giving a measure of the sound intensity at that location. By scanning the entire surface of the component, and integrating over its area, the sound power for that component can be established. This technique can be applied equally well to either a running or a stationary engine. In the case of a stationary engine the structure is excited by a force representing as closely as possible that of the operating engine. This approach is well suited for investigating the effects of various environmental factors, such as temperature, on the performance of the damping treatment, as will be illustrated for the valve cover case.

The experimental setup consisted on an engine head with a valve cover attached to it as in normal installation. The head was installed on soft mounts inside an environmental chamber and then excited with a shaker. The force was selected to be random, over a frequency range considered to be important on the basis of earlier vibration response measurements made on the head while the engine was operating. Although an attempt was made to simulate the relative amplitude and frequency contents of the experimental force, no attempt was made to match its magnitude with that of the operating engine. The measured noise and vibration data resulting from this force will be shown in normalized form. The vibration of the valve cover was measured by an accelerometer mounted on its surface while its noise contribution was measured by scanning its surface in accordance with the acoustic intensity technique. Figures 6.85 and 6.86 illustrate typical valve cover noise and vibration response spectra resulting from the input applied force. Although there is a strong indication, as can be seen from these figures, that the noise of the valve cover is caused by its vibration, the nature of such a vibration (e.g., whether it is of the forced and/or resonant type) cannot be fully determined at this point without performing a number of additional tests.

To gain more understanding of the nature of the valve cover vibration, it is necessary to determine its dynamic behavior over the frequency range of interest. This is usually accomplished by applying a known exciting force to the valve cover and measuring its response. This can be done by a number of

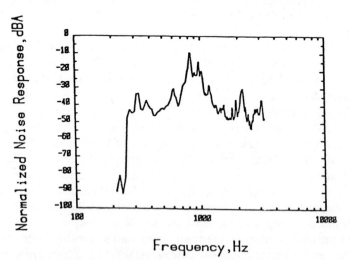

**FIGURE 6.85.** Noise contribution of the undamped valve cover.

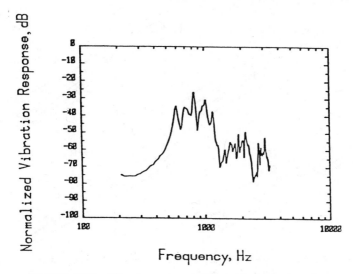

**FIGURE 6.86.**   Vibration response of the undamped valve cover.

different methods [6.15], all of which generate driving and transfer point frequency response plots. Figure 6.87 illustrates a typical driving point frequency response plot generated on the top side of the valve cover using the impact technique. It can be seen from this figure that there are a number of resonances, starting at approximately 600 Hz, with low damping values, that is, having modal loss factors of the order of 0.02 or less.

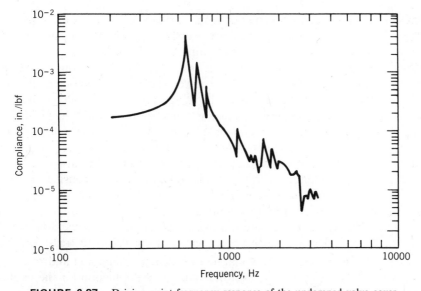

**FIGURE 6.87.**   Driving point frequency response of the undamped valve cover.

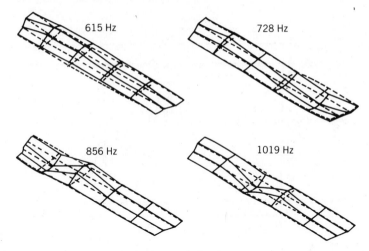

**FIGURE 6.88.** Typical mode shapes of vibration of the undamped valve cover.

To characterize further the dynamic behavior of the valve cover, it is also necessary to measure the mode shapes of vibration for the resonances of interest. To obtain the mode shapes, a number of frequency plots are generated, similar to the one shown in Figure 6.87 but at a number of different points on the valve cover. By comparing the relative amplitude and phase angle between all points for each resonance, it is possible to obtain the deformation patterns for the valve cover for a number of resonances, as illustrated in Figure 6.88.

**Design of the damping treatment.** Since the valve cover is seen to have a number of resonances, which are usually excited by the engine forces, it is necessary to follow the design procedure for at least the most important resonances. For practical purposes it is usually sufficient to consider resonances typical of those occurring at the lower and upper portions of the frequency range. Thus, if the results of the design procedure are acceptable for both extremes, the resonances within this range should be well handled.

As with all applications of damping treatments to solve noise and vibration problems, it is usually necessary to consider the effects of a given treatment not only in increasing the modal damping of the structure but also as to how it affects its other modal properties. This is very important to consider because it is usually difficult to increase significantly the modal damping of a structure without changing some of its other modal properties.

For the case where a structure is excited at its support with random vibration, the criterion measuring the effectiveness of the damping treatment (see Section 1.3.3.) is $\sqrt{\eta/\omega}$, where $\eta$ and $\omega$ are the modal damping and resonant frequency, respectively. Thus, to get maximum noise reduction for the treated cover, it is necessary to increase its damping as much as possible while keeping its resonant frequency as low as possible. This implies that the treatment has to be operating in a range where the damping material is relatively soft.

Although many modes of vibration are often considered, the design procedures in this chapter will be limited to those associated with the fundamental mode. To design the damping treatment, it is necessary to obtain the following information relative to the operating environment and dynamic response of the valve cover:

1.  The temperature range of operation for the valve cover, which was approximately 50° C(122° F) to 120° C(248° F).
2.  The frequency of the mode of vibration to be damped, which was 615 Hz.
3.  The mode shape of vibration of the resonance to be damped. This can be simplified by representing it in terms of an equivalent simply supported beam or plate. In this approximate case a semiwavelength of vibration of 76.2 mm (3 in.) was used.
4.  The initial modal damping in the system, which corresponds to a loss factor of approximately 0.02.
5.  The thickness of the valve cover in production, which was 1.6 mm (0.063 in.).
6.  The allowable limits for reducing the fundamental resonant frequency of the cover without causing any durability problems.

Having established the information, the damped cover can now be designed. First, a damping material is selected that has good damping capabilities over the operating temperature range of the valve cover. The material had to be resistant to oil and long-term exposure to heat. Second, if the treatment is going to be manufactured using existing tools, then the total thickness of the damped valve cover should be the same as that of the original undamped valve cover. This implies that the damping material needs to be sandwiched between two steel layers, each having a thickness equal to slightly less than half of the original undamped steel cover. Third, the thickness of the damping material is computed, as in Section 6.5, to generate the desired modal damping and natural frequency.

Figures 6.89 and 6.90 illustrate the resulting damping performance and natural frequency of a sandwich configuration utilizing the material properties of Figure 6.91 and a semiwavelength of vibration of 76.2 mm (3 in.). It can be seen from these figures that the treatment performed well in achieving a high loss factor and a low natural frequency over a wide temperature range, exceeding the operating range of the valve cover.

**Fabrication and evaluation of the damping treatment.** Once the damping treatment has been designed, it must be fabricated in a way suitable for production. The damping material was placed between two steel layers

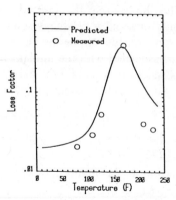

**FIGURE 6.89.** Variation of the laminated valve cover damping with temperature.

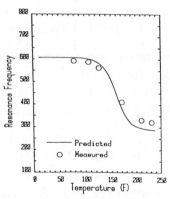

**FIGURE 6.90.** Variation of the laminated valve cover frequency with temperature.

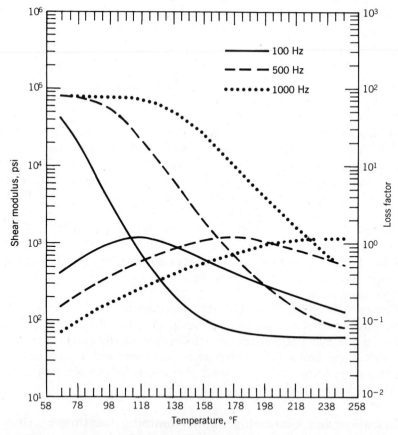

**FIGURE 6.91.** Variation of the properties of the damping material with temperature and frequency.

using a special adhesive layer for improving its bonding properties. Thus the system is actually a five-layer system consisting of two steel layers, two adhesive layers, and a damping material in the middle. The adhesive layer thickness was kept to a minimum so that it did not interfere with the damping layer performance, and was much smaller than that of the damping layer. A specially selected steel alloy was used which could be deep-drawn during the stamping process. Such a system has similar stiffness properties, at room temperature, as those of a solid steel construction, as seen in Figure 6.90, which shows the variation of the resonant frequency with temperature.

After assembly the sandwich treatment can be formed into the shape of the valve cover. The damped valve cover can then be evaluated for its damping performance and noise reduction. Figure 6.92 represents a driving point frequency response plot on the valve cover at 65° C(149° F), along with the results of Figure 6.87. It is seen that good damping has been achieved at the desired temperature. The measured loss factor at this temperature, for the fundamental mode of vibration, is approximately 0.4, which is in good agreement with the predicted values of Figure 6.89.

The next step is to measure the noise contribution of the damped valve cover and compare it with the original solid steel one. Figures 6.93 and 6.94 illustrate the resultant vibration and noise reductions achieved with the damped valve cover, at 150° F, as compared with the original undamped case. It is seen that both the vibration and the noise levels have been significantly

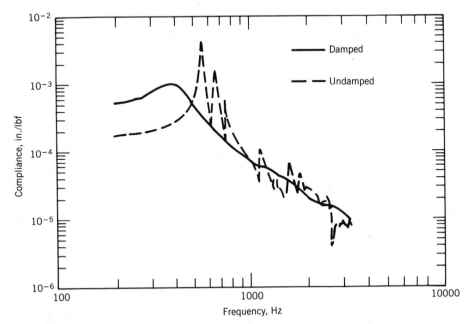

**FIGURE 6.92.** Comparison of the driving point frequency response for the damped and undamped covers.

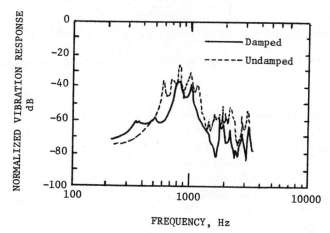

**FIGURE 6.93.** Comparison of the vibration response for the damped and undamped covers.

reduced by the incorporation of damping. The previous plots were repeated for a number of different temperatures, and the overall noise contribution of the cover is shown in Figure 6.95 as a function of temperature. Superimposed on the measured data of Figure 6.95 are the variations of the term $\sqrt{\eta/\omega}$, with temperature, for the fundamental mode of vibration of the valve cover. These parameters were calculated from the modal information generated at different temperatures, giving both the damping and natural frequency. It is interesting to note that this simple criterion for the damping effectiveness of the treatment is reasonably effective in correlating with the measured data. At low temperatures the damping of the system is low, and its stiffness is high so that it is quite similar to the solid steel construction. At very high temperatures the material is very soft with low damping capabilities, and thus its effectiveness is

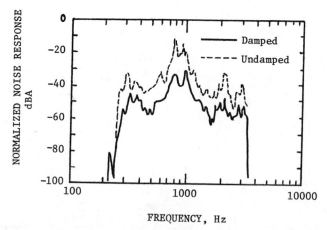

**FIGURE 6.94.** Comparison of the noise contribution of the damped and undamped covers.

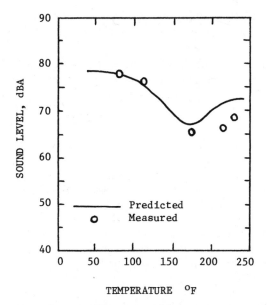

**FIGURE 6.95.** Variation of the noise contribution of the damped valve cover with temperature.

due to the reduced resonant frequency. In the transition region of the material the damping is extremely high, and the resonant frequency is somewhat lower than for the solid steel one so that the effectiveness of the treatment is greatest.

**Practical considerations.** The application of damping treatments to solve noise and vibration problems is a frequently misunderstood subject. It can be seen from the foregoing examples that extreme care must be taken to define the problem, design the treatment, fabricate it, and apply it to the structure. This process is necessarily lengthy and comprehensive; hence the chance for success of a "hit and miss" application of these techniques is very small. Also it should be emphasized that optimizing the treatment for only its damping performance without considering the effects of other modal properties will usually result in insufficient and unacceptable noise reductions. It is this necessity, along with the additional cost requirements, that has often led to situations where damping was considered only as a last resort. However, when used properly, damping can play an important role as one of the several necessary techniques needed to solve fully noise and vibration problems.

For reducing the noise contribution of sheet metal components in engine structures, such as oil pans, valve covers, and timing gear covers, where damping is not designed into the system, the alternative is to build enclosures around such components or to isolate them. However, as mentioned before, many isolation systems lead to undesirable leakage at the joints. The enclosure approach is not an attractive solution because it adds to the product weight, and cost, and leads to handling and serviceability problems. The damped sheet metal components, when properly stamped in a sandwiched configuration, avoid all of these shortcomings, except for the increase in the material cost which is a small portion of the total product cost. It is in situations like these that the damping approach provides an attractive solution because it is practical and cost-effective when compared with other alternatives.

## REFERENCES

6.1. D. Ross, E. E. Ungar, and E. M. Kerwin, Jr., "Damping of plate flexural vibrations by means of viscoelastic laminate," *Structural Damping*, ASME, New York, pp. 49–88, 1959.

6.2. H. Oberst and K. Frankenfeld, "Uber die Dampfung der Biegeschwingungen dunner Bleche durch festhaftende Belage," *Acustica*, **2**, 181–194 (1952).

6.3. G. E. Warnaka, "Effect of adhesive stiffness on extensional damping," ASME, 1967.

6.4. D. J. Mead, "The Effect of Certain Damping Treatments on the Response of Idealized Aeroplane Structures Excited by Noise," Air Force Materials Lab., AFML-TR-65-284, WPAFB, August 1965.

6.5. L. Bohn, F. Linhardt, and H. Oberst, "Progress in the development of vibration damping materials," WADC—Univ. of Minnesota Conference on Acoustical Fatigue, March 1961.

6.6. F. S. Owens, "Elastomers for damping over wide temperature ranges," *Shock Vib. Bull.*, **36**, Pt 4, 25–35 (1967).

6.7. D. I. G. Jones, "Effect of free layer damping on response of stiffened plate structures," *Shock Vib. Bull.*, **41**, Pt. 2, 105–120 (December 1970).

6.8. D. I. G. Jones, "Design of constrained layer treatments for broad temperature damping," *Shock Vib. Bull.*, **44**, Pt. 5, 1–12 (1974).

6.9. D. I. G. Jones and W. J. Trapp, "Influence of additive damping on resonance fatigue of structures," *J. Sound Vib.*, **17**(2), 157–185 (1971).

6.10. L. C. Rogers and A. D. Nashif, "Computerized processing and empirical representation of viscoelastic material property data and preliminary constrained layer damping treatment design," *Shock Vib. Bull.*, **48**(2), 23–37 (1978).

6.11. A. D. Nashif and T. Nicholas, "Vibration control by a multiple-layered damping treatment," *Shock Vib. Bull.*, **41**(2), 121–131 (1970).

6.12. D. I. G. Jones, A. D. Nashif, and M. L. Parin, "Parametric study of multiple-layer damping treatments on beams," *J. Sound Vib.*, **29**(4), 423–434 (1973).

6.13. D. L. Knighton, "The design of multiple constrained layer damping treatments," Paper presented at ASA Meeting, Cambridge, Mass., 1979.

6.14. J. J. DeFelice and A. D. Nashif, "Damping of an engine exhaust stack," *Shock Vib. Bull.*, **48**(2) 75–84 (1978).

6.15. D. E. Baxa (ed.), *Noise Control in Internal Combustion Engines,* Wiley, New York, 1982.

6.16. J. P. Henderson, "Damping applications in Aero-propulsion systems," ASME Publication AMD-Vol. 38, *Damping Applications for Vibration Control,* ed. P. J. Torvik, 1980.

# 7

# DESIGN
# DATA SHEETS

## 7.1. CAUTION

Since commercial viscoelastic materials are subject to batch by batch variation, and manufacturers may change composition or processing from time to time, the properties displayed in the Data Sheets cannot be guaranteed. They accurately depict the properties of the particular samples tested, and can be used as a guide for selecting materials for specific applications, but each user is responsible for verifying or otherwise establishing that the desired properties, both physical and damping related, are retained in each batch being considered for application in engineering practice. Table 7.1 lists some manufacturers of damping materials, along with relevant data sheets.

TABLE 7.1. Table of Manufacturers, and Materials, (1983)

| Manufacturer or Source | Material(s) | Data Sheets | | |
|---|---|---|---|---|
| Antiphon | Antiphon-13 | 001-A | 001-B | 001-C |
| Barry Controls | H-326 | 003-A | 003-B | 003-C |
| Blachford | Aquaplas | 002-A | 002-B | |
| Chicago Vitreous | CV325 | 004-A | 004-B | 004-C |
| Corning Glass | 0010 | 005-A | 005-B | 005-C |
| Dow Corning | Sylgard 188 | 006-A | 006-B | |
| EAR Corporation | C-1002 | 007-A | 007-B | 007-C |
| | C-2003 | 008-A | 008-B | 008-C |
| Fel-Pro | Nitrile rubber | 016-A | 016-B | |
| General Electric | G. E. SMRD | 022-A | 022-B | |
| Lord Corporation | BTR | 009-A | 009-B | 009-C |
| | LD-400 | 010-A | 010-B | 010-C |
| Shell Chemical | Kraton | 024-A | 024-B | |
| Soundcoat, Inc. | DYAD 601 | 012-A | 012-B | 012-C |
| | DYAD 606 | 013-A | 013-B | 013-C |
| | DYAD 609 | 014-A | 014-B | 014-C |
| | Soundcoat N | 015-A | 015-B | |
| 3M Company | ISD-110 | 017-A | 017-B | 017-C |
| | ISD-112 | 018-A | 018-B | |
| | ISD-113 | 019-A | 019-B | 019-C |
| | 468 | 020-A | 020-B | 020-C |
| | ISD-830 | 021-A | 021-B | |
| | EC-2216 w/graphite | 023-A | 023-B | |

## 7.2.  DATA SHEETS GIVING COMPLEX MODULUS PROPERTIES AS FUNCTION OF TEMPERATURE AND FREQUENCY

**Data Sheet 001.   Damping properties of Antiphon-13.**

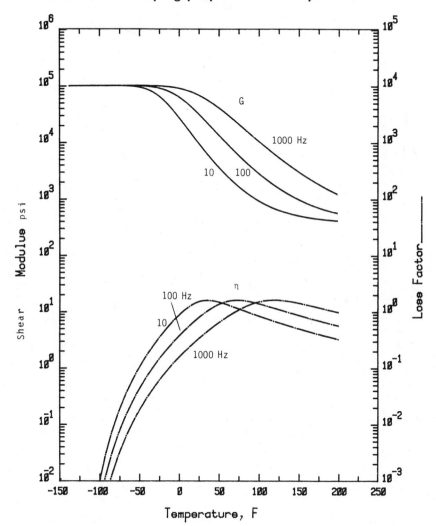

**001A.**  Damping properties with temperature.

**Data Sheet 001.   Damping properties of Antiphon-13 (continued).**

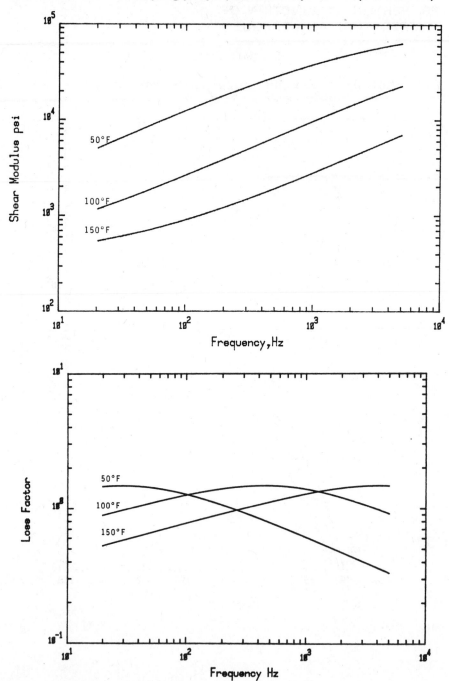

**001B.**   Damping properties with frequency.

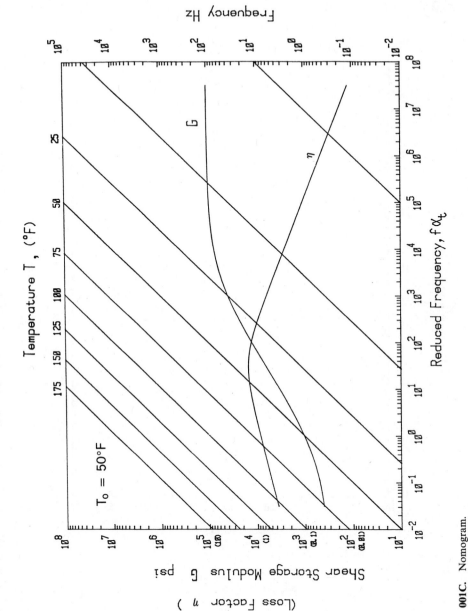

**001C.** Nomogram.

**367**

## Data Sheet 002. Damping properties of Blachford Aquaplas.*

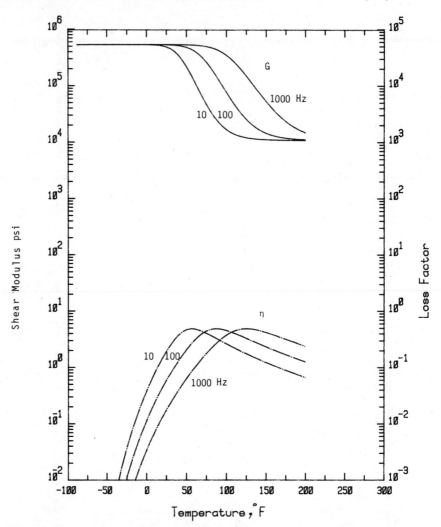

* Extensional damping treatment ($E = 3\,G$)

**002A.** Damping properties with temperature.

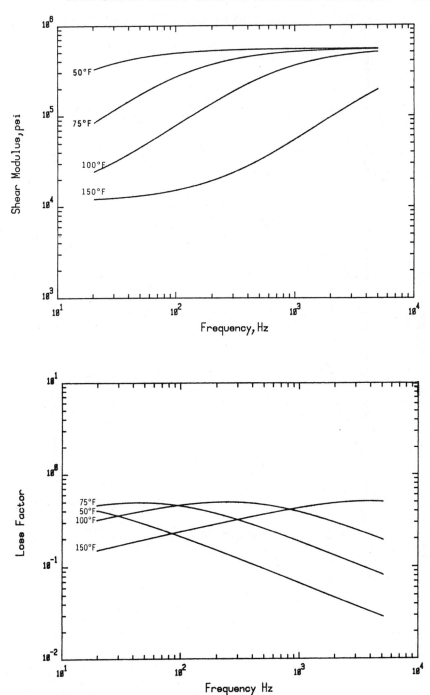

**002B.**  Damping properties with frequency.

## Data Sheet 003. Damping properties of Barry Controls H-326.

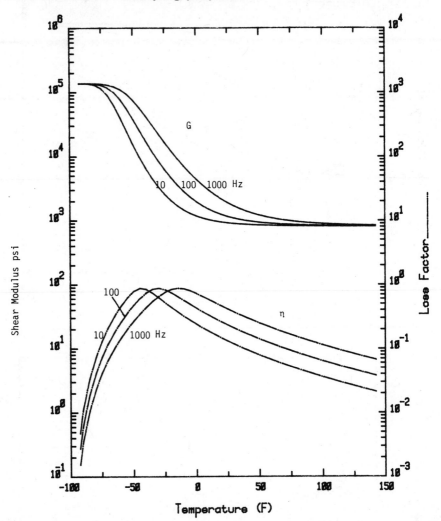

**003A.** Damping properties with temperature.

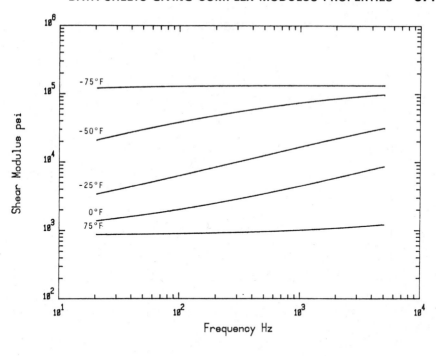

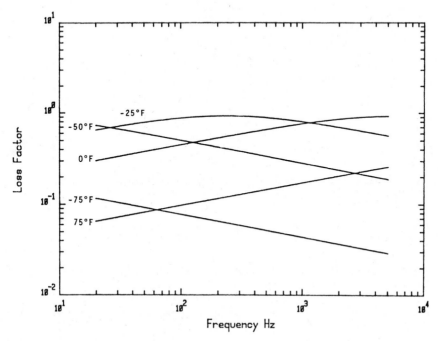

**003B.** Damping properties with frequency.

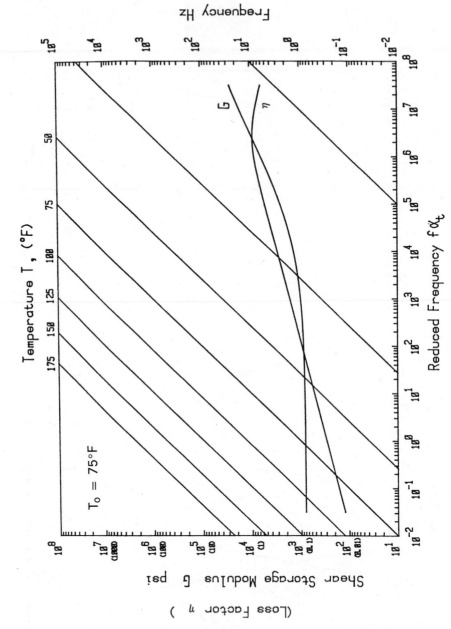

**003C.** Nomogram.

372

**Data Sheet 004.    Damping properties of Chicago Vitreous 325.**

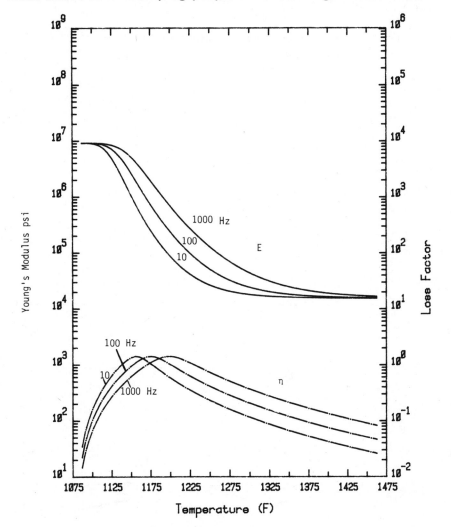

**004A.**    Damping properties with temperature.

**Data Sheet 004. Damping properties of Chicago Vitreous 325 (continued).**

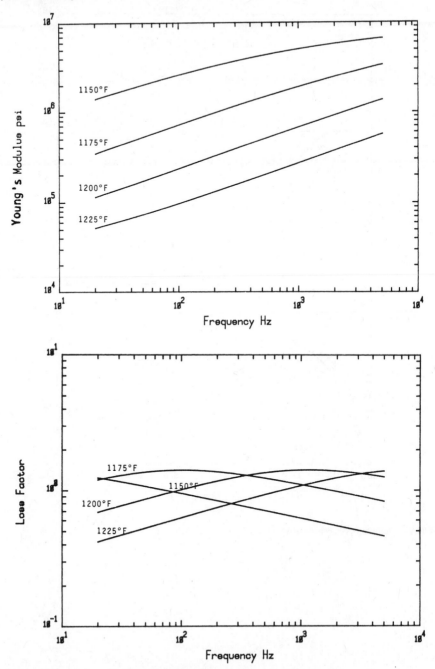

**004B.** Damping properties with frequency.

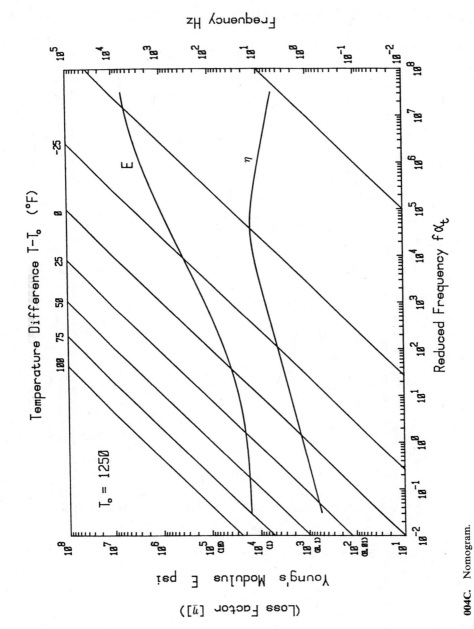

004C. Nomogram.

**375**

## Data Sheet 005. Damping properties of Corning 0010.

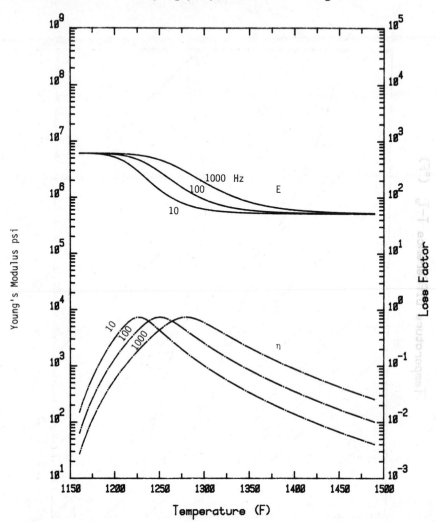

**005A.** Damping properties with temperature.

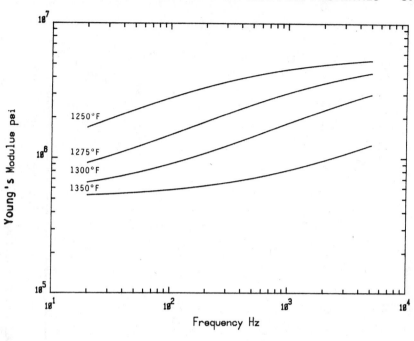

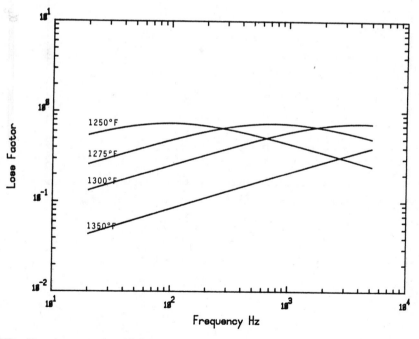

**005B.**  Damping properties with frequency.

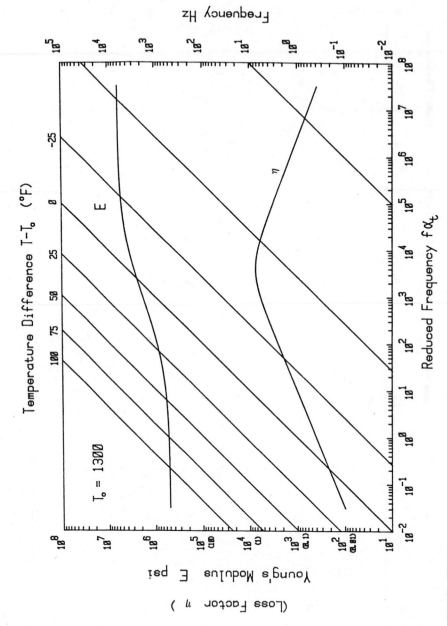

005C. Nomogram.

378

**Data Sheet 006.   Damping properties of Dow Corning Sylgard 188.**

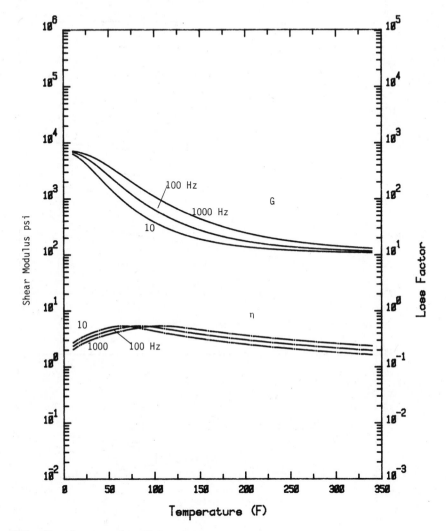

**006A.**   Damping properties with temperature.

**Data Sheet 006. Damping properties of Dow Corning Sylgard 188 (continued).**

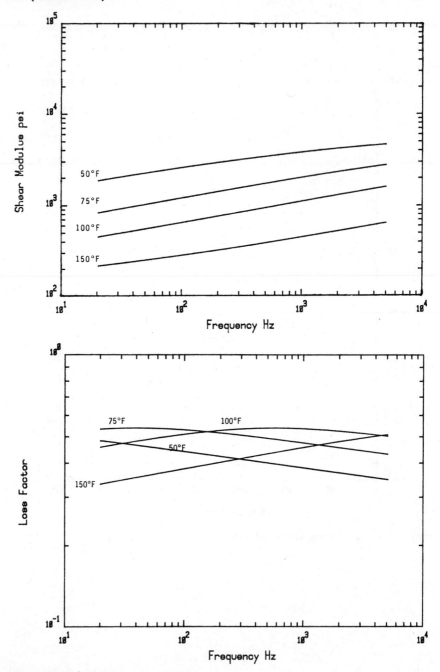

**006B.** Damping properties with frequency.

**Data Sheet 007.    Damping properties of EAR C-1002.**

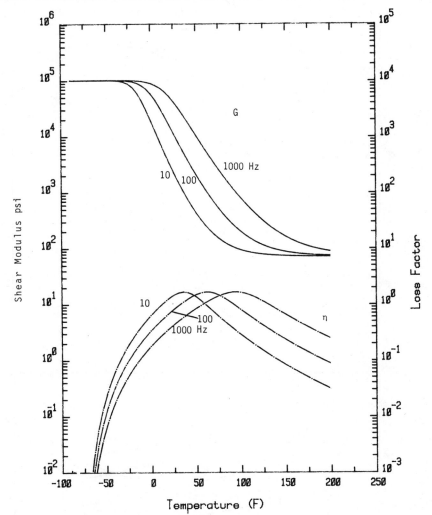

**007A.**   Damping properties of temperature.

## Data Sheet 007.  Damping properties of EAR C-1002 (continued).

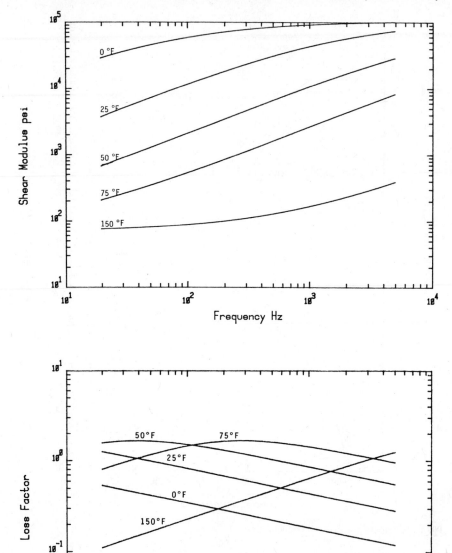

**007B.** Damping properties of frequency.

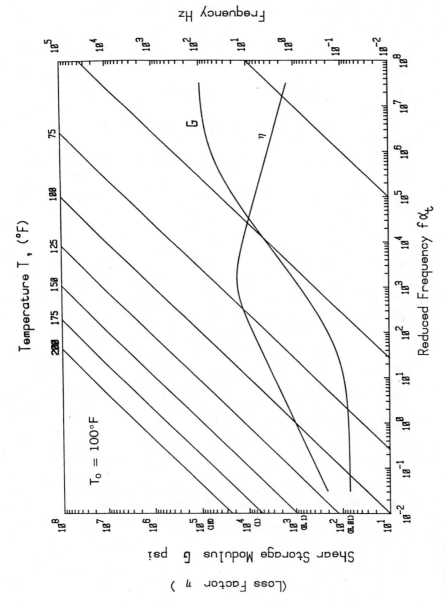

007C. Nomogram.

**383**

## Data Sheet 008.   Damping properties of EAR C-2003.

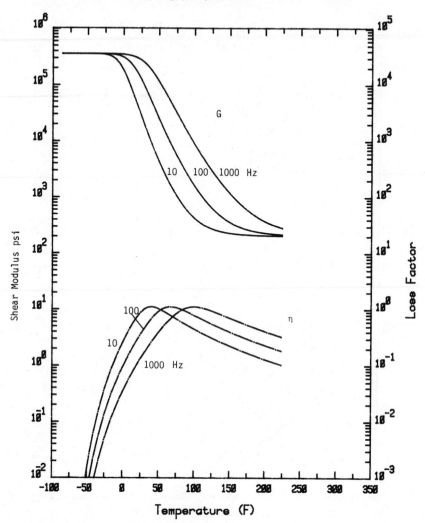

**008A.**   Damping properties with temperature.

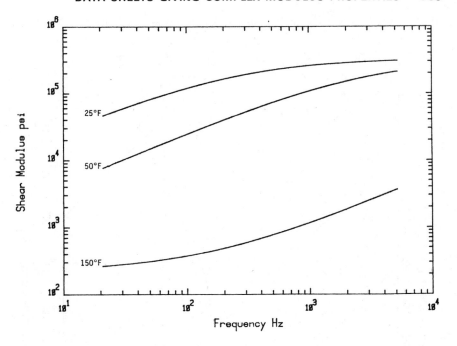

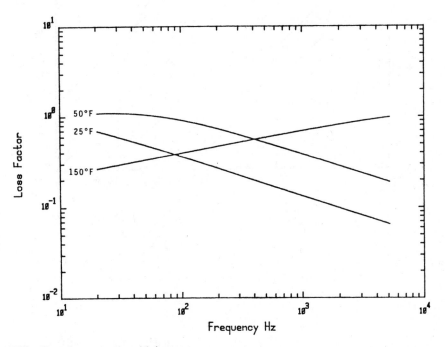

**008B.**  Damping properties with frequency.

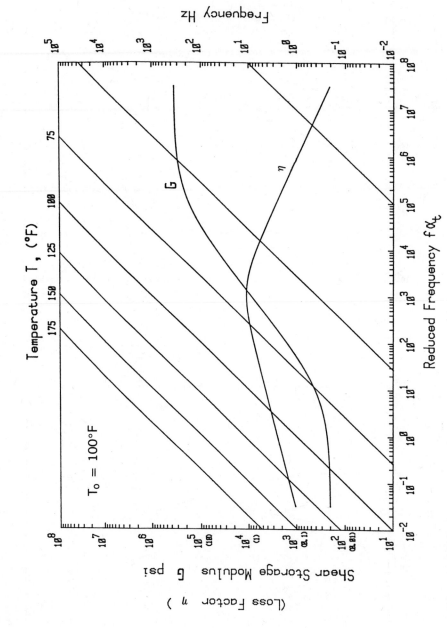

**008C.** Nomogram.

386

**Data Sheet 009.  Damping properties of Lord BTR.**

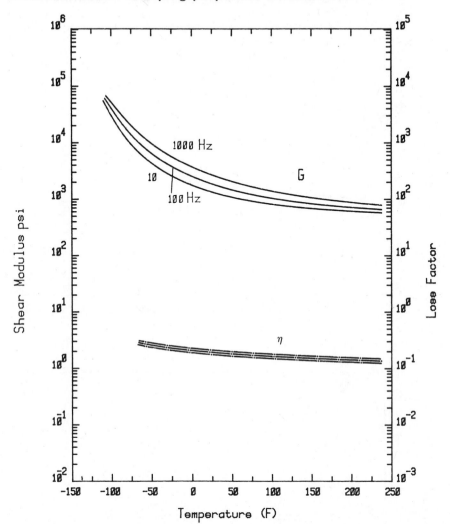

**009A.**   Damping properties with temperature.

## Data Sheet 009.  Damping properties of Lord BTR (continued).

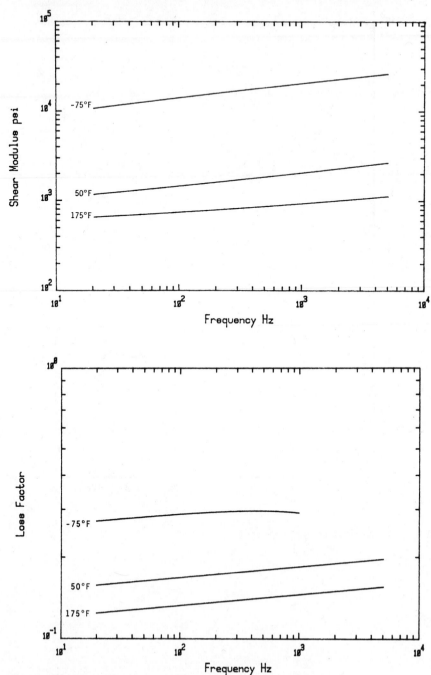

**009B.** Damping properties with frequency.

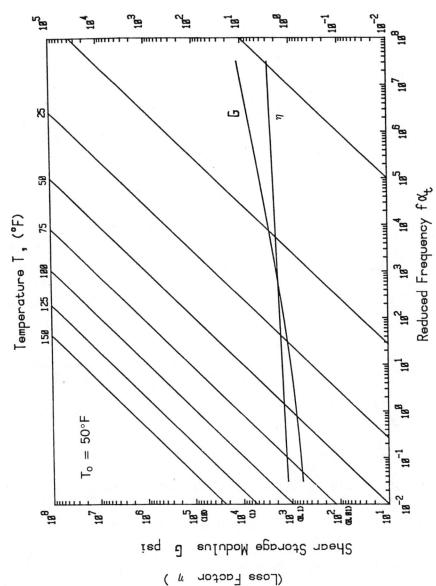

009C. Nomogram.

389

## Data Sheet 010.   Damping properties of Lord LD-400.*

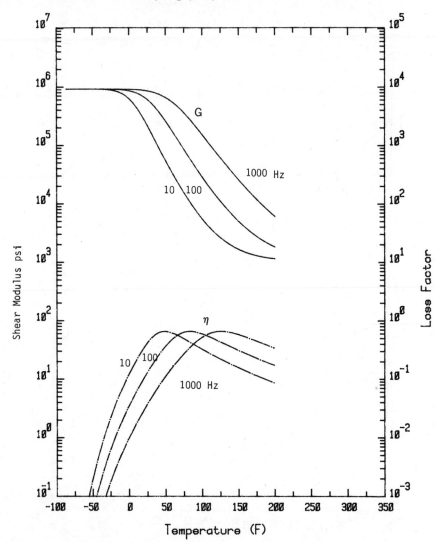

* Extensional damping treatment ($E = 3G$)

**010A.**   Damping properties with temperature.

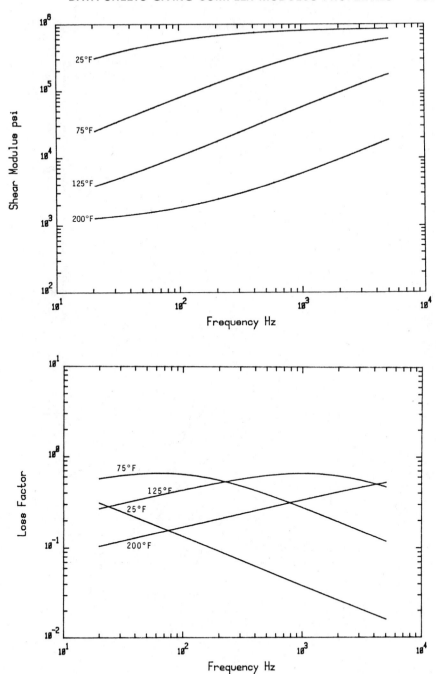

**010B.**   Damping properties with frequency.

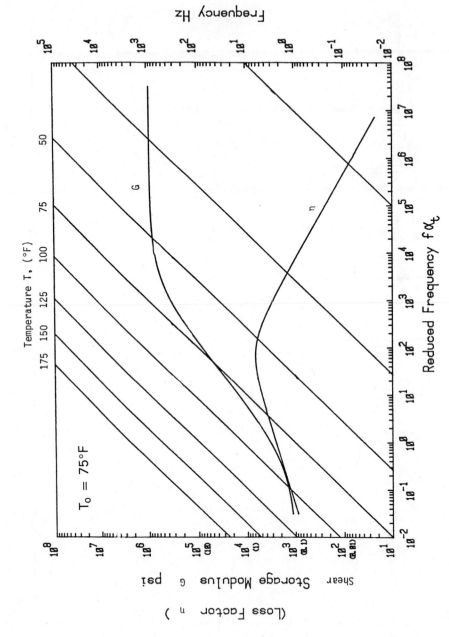

**010C.** Nomogram.

392

## Data Sheet 011.   Damping properties of Polyisobutylene (PIB).

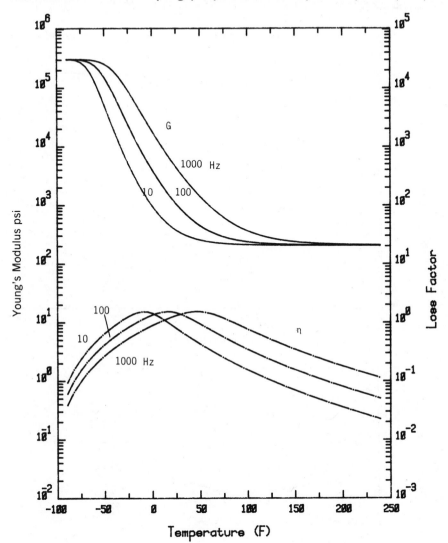

**011A.**   Damping properties with temperature.

## Data Sheet 011. Damping properties of Polyisobutylene (PIB) (continued).

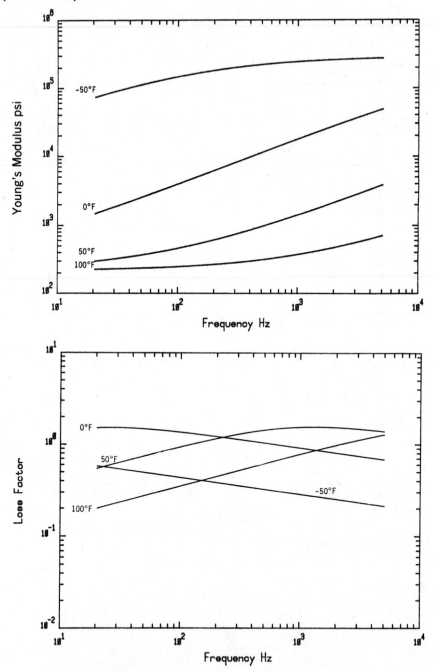

**011B.** Damping properties with frequency.

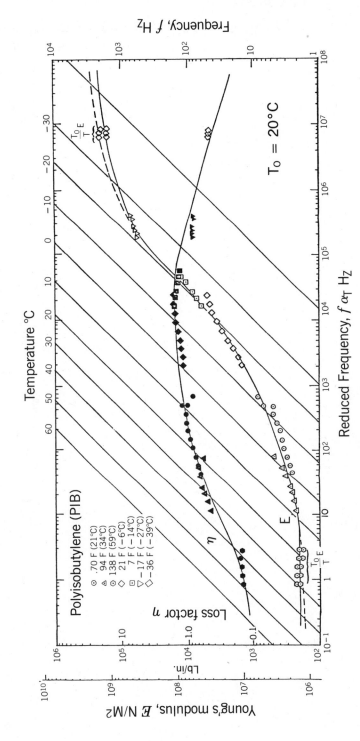

011C.  Nomogram.

**395**

## Data Sheet 012. Damping properties of Soundcoat DYAD 601.

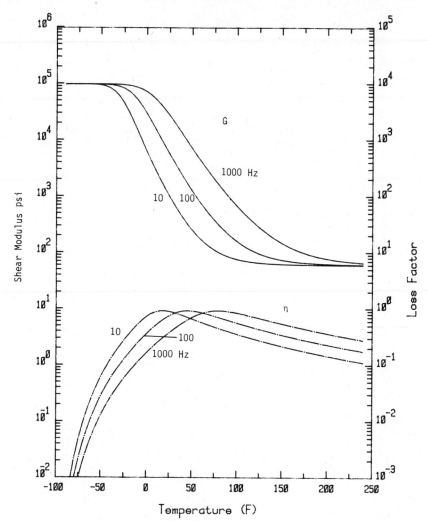

**012A.** Damping properties with temperature.

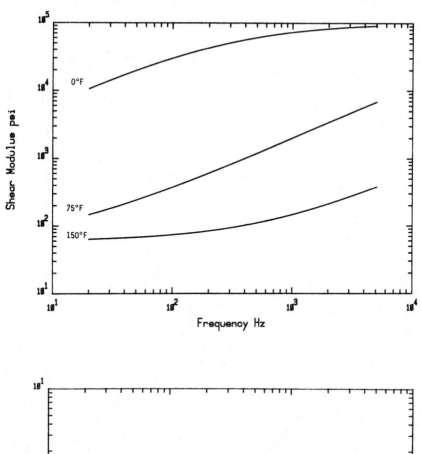

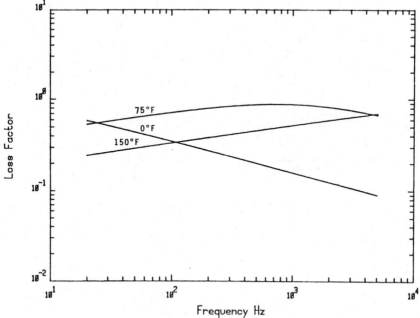

**012B.**   Damping properties with frequency.

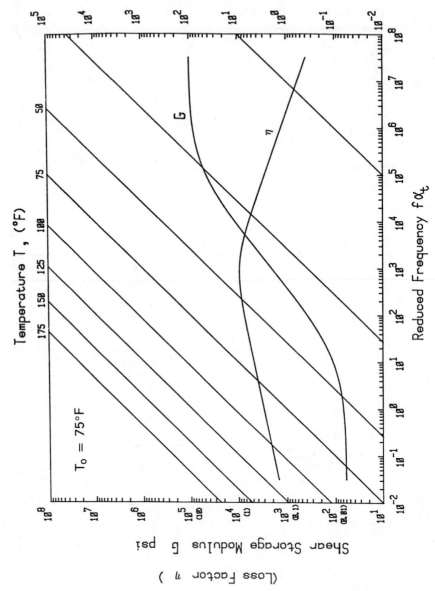

012C. Nomogram.

398

## Data Sheet 013.  Damping properties of Soundcoat DYAD 606.

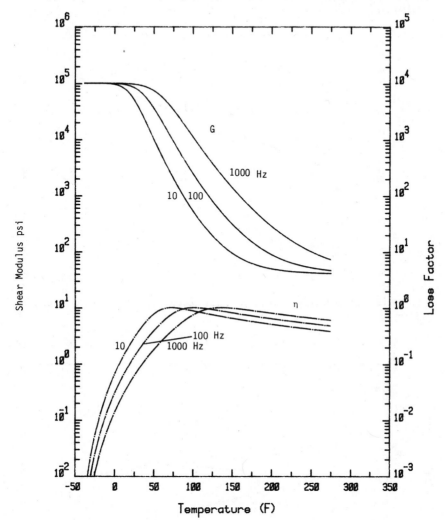

**013A.**  Damping properties of temperature.

**Data Sheet 013. Damping properties of Soundcoat DYAD 606 (continued).**

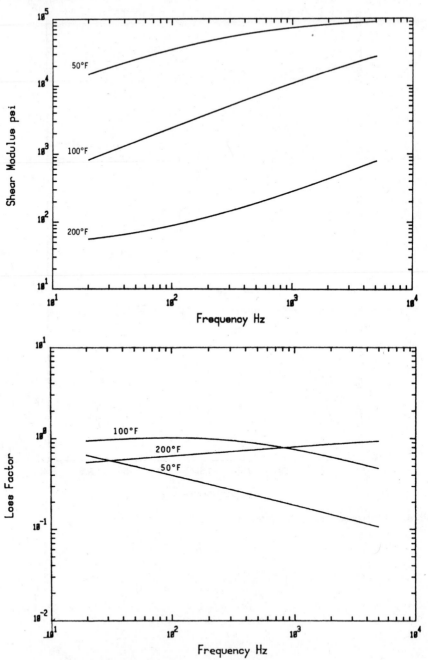

**013B.** Damping properties of frequency.

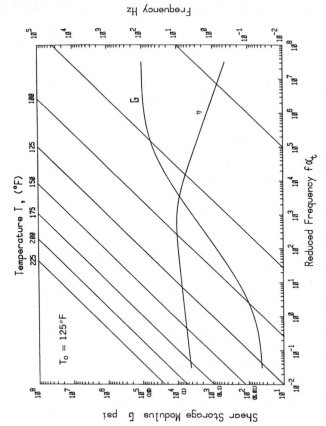

013C. Nomogram.

## Data Sheet 014.   Damping properties of Soundcoat DYAD 609.

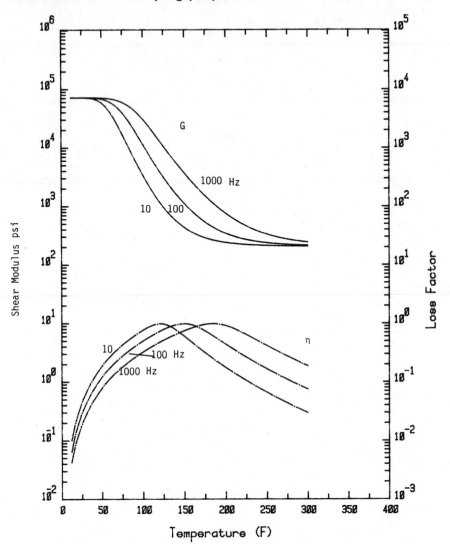

**014A.**   Damping properties of temperature.

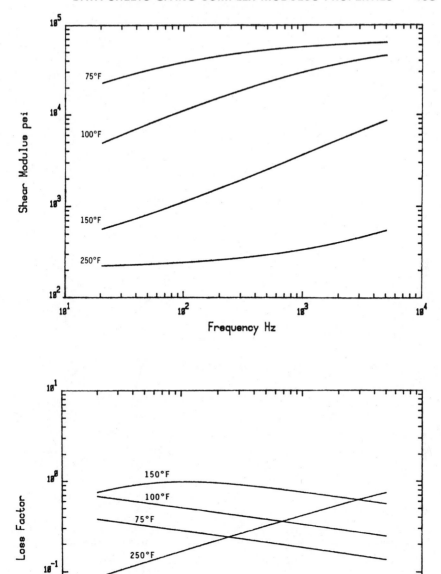

**014B.**  Damping properties with frequency.

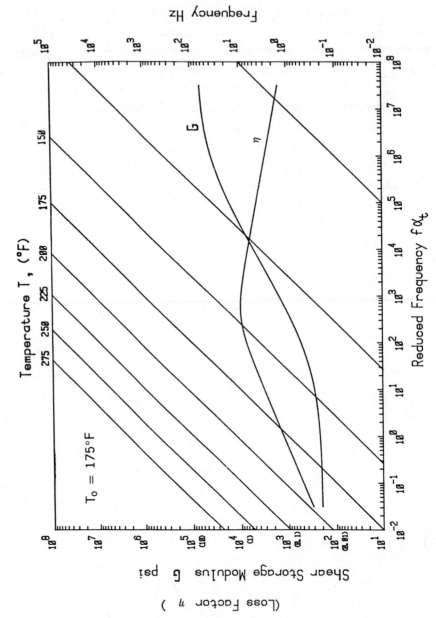

014C. Nomogram.

404

Data Sheet 015. Damping properties of Soundcoat N.

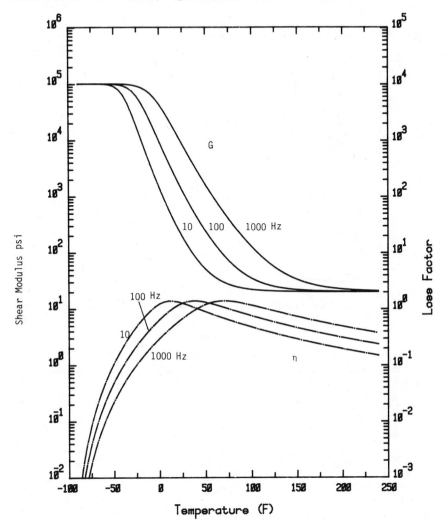

**015A.** Damping properties with temperature.

## Data Sheet 015. Damping properties of Soundcoat N (continued).

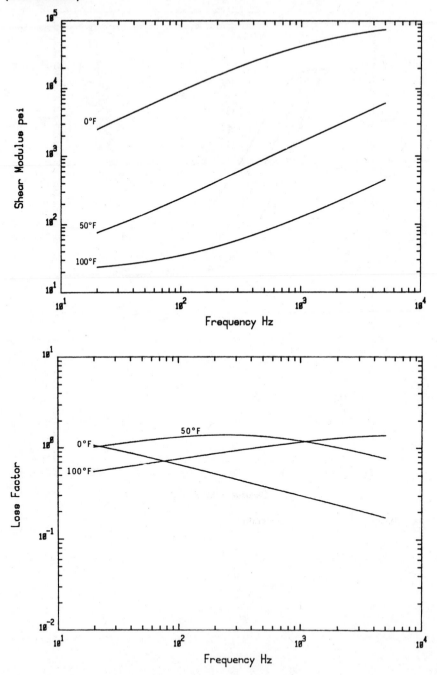

**015B.** Damping properties with frequency.

## Data Sheet 016.  Damping properties of Fel-Pro Nitrile Rubber.

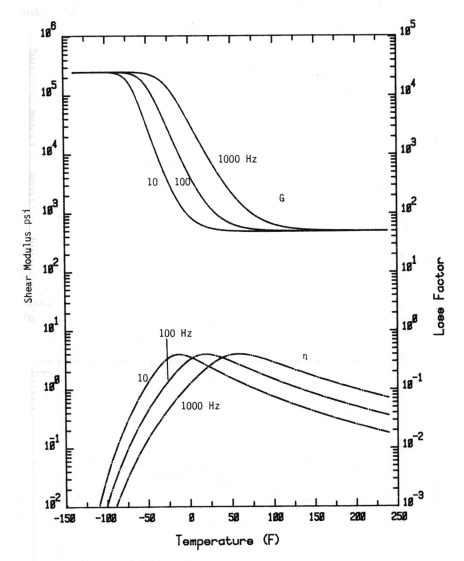

**016A.**  Damping properties with temperature.

## Data Sheet 016. Damping properties of Fel-Pro Nitrile Rubber (continued).

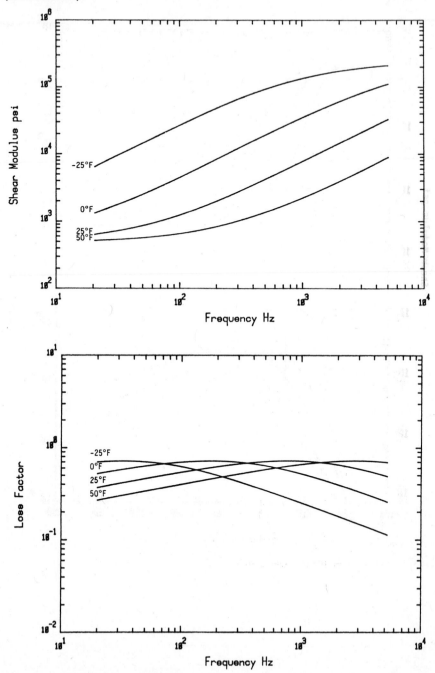

**016B.** Damping properties with frequency.

## Data Sheet 017.   Damping properties of 3M ISD-110.

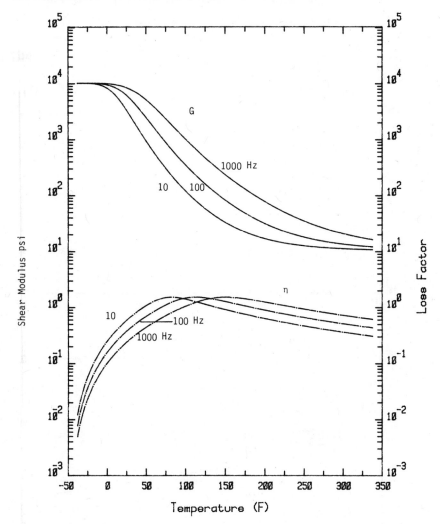

**017A.**   Damping properties with temperature.

## Data Sheet 017.   Damping properties of 3M ISD-110 (continued).

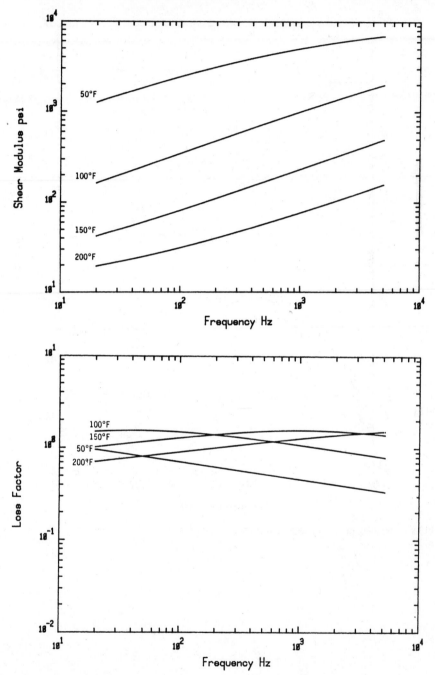

**017B.**   Damping properties with frequency.

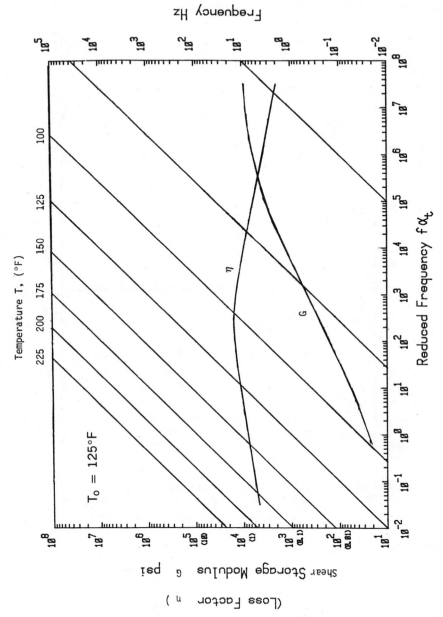

017C. Nomogram.

**411**

## Data Sheet 018.   Damping properties of 3M ISD-112.

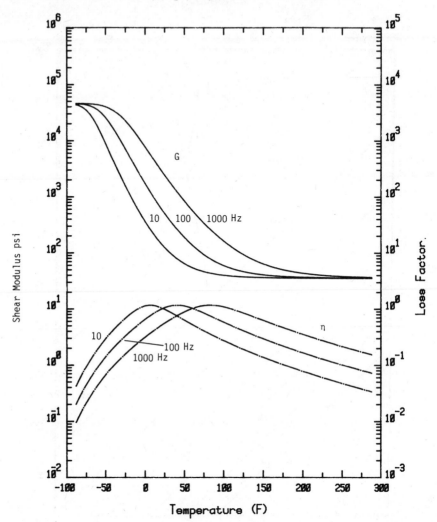

**018A.**   Damping properties with temperature.

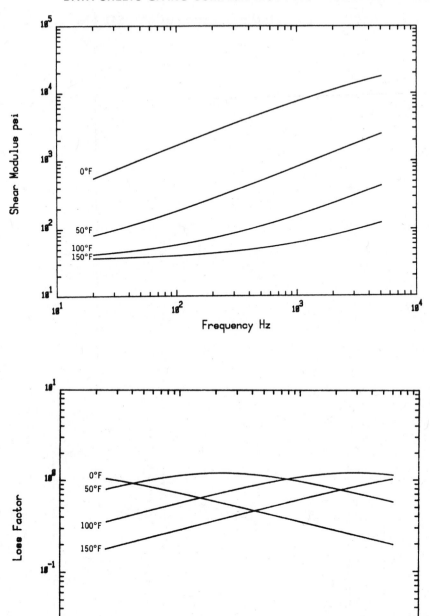

**018B.** Damping properties with frequency.

## Data Sheet 019.   Damping properties of 3M ISD-113.

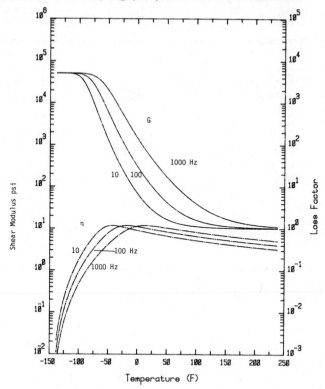

**019A.**   Damping properties with temperature.

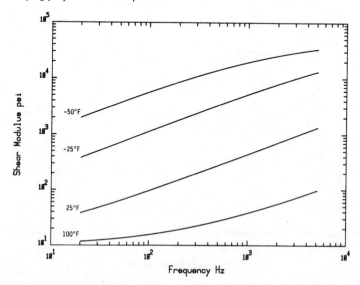

**019B.**   Damping properties with frequency.

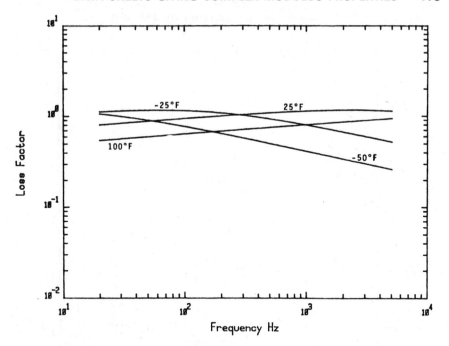

**019B.** (continued)

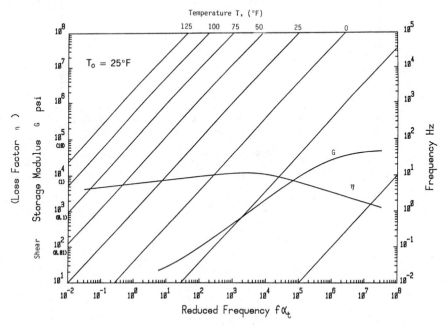

**019C.** Nomograms.

## Data Sheet 020.  Damping properties of 3M 468.

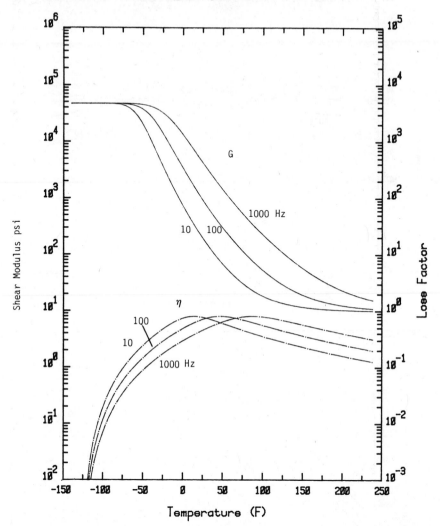

**020A.**  Damping properties with temperature.

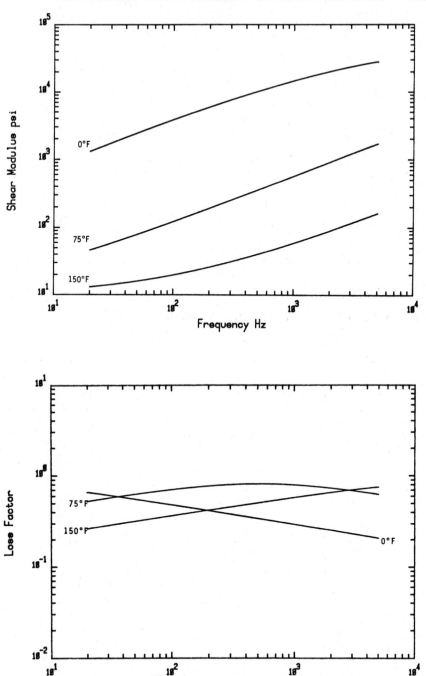

**020B.** Damping properties with frequency.

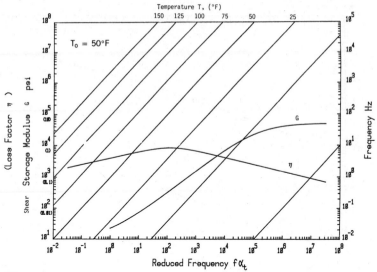

**020C.** Nomogram.

## Data Sheet 021. Damping properties of 3M ISD-830.

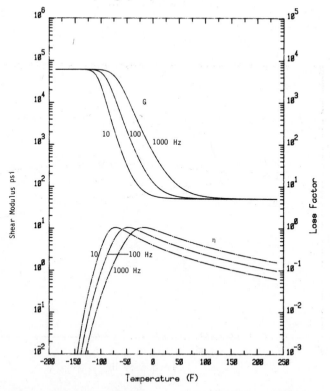

**021A.** Damping properties with temperature.

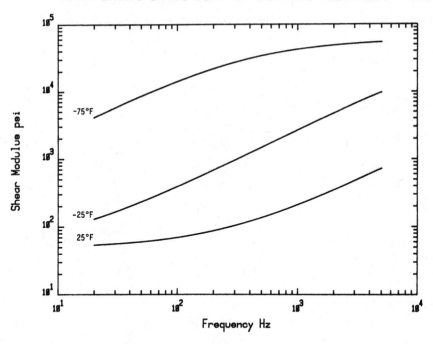

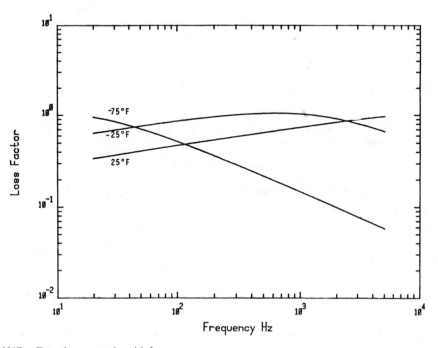

**021B.**  Damping properties with frequency.

## Data Sheet 022.   Damping properties of G.E. SMRD.

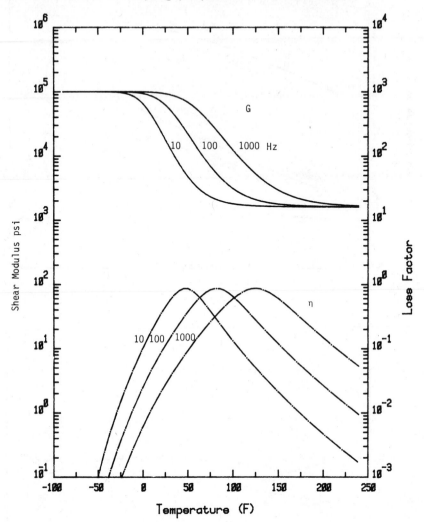

**022A.**   Damping properties with temperature.

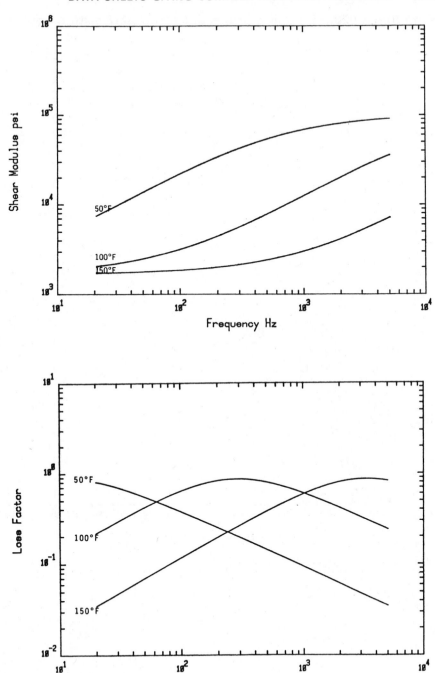

**022B.**  Damping properties with frequency.

## Data Sheet 023.  Damping properties of EC 2216 with graphite.

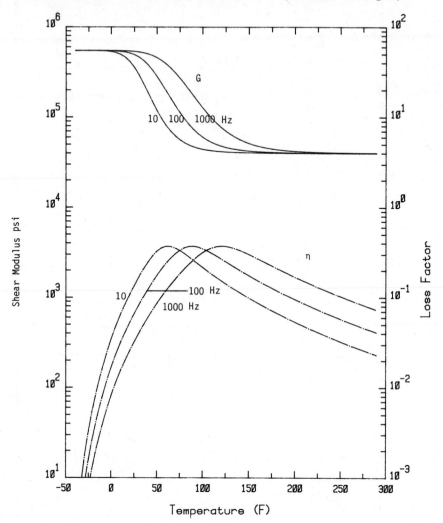

**023A.**  Damping properties with temperature.

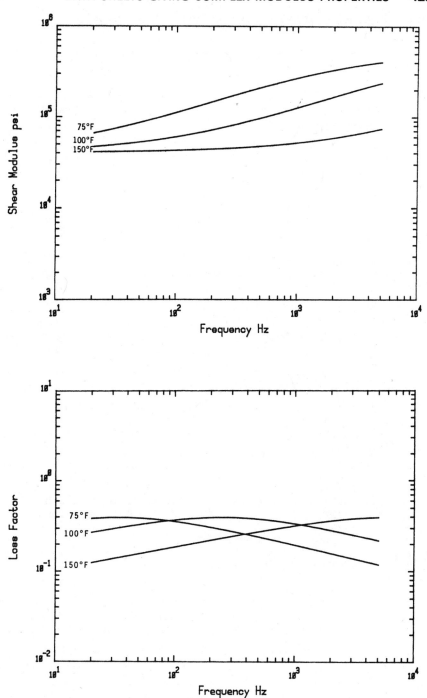

**023B.**  Damping properties with frequency.

## Data Sheet 024.  Damping properties of Shell Kraton.

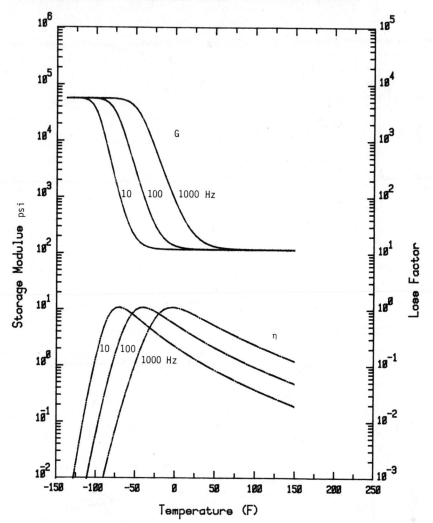

**024A.**  Damping properties with temperature.

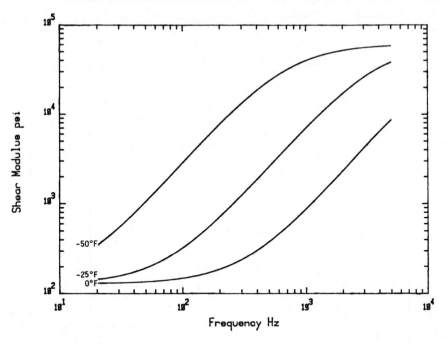

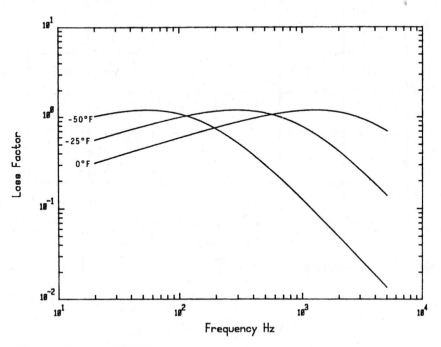

**024B.** Damping properties with frequency.

## Data Sheet 025. Complex modulus properties of Paracril-BJ with 0 PHR C.

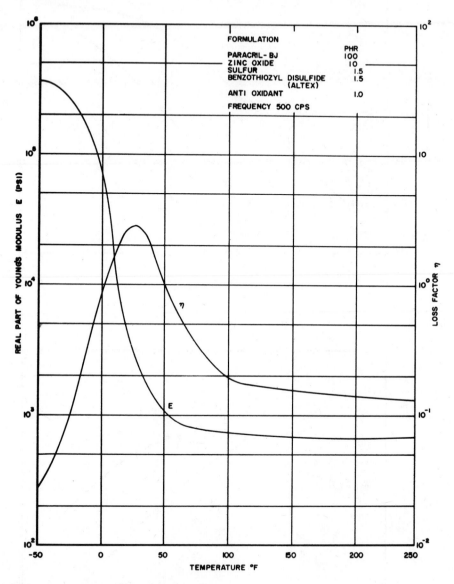

**025A.** Damping properties with temperature.

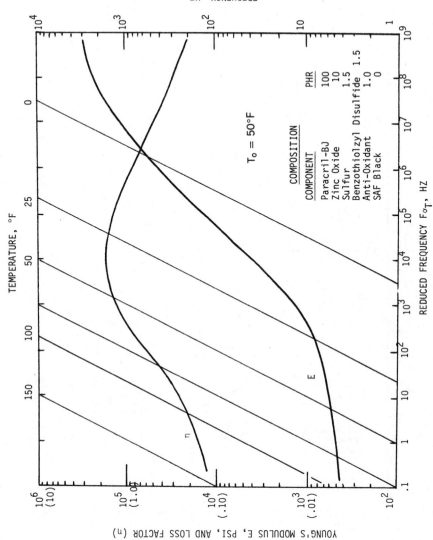

**025B.** Nomogram.

**427**

## Data Sheet 026.   Complex modulus properties of Paracril-BJ with 25 PHR C.

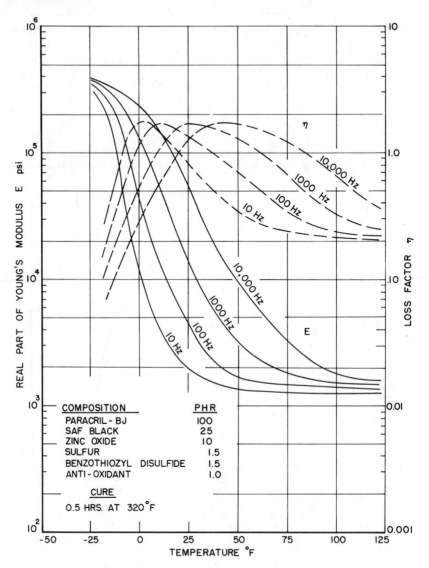

**026A.**   Damping properties with temperature.

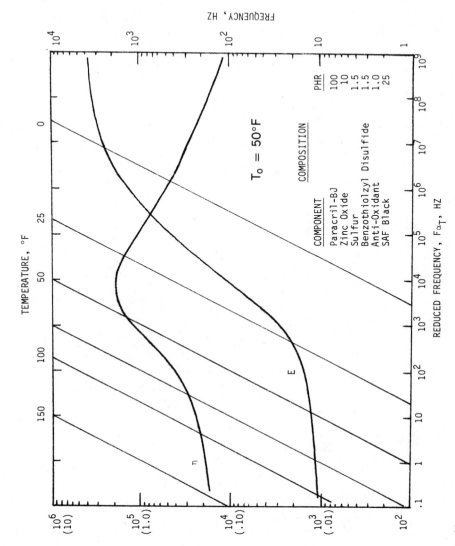

**026B.** Nomogram.

## Data Sheet 027. Complex modulus properties of Paracril-BJ with 50 PHR C.

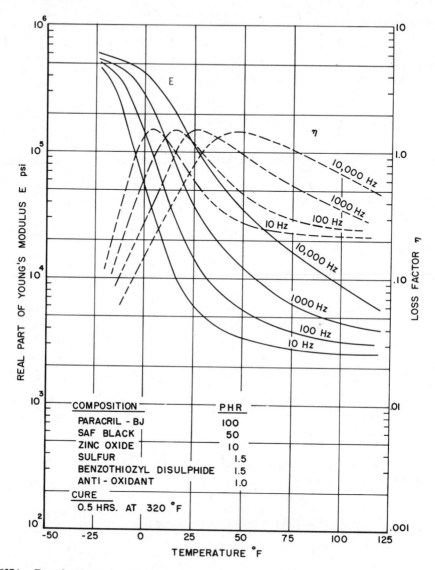

| COMPOSITION | PHR |
|---|---|
| PARACRIL - BJ | 100 |
| SAF BLACK | 50 |
| ZINC OXIDE | 10 |
| SULFUR | 1.5 |
| BENZOTHIOZYL DISULPHIDE | 1.5 |
| ANTI - OXIDANT | 1.0 |

CURE
0.5 HRS. AT 320 °F

**027A.** Damping properties with temperature.

FREQUENCY, HZ

$T_0 = 50\,°F$

COMPOSITION

| COMPONENT | PHR |
|---|---|
| Paracril-BJ | 100 |
| Zinc Oxide | 10 |
| Sulfur | 1.5 |
| Benzothiolzyl Disulfide | 1.5 |
| Anti-Oxidant | 1.0 |
| SAF Black | 50 |

TEMPERATURE, °F

REDUCED FREQUENCY $F\alpha_T$, HZ

YOUNG'S MODULUS E, PSI, AND LOSS FACTOR ($\eta$)

027B.  Nomogram.

## Data Sheet 028. Complex modulus properties of Paracril-D.

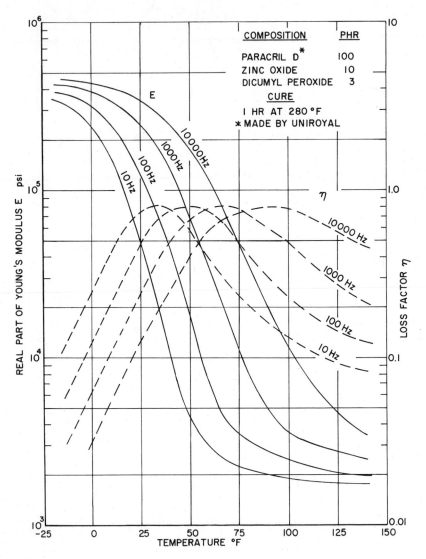

**028A.**  Damping properties with temperature.

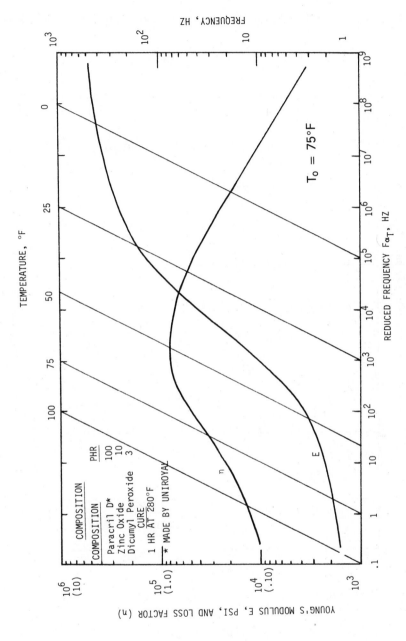

**028B.** Nomogram.

TEMPERATURE, °F

FREQUENCY, HZ

REDUCED FREQUENCY $F\alpha_T$, HZ

YOUNG'S MODULUS E, PSI, AND LOSS FACTOR ($\eta$)

$T_o = 75°F$

COMPOSITION

| COMPOSITION | PHR |
|---|---|
| Paracril D* | 100 |
| Zinc Oxide | 10 |
| Dicumyl Peroxide | 3 |

CURE
1 HR AT 280°F
* MADE BY UNIROYAL

Data Sheet 029. Complex modulus properties of Viton-B.

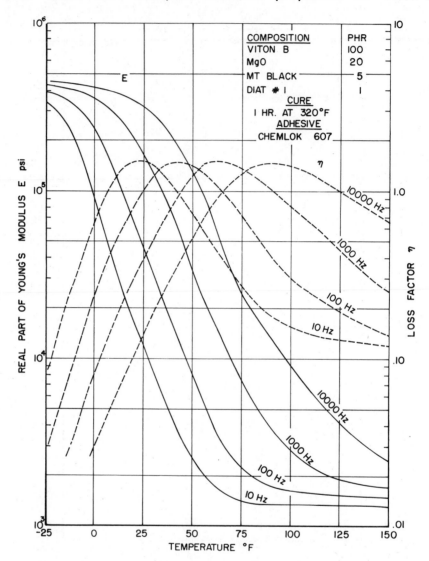

| COMPOSITION | PHR |
|---|---|
| VITON B | 100 |
| MgO | 20 |
| MT BLACK | 5 |
| DIAT # 1 | 1 |

CURE
1 HR. AT 320°F
ADHESIVE
CHEMLOK 607

**029A.** Damping properties with temperature.

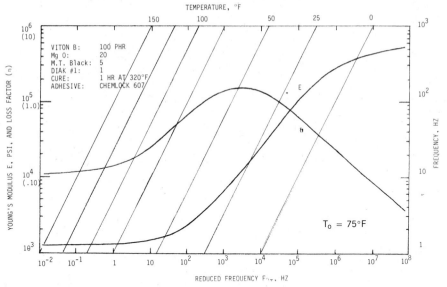

**029B.** Nomogram.

## Data Sheet 030. Complex modulus properties of GE RTV-630 Silicone (10:1) polymer.

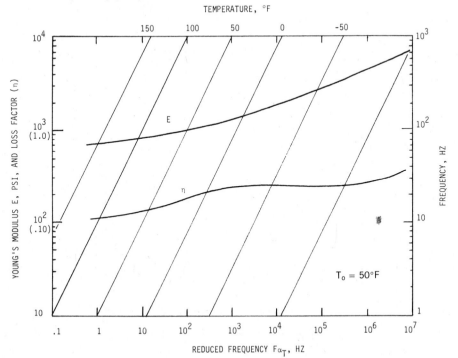

**030.** Nomogram.

## Data Sheet 031. Complex modulus properties of polymer blend.

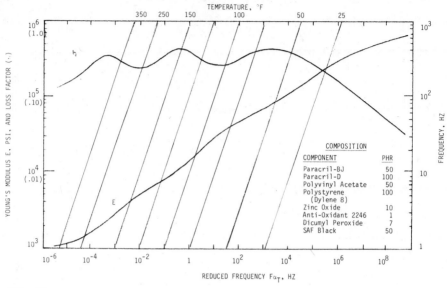

**031.** Nomogram.

## Data Sheet 032. Complex modulus properties of VPCO-15080 (Farbwercke Hoechst AG) [5.30].

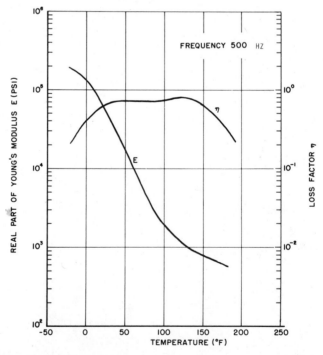

**032.** Damping properties with temperature.

# Data Sheet 033. Complex modulus properties of Styrene-Butadiene Rubber

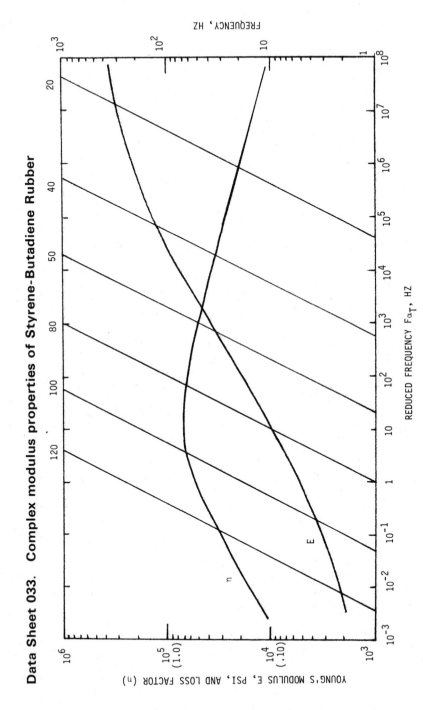

033. Nomogram.

## Data Sheet 034.  Complex modulus properties of 3M-467.

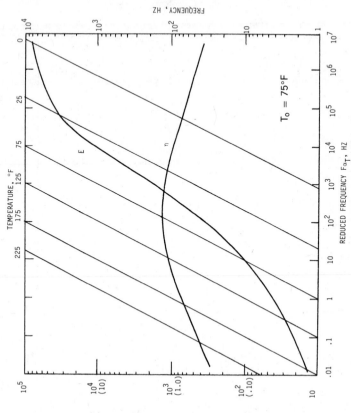

034.  Nomogram.

**Data Sheet 035. Complex modulus properties of 3M-428 adhesive.**

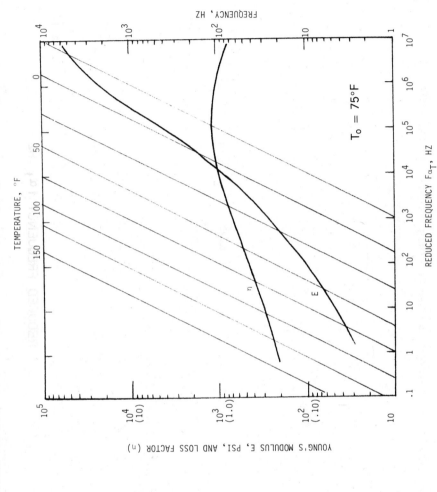

**035.** Nomogram.

**Data Sheet 036. Complex modulus properties of Kraton 107.**

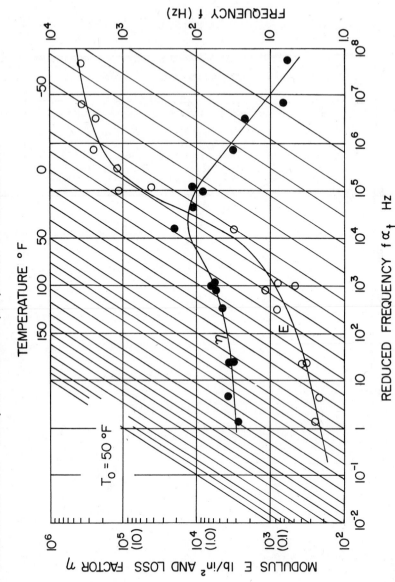

036. Nomogram.

# INDEX

**441**